CRC
Handbook
of
Lichenology

Volume I

Editor

Margalith Galun, Ph.D.
Professor
Department of Botany
The George S. Wise Faculty of Life Sciences
Tel Aviv University
Tel Aviv, Israel

CRC Press
Taylor & Francis Group
Boca Raton London New York

CRC Press is an imprint of the
Taylor & Francis Group, an **informa** business

CRC Press
Taylor & Francis Group
6000 Broken Sound Parkway NW, Suite 300
Boca Raton, FL 33487-2742

Reissued 2019 by CRC Press

© 1988 by Taylor & Francis Group, LLC
CRC Press is an imprint of Taylor & Francis Group, an Informa business

No claim to original U.S. Government works

A Library of Congress record exists under LC control number:

Publisher's Note
The publisher has gone to great lengths to ensure the quality of this reprint but points out that some imperfections in the original copies may be apparent.

Disclaimer
The publisher has made every effort to trace copyright holders and welcomes correspondence from those they have been unable to contact.

ISBN 13: 978-0-367-26163-4 (hbk)
ISBN 13: 978-0-367-26165-8 (pbk)
ISBN 13: 978-0-429-29178-4 (ebk)

Visit the Taylor & Francis Web site at http://www.taylorandfrancis.com and the
CRC Press Web site at http://www.crcpress.com

PREFACE

Lichens are among the most widely distributed eukaryotic organisms in the world. There are, to date, about 13,500 lichen species known, which accounts for approximately 20% of all the fungi described. All are the result of a symbiotic association between two unrelated organisms — a fungus and an alga (or cyanobacterium) — which, when fully integrated, form a new biological entity with very little resemblance to either one of its components.

Despite their wide ecological amplitude and abundance, also in extreme environments where often other plants cannot exist, lichens have received relatively little attention by plant physiologists. It was not until 1984 that the first international conference on lichen physiology and cell biology was held (organized by D. H. Brown, in Bristol). The second conference, devoted in part to lichen physiology, took place in 1986 (organized by E. Peveling, in Münster).

I have attempted to invite as many as possible lichenologists who have made significant contributions to lichen research, to contribute chapters to this book. Each author has been encouraged to approach his subject in his own way, in order to communicate the author's enthusiam to the reader. Naturally, in a book where each chapter deals with its own topic, there is bound to be some overlap of subject matter. Thus, people who are interested in a specific subject and read only one of the chapters, need to have some background pertinent to that subject. The reader will also find some contradictory statements in the various chapters, mainly on assumptive aspects, which I did not find necessary to unify, in order to stimulate further investigations.

Topics which have not been comprehensively surveyed in other treatises or are scattered in different journals, such as Nitrogen Metabolism (Chapter VI.B), Enzymology (Chapter VI.C), and Ecophysiological Relationships in Different Climatic Regions (Chapter VII.B.2), are here given more substantial scope. The Algal Partner (Chapter II.B) is much more extended that The Fungal Partner (Chapter II.A), because the systematic status of the photobionts has not been treated before in an integrated manner.

The terms "phycobiont" for the algal symbiont and "mycobiont" for the fungal symbiont of lichens have been introduced by G. S. Scott in 1957 [Lichen Terminology, *Nature (London)*, 179, 486, 1957]. After the "blue-green algae" were classified as "cyanobacteria" it was proposed (Ahmadjian, V., Holobionts Have More Parts, *International Lichenological Newsletter*, 15, 19, 1982) that "photobiont" should be used for the photosynthetic partner(s) of the lichens and more specifically "cyanobiont" for the cyanobacterial symbionts and "phycobiont" for the eukaryotic algal symbionts. This terminology has been adopted here, except for Chapter II.B (Tschermak-Woess) where, upon the author's request, the previous terms "phycobiont" and blue-green algae (Cyanophyceae) appear.

The names of the lichens cited in the various chapters are always those of the original articles. This may, in many cases, not coincide with the modernized system by Haffelner (Chapter X).

It is my hope that this book will bring lichenology to the notice of biology researchers, teachers, and students, as an important phenomenon in the mainstream of biology.

I greatly appreciate the efforts of the contributors and am especially indebted to Dr. Paul Bubrick, who encouraged me to launch this endeavor and who was very helpful in the editing of several chapters. I also wish to thank Dr. Leslie Jacobson, who read and commented on the chapters I contributed to this book, and Ms. Ruth Direktor and Ms. Henriette van Praag for their skillful secretarial assistance.

Finally, I am deeply grateful for the help, forbearance, and encouragement of my family.

Margalith Galun

THE EDITOR

Margalith Galun, Ph.D. is Professor of Botany at The George S. Wise Faculty of Life Sciences of Tel Aviv University. She received her M.Sc. and Ph.D. degrees from The Hebrew University of Jerusalem. Dr. Galun is a member of several professional and scientific organizations and in some she holds executive positions, such as Vice-President of the International Association for Lichenology, while in others she serves on the Executive Committee, such as that of the International Mycological Association and of the Israel National Collections of Natural History. She is the editor-in-chief of the journal *Symbiosis* and is a member of the editorial board of the journals *Israel Journal of Botany, Endocytobiosis and Cell Research,* and *Lichen Physiology and Biochemistry.* Dr. Galun has been the recipient of many research grants and is the author or co-author of two books and of about 90 research articles, including reviews, symposia, and chapters in books. Her current major research interests relate to the interaction between symbionts of plant symbiotic systems.

CONTRIBUTORS

Volume I

André Bellemère, Ph.D.
Professor
Department of Mycology
Ecole Normale Superieure
Saint Cloud, France

Carlos Vicente, Ph.D.
Professor and Head
Faculty of Biology
Department of Plant Physiology
Complutense University
Madrid, Spain

Margalith Galun, Ph.D.
Professor
Department of Botany
The George S. Wise Faculty of Life
  Sciences
Tel-Aviv University
Tel-Aviv, Israel

David L. Hawksworth, Ph.D., D.Sc.
Professor and Director
CAB International Mycological Institute
Surrey, England

Hans Martin Jahns, Ph.D.
Professor
Botanisches Institut
Universität Düsseldorf
Düsseldorf, West Germany

María Estrella Legaz, Ph.D.
Professor
Faculty of Biology
Department of Plant Physiology
Complutense University
Madrid, Spain

Marie Agnes Letrouit-Galinou, Ph.D.
Director of Research
Department of Cryptogamy
Universite Pierre & Marie Curie
Paris, France

Jacob Lorch, Ph.D.
Professor
Department of Botany
The Hebrew University
Jerusalem, Israel

Amar Nath Rai, Ph.D.
Professor and Head
Department of Biochemistry
School of Life Sciences
North-Eastern Hill University
Shillong, Meghalaya, India

Elisabeth Tschermak-Woess, Ph.D.
Professor
Department of Botany
University of Vienna
Vienna, Austria

TABLE OF CONTENTS

Volume I

TABLE OF CONTENTS

Volume II

TABLE OF CONTENTS

Volume III

Section I: Introduction

Chapter I

THE TRUE NATURE OF LICHENS — A HISTORICAL SURVEY

Jacob Lorch

"Les Lichens gélatineux ne seroient-ils pas des individus de Nostoc qui auroient changé de forme?"

E. P. Ventenat, *Tableau du Règne Végétal
selon la méthode du Jussieu*, Paris, 1799.

PRELUDE

Upon opening a handbook, one expects to be presented with an outline of the great periods in the history of the subject at hand. For lichenology, these great periods have been sketched in great detail by Krempelhuber in his two-volume *Geschichte und Litteratur der Lichenologie bis 1865*.[1] As they have been widely copied and adopted and are readily available,[2] the author of this chapter, in agreement with the editor, will confine himself to the events of those historic years, full of toil and strife, which saw the emergence of the Schwendenerian theory and its slow yet inexorable progress and victory.

True, that was not the first time that lichenology was in turmoil. Decades earlier, Acharius — whose system of lichens had become very widely accepted — was challenged and threatened by doubts and distrust sown by Meyer and Wallroth. Indeed, the resulting tendency to blur the definitions of species, even to the lumping of all lichens with a merely superficial resemblance in one species, brought lichenology into disrepute. That crisis was resolved in 1831 by E. Fries, who smoothed the way for the profusion of lichenological studies in the years to come.[3]

Whereas the Acharius-Fries interlude involved divergent views on matters of systematics, the events which will be dealt with here concern the basic nature of lichens and their proper place in the order of living beings. Anatomical studies, with emphasis on growth and development, provided the core of the story. Work on physiology — elementary as it was — and studies on the absorption of light provided supplementary evidence. In any case, all the leading taxonomists looked askance at the theorizing of "outsiders".

The story of the Schwendenerian theory surely deserves to rank among the most fascinating cases in the history of science. Its message to working lichenologists transcends the limits of a drama culled from the pages of old journals. Though the theory under scrutiny is generally ascribed to Simon Schwendener, it has occasionally been named the "Bornet-Schwendener Theory"[4] or the "de Bary-Schwendener Theory".[5] At least one of its antagonists loved to refer to "Schwendenerism", which yielded two warring camps, "Schwendenerians" and "Anti-Schwendenerians".[6]

LARGO — 1848

One of the very first studies of lichens which pays major attention to their growth and reproduction — besides and beyond their anatomy — was von Flotow's study of 1850, the subtitle of which mentioned that it was "extracted from a letter to Fries in Upsala, September 1848":[7]

"Ever since I have come to possess a good microscope of Schiek's, the lichens have become twice as dear to me . . . . I made my first microscopic studies in 1838, on the *Verrucarias*".[7] After explaining that he did not wish to go more deeply into a study of algae — so as not to come to loggerheads with one of his contemporaries — von Flotow proceeds to describe some of his results concerning *Ephebe*.

"From the lichenological point of view it is certain that *Gloeocapsa (Protococcus* [sic!] Ag.) *sanguinea, rubicunda, atrata*, etc. are different conditions of the gonidia of *Ephebe pubescens* and are related to it as *Lepra viridis* is related to *Parmelia parietina*." "These gonidia of *Ephebe* are dispersed in nature as abundantly, nay, even more abundantly than *Lepra viridis*. In our montane valleys they produce a reddish-black covering of entire slopes and rocky escarpments, visible from several thousand steps away. But they also penetrate between other lichens and mosses, where they appear as numerous, low forms which are more or less developed into *Thermutis* or *Ephebe*. Likewise, the '*corpuscula fungosa* Fik.' in the genus *Stereocaulon* belongs to these gonidia of *Ephebe* and I tend to believe that the mode and manner in which the stem of *Stereocaulon* relates to its parasite — in that it furthers or restricts its development — affords some evidence of its own nature. Therefore, I do not consider Florke's hints as altogether rejectionable."[8]

Another pioneer, a lecturer on botany and vegetable physiology at the Bristol Medical School, was G. H. K. Thwaites. In 1849, he published a short contribution in *The Annals and Magazine of Natural History*. This journal had as its motto:

> " . . . The sylvan powers obey or summon . . .
> The Dryads come . . .
> The Nymphs come not empty-handed
> But scatter round ten thousand forms minute
> of velvet moss or lichen . . .
> All, all to us unlock their secret stores
> And pay their cheerful tribute."

Thwaites seemingly takes this quite literally:

" . . . I am able to state with confidence what is the true character of *gonidia* . . . When examined more carefully, (the *thallus* of *Collema*) is found to consist of numerous *Nostoc*-like vesicles cohering closely, among which ramify anastomosing filaments. The cellular cuticle which invests the *thallus* of some species of *Collema* is but a modification of those anastomosing filaments."[9]

After going into some more anatomical details, Thwaites says:

"From the above it is clear that the *gonidia* of a lichen are the analogues as regards their function of the *Nostoc*-vesicles of *Collema* and this view enables us to understand what previously appeared an anomalous character in these organs. The *gonidia* are in fact the *essential* part of the whole structure and can scarcely be considered as *gemmae* except when under certain circumstances they put on that character, just as ordinary cells do in other plants. The other elements of the lichen-thallus may without difficulty be believed to represent the modifications of the anastomosing filaments of *Collema*, which no doubt they are."[9]

Thwaites is aware of the more general implications:

"So true is it, that in the smallest natural groups of organized structures the same great principles are to be discovered, when carefully sought for, which exhibit themselves so obviously in the larger divisions of the Kingdom of Nature."[10]

Louis René Tulasne, one of the two brothers who made for themselves an immortal name in the discovery of the life cycles of several groups of fungi, worked on the lichens as well. Starting from the studies of Itzigsohn and von Flotow, he proceeded to examine certain black spots "which, among the various productions born from the thallus of lichens, are not the least deserving of serious examination".[11] Itzigsohn had interpreted them as antheridia, like those of mosses or liverworts, "corpuscles which contain animalcules which are able to move".[11] However, even by using all the means recommended by Itzigsohn, Tulasne failed to confirm this.

This episode, though it does not directly relate to the alga-lichen hypothesis, is mentioned here because Tulasne, there and then, introduced for these puntiform conceptacles the term "spermogonie", adding: "One finds oneself pushed to regard them as strangers to the lichen, like parasites on their thallus, analogous to *Septoria* or *Phyllosticta* or other small fungi which live on decaying leaves, the more so as the latter show an organisation almost identical with that described".[12]

This analogy between lichens and parasitized leaves of higher plants must have been charismatic in its way. Indeed, the examination of numerous crustaceous and foliaceous lichens "revealed, in a similar manner, that the 'corps itzigsohniens' or spermogonies found in them do not at all appertain to them. Nor can one doubt that they are unique organs until now unfairly neglected by the lichenologists".[12]

After briefly grappling with whatever was known at the time about mosses, ferns, Salviniaceae, and algae, Tulasne suggested that the study of lichens may not in itself resolve the problems of the nature and function of "spermaties".

It was this doubt that caused Tulasne to extend his studies to the class of fungi. "The relevant results together with those obtained from observations exclusively devoted to lichens will — if I am not wrong — show that the latter, though they have been called 'aerial algae', are linked with fungi by an affinity much more close than what is generally believed".[13]

In the second part of his study, Tulasne only once mentions lichens in passing, being mostly preoccupied with fungi. Yet he concludes: "Without prejudging the role of the spermatia in any way, one cannot fail to remark that they precede the formation of 'endotheque' spores in the same manner as the antheridia of ferns or of *Equisetum* precede the development of their 'seminiferous capsules' (i.e., the sporangia)".[14]

The following year, on December 13, 1852, Tulasne presented to the Académie des Sciences numerous new observations on the reproduction in fungi. He opened his report by pointing out that "the organic types (which exist) in that vast class of vegetables are infinitely more diverse than in the lichens. This makes it naturally evident that the problem of their sexuality, if they have sexuality, poses a much more laborious task".[15] "In fact, when one reflects upon the numerous means of propagation which the Creator has bestowed upon algae, lichens and mosses, one should have no difficulty in this respect."[16]

The rich and varied crop of unheard-of and undreamt-of discoveries of reproductive organs in algae and fungi gleaned in the fifties of the last century proved a strain on the eyes and the minds of contemporary botanists and biologists, as well as on their character. H. Itzigsohn, mentioned earlier, was one of those bent on deciphering the secrets of lichens. A doctor of medicine, Itzigsohn spent his leisure hours at his microscope, in an attempt to unravel the most discrete facts of reproduction. For instance, after immersing and softening *Andraea grimsulina* in water he discovered wonderfully clear specimens of *Sirosiphon sauteri*. Pursuing his observations, he saw that, far from persisting at that stage of vegetation, they slowly grew into what Kützing had described as *Stigonema pannosum*. "The same thing that is figured as an alga by Weber and Mohr, Dillwyn, Agardh, etc., is treated by the lichenologists as *Collema, Parmelia, Cornicularia, Lichen*. One merely has to compare this with the lichens in Rabenhorst's 'Kryptogamenflora'. Currently, it is often referred to as *Ephebe pubescens* Fries. And if indeed . . . these plantlets produce apothecia in the manner of *Collema* then of rights the lichenologists claim it — as *Ephebe, Thermutis* or *Collema* — as theirs. Hence we must concede that *the definitive form of this plant pertains to the lichens* [author's emphasis]."[17]

"If we now raise the question whether Kützing was right when in his phycological writing he presented these filaments as *Stigonema*, the answer — however paradoxical it may sound — is definitely in the affirmative. Younger stages of *Stigonema* are totally identical with *Sirosiphon*, as I have shown . . . and can affirm with complete confidence. However, if one were to claim that if so neither can *Sirosiphon* be a genus of algae, but is merely a point of transition in the development towards the lichens, perhaps equivalent to the protonematous formations of mosses, this would be contradicted by my own very comprehensive researches on the Nostocineae.''[18]

Itzigsohn's highly creative view of nature is marked by the assumption of a very high degree of plasticity. Small wonder that in his eyes *Sirosiphon silvestris* seems originally to germinate from the gonidia of lichens ''which have failed their purpose; though my researches on this point have not been concluded''.[19]

''As for the 'Collemen', their gonidial filaments are not merely superficially similar but 'totally identical' with *Nostoc* filaments . . . .''[19] Their manner of breaking up at the interstitial cells is but one more proof of this. Later, in the same journal and the same year, H. Itzigsohn returns to the fray: ''How does *Collema* relate to *Nostoc* and the Nostocineae?''[20]

Instinct (or rather ''ein instinktartiger Takt'') led him to the study of young lichen thalli, especially those which develop from isolated gonidial clusters: ''Everything that has been interpreted as unicellular algae are merely developmental stages of other algae or lichens. How much longer will the spell of unicellular algae mislead plant physiology! After seven years' studies of algae I have never once seen a unicellular alga! . . . .''[20]

''To come back to our gonidia, these develop by repeated division either into the so-called *Protococcus* form, familiar to the lichenologists as *Lepraria*, or in very humid localities . . . into *Ulothrix*; in even more unique, as yet unknown conditions, . . . perhaps into *Prasiola*. All this happens as long as the gonidia are isolated, i.e., without the influence of male organs.''[20]

The rest of the article deals with what Itzigsohn calls ''das spermatische Gewimmel'', literally ''the spermatic chaos'', i.e., presumed male sexual cells, ''not as motile as those of higher cryptogams, but sufficiently motile not to be mistaken for the dance of atoms''.[20] The fecundation of the female gonidia is followed by the elongation of the spermatia and, simultaneously, by the first indications of bifurcated branches (growing from the spermatia). In due course, ''a solid body is formed, i.e., the complete young thallus with a fibrous layer formed from the spermatia, and a gonidial layer developed from the gonidia''.[21]

Some more details lead the author to the conclusion that ''there can be no talk about an alga which may turn into a lichen. What appears to be a transition from *Nostoc* to *Collema* is to be searched for merely in the fertilization of a soredium — for that is what the Nostoc-globe or Nostoc-mass represents''.[22]

Is it not conceivable that but for the unfortunate term ''gonidia'' all of these discoveries would not have been made?

Another dimension — and a highly relevant dimension at that — of the contemporary scientific scene is evident in a brief report by A. de Bary to the readers of *Flora or General Botanical Journal* on ''The latest works, in particular those of Pasteur,''[23] on the formation (''Entstehung'') and growth (''Vegetation'') of the lower fungi.

In this balanced report on Pasteur's denial of spontaneous generation, de Bary mentions the objections made in France (in particular by Pouchet) and in Germany (where Nägeli, in his voluminous work on starch-grains, argued that what had been proven in individual instances, i.e., the absence of spontaneous generation, yet left the problem in general undecided).

''The possibility — so de Bary — of the formation of cells, as the beginnings of low organisms (produced from) living or disintegrating organic substance yet remains conceivable, once the difficulty which consists in the absence of any well-founded conception about the transition of an organic body into an organized one, is overcome.''[23]

This may be an adequate juncture to introduce one of the central actors in our drama. Surely the most prolific writer ever on lichen taxonomy, with hundreds of works and numerous new taxa to his credit, William Nylander was the staunchest supporter of the autonomy of lichens. He died almost synchronously with the century, in 1899.

A necrologue by father Hué refers to the dramatic effect of the events treated in this chapter on the life of Nylander. After relinquishing his chair of botany at the University of Helsingfors in 1863, Nylander moved to Paris, but after Bornet published in 1873 his *Recherches*, which corroborated Schwendener's views, Nylander's life was poisoned by his obsession with what he called "the autonomy of the lichens".[24]

"He had always consistently rejected the theory of symbiosis, insisting that the gonidia are the offspring of germinating spores. This problem made him leave from one day to the next the laboratory of the Museum, where he had worked for many years. (He never went back.) It became a nightmare which relentlessly haunted his existence, and which made him regard as personal enemies all those who thought differently. He always encouraged beginners in lichenography, but once they got settled down he never failed to press them into adding a note on the autonomy of lichens, with himself providing the arguments. Some, like Crombie and Richards, accepted. With the others, he came to loggerheads. He passed his last years, and died, in total isolation."[24]

Nylander's faith seems to have been first, and most clearly, stated in a brief communication submitted by the secretary of the Konglige Vetenskaps-Akademie in Stockholm at its session of October 1, 1855:

"Lichens have long been presented in the botanic system as a subdivision of the algae, and have consequently been called *Algae terrestres*, Aerophyceae, etc., but lately a more careful study of their organisation has shown that they are far more closely related to the fungi. In particular, the groups *Licheneae* and *Collemeae* have an apparent affinity with certain algae. But on the other hand the *Graphideae* and the *Verrucariae* are a real transition to several genera of fungi, both among the Pyrenomycetes and the Discomycetes."

"*Scytonema* stands in relation to *Gonionema* just as *Sirosiphon* stands to *Ephebe* . . . . It is more difficult to assume the transition of *Nostoc* to *Collema*. But all their ambiguous formations are consistently sterile or at the least lack any fructification which has been clearly seen, and described with suitable care."[25]

All considered: "In the present state of science it is impossible to draw an absolute border between lichens and fungi. They make up two separate groups of plants which blend with each other through certain forms of incomplete organisation . . . ."[26]

The taxonomy of lichens also was the chief concern of Speerschneider (1855), who used his microscope in order to determine the proper place of *Ramalina* — his favorite genus at the time — in the Natural System.

"Any fracture . . . will at once make possible the release of the enmeshed gonidial cells. This release is of no mean importance for the reproduction of many lichens . . . the massed multiplication of which, linked as it is with a rather sparse appearance of apothecia, respectively spores, seems somewhat strange. But when one considers that the reproduction of lichens — as now appears beyond all doubt — may take place by means of the gonidial cells, the importance of the soredia discussed (in this study) is evident enough. In order to establish this importance by experiment I have sown these mealy soredial masses. Though up to now I have not obtained a rigorously convincing result, I did become convinced that such a result will sooner or later be obtained by me."[27]

Bornet (1856) is also largely guided by taxonomical interests:

"Among the algae which are today referred to the tribes Palmellées and Scytonemées, there are some which will without doubt be ranged with the lichens, once their reproductive apparatus shall become known. Such is the case of *Ephebe pubescens* Fr., the fructiferous organs of which I have described. I shall also mention the very interesting genus which Nylander has named *Gonionema*, whose fructification is that of a lichen, but whose thallus is composed of filaments identical with those of *Scytonema*. Undoubtedly this is a fertile field for discoveries (to be made by) an attentive and persevering observer.

Last winter, at Cannes, I have found three lichens which belong to this category, i.e., which have the thallus of algae and fructifications proper to lichens . . . "*Spilonema*, gen. nov.; *Synalissa conferta*, n.sp.; *Synalissa micrococca*."[28]

On the second of these, Bornet has this to say: "The first state of this lichen did not appear to me as distinct from the algae known under the name of *Gloeocapsa Magma* Kutz., a tiny 'Palmellee' which is perhaps merely the primordial state of diverse lichens, *Sirosiphon*, etc., ... One readily finds on one and the same rock all transitions between well-developed specimens of this lichen and a brown-reddish powdery layer made up of thick-walled globules of red colour, solitary or in groups of 2, 4, 8, etc., which contain green matter and agree totally with the above-mentioned alga."[28]

Bayrhoffer (1857) attempts to fill a gap in the knowledge of *Thrombium Nostoc*, in view of the fact that Wallroth — who had earlier observed apothecia on *Nostoc* — had failed to describe their development and spores: "This description will in many respects throw an explicatory light on many other lichens".[29] In his exceedingly fanciful study, Bayrhoffer provides descriptions of what he thought were the male and female layers of the lichen thallus. Thus, the female gonidia each have a nucleus, equated with the yolk of an egg, etc.[29] (Specimens of the related species, *Thrombium bacillare*, collected in Lorch, failed to yield ripe spores, but otherwise agreed with the former.[30]) Years later, one of Schwendener's opponents could think of no more terrible condemnation of the theory of the double nature of lichens, than to compare its merits with those of Bayrhoffer's theories!

G. Thuret (1857) also occupies himself with *Nostoc*. His own observations, about 13 years earlier, on the curious phenomena which accompany the reproduction of an aquatic *Nostoc* do not — he points out — seem to have occupied anyone ever since. Over several years, Thuret repeated those observations on a terrestial species common in the vicinity of Cherbourg. These confirmed the results of his earlier observations on *Nostoc mougeottii*. In spite of their divergent environments, the two species showed exactly the same phenomena. In fact, the new paper is "justified merely by the wish to profit from the high quality of a new draughtsman, Monsieur Riocreux".[31]

The year 1857 also saw Simon Schwendener graduating from the University of Zürich and move with his much-admired teacher, C. Nägeli, from there to München. That same year, he began his observations of the lichen thallus, motivated by the paucity of anatomical data on lichens. His first report, dealing with the arbuscular lichens, saw light in 1860, the year of his appointment as Privatdocent. Schwendener did not wish to delay publication until his study of the foliose lichens was completed and moreover — as he says — he had not given up hope to follow in detail the development of a thallus from a spore.

By boiling the thallus in a dilute solution of KOH or by its prolonged heating in a concentrated solution, the branching hyphae could be observed in some detail. Thus, close to the tip of the thallus — according to Schwendener — there begins the formation of the green spherical gonidia. He goes on to describe the production of gonidia from articulated cells ("Gliederzellen") by the formation of branches which divide into two: a stalk-cell and a spherical apical cell. These cells often exude oily drops. Each gonidium divides at first into two, and then again by partitions perpendicular to each other. Following division, the parts become spherical. The stalk-cell, too, sometimes divides into two to three cells, which do not develop further. However, in other cases there are formed at the juncture of stalk and gonidium one to two protuberances which penetrate into the gonidium, between its cells, and branch there. They then grow outwards to the surface. The branches adhere to the gonidium and continue to grow and divide.[32]

This brief outline of some salient points in Schwendener's study of 1860 leaves no doubt that at that time Schwendener's theory had not crossed its author's mind. Surely, even a slight suspicion in the line of his later theory would have been reflected in a more cautious vein in his description of growth in lichens. Which brings us to the textbook which best sums up the situation prior to the "affair", the outbreak of the Algo-Lichenic Theory.

ALLEGRO CON BRIO — 1865

In his foreword (December 1865), de Bary explains that in his *Morphologie und Physiologie der Pilze, Flechten und Myxomyceten* he follows Koerber's system, "without any wish to join forces with one or other of the parties of lichenographers, but merely because Koerber's books are relatively comprehensive, and generally available".[33]

Chapters 8 to 11, on the lichens, are sandwiched between the four sections (234 pp) on fungi and the discussion (20 pp) of the Myxomycetes. As to the fibrous tissue or pseudoparenchyma of the lichens, de Bary accepts the views of Schleiden and Schacht who have hinted at the similarity between this and certain structures in mushrooms. "Speerschneider and above all Schwendener have proved this beyond doubt."[34]

de Bary goes on:

"To the hyphae of the lichen thallus there are added a second kind of form-elements: rounded or elongated, green or bluish-green cells. Though these have in many cases been shown to be initially produced from the hyphae, they subsequently show a development — independent to a certain degree — and a unique mode of reproduction, mostly by dividing in the three dimensions of space. Since Wallroth, these cells are known as gonidia, an expression which insofar as it is accepted literally and in the sense of its author, as 'Brutzellen', i.e., reproductive organs, is certainly an unfortunate choice."[35]

As far as the development of gonidia is concerned, de Bary follows Bayrhoffer who first claimed to have shown how the green gonidium is produced by the swelling of a short lateral branch of a hypha, followed by its detachment as a spherical cell which turns green.[36] de Bary illustrates this process in his Figure 88, a and g, and after some more details proceeds to the mode of reproduction of the blue-green gonidia, relying on Schwendener's observations "which I have repeated for most, if not for all, stages: gonidia reproduce by repeated division into two, through the formation of a cross-wall. The first cross-wall passes through the juncture with the stalk-cell (Figure 88 f); the two following divisions proceed in such a manner as to produce four cells ('Theilzellen') arranged in a near-tetrahedron and only rarely in one plane."[37]

As for the blue-green gonidia, "following Schwendener, these — like the green gonidia — are formed as the terminal cells of lateral hyphal branches, and divide like them, but for the less rapid formation of the cross-walls . . . ."[37] The hyaline envelope of groups of those gonidia reminds de Bary of "cell-families" of Nostocaceae and Chroococcaceae. The similarity with these latter organisms becomes very complete by virtue of the nature of the contents of the gonidia. Phycochrome bestows on the uniform "Protoplasmakörper" a clear or dirty blue-green color. "This color has not as yet been examined in detail, but its designation as phycochrome is justified by the complete similarity to that of the algae referred to."[38]

Further on (Reference 38, p. 295) de Bary comes back to the relationship between the hyphae and the gonidia. After abstracting the views of Cassini, Itzigsohn, Bayrhoffer, and Wallroth, de Bary makes the following open-minded comment:

"As mentioned above, the genetic connection of the hyphae with the gonidia is dubious. But, even if it were clear that such a connection never exists, the substantiated presumption ('begrundete [sic!] Vermuthung') of a genetic relationship between *Collema* and forms of *Nostoc* will have to persist. This, for the following reasons: The 'Gonidienschnüre' (gonidial threads) of *Collema* do not merely resemble the threads of *Nostoc*, but are identical with them in every essential character. The same is true for the surrounding gel. If one imagines what *Collema* would be like with its hyphae removed, the shape of the thallus will, indeed, be different from that of *Nostoc* but there will no longer be any structural difference, in many cases not even to a degree that separates individual forms or species of *Nostoc*. Secondly, one frequently observes on the thallus of indubitable *Collema*, besides its spherical buds which resemble soredia, individual threads of gonidia in sharply defined gelatinous envelopes, without hyphae. These are identical in every respect with young *Nostoc* threads and I believe I have not erred in my observation that they are in fact (derived) from the thallus of *Collema*."[39]

de Bary states with certainty that between *Ephebe* and its relatives there exists a genetic relation:

"If one imagines the hyphae as removed, then the thallus of *Ephebe pubescens*, *Spilonema* — in particular its thinner branches — represent typical forms of the genus *Sirosiphon* . . . . I have in my possession a preparation (stained with Kali) in which a strong sprout of *Ephebe* has grown into a thread of *Sirosiphon* with 32 members. The fact that it is not merely attached, but represents a true branch, is most reliably indicated by the circumstance that several hyphae of the main branch penetrate into it, to the fourth member."[40]

"All considered, it is surely beyond doubt that a large part of the Nostocaceae and Chroococcaceae are genetically related to the gelatinous lichens *Ephebe*, etc. In what way? That remains to be investigated. If I may briefly allude to my subjective opinion — the detailed motivation of which would lead us too far — two suppositions appear to be justified: Either the lichens referred to are the fully developed, fructifying phases of plants which have until now been included in the algae as Nostocaceae or Chroococcaceae. Or the latter are typical algae which assume the form of *Collemas*, *Ephebes*, etc., because certain parasitic ascomycetes penetrate into them, their hyphae spread in the continuously growing thallus, and repeatedly adhere to its phycochrome-bearing cells. In the latter case, the plants referred to would be pseudo-lichens, comparable with phanerogams similarly deformed by parasitic fungi."[40]

Doubts or no doubts, the gonidia continued to lure and fascinate all those who came within their reach, much like the dug-in larvae of ant-lions patiently awaiting their victims at the bottom of their cone-shaped pits, often with fatal consequences.

In 1867, F. Cohn published his latest studies on the physiology and systematics of "Oscillarien" and "Florideen". On the basis of his analysis of their colors he concluded that the Ascomycetes should be included in the lichens:

"In the class 'Algae' two main types are united which, starting from homologous low forms, diverge in their higher developmental stages. They are most readily distinguished by the absence or presence of motile cells . . . ."[41]

"The first series begins with Chroococcaceae, (which include *Bacteria*, *Oscillatoria*, *Vibrions*). The Nostocaceae, Rivulariae, Scytonemae are linked through *Banoia* and *Goniotrichum* with the Florideae. With the Collemaceae as intermediaries, they lead to the Licheneae (including Ascomyceteae)."[41]

Later in the same year — and volume — Askenazy published the last part of an extensive report on chlorophyll and accompanying dyes. His study deals with solutions prepared from a variety of algae and lichens and includes a selection of the absorption spectra obtained. Thus, Askenazy notes that in *Anthoceros* ("unfortunately too rare in the vicinity of Heidelberg for a detailed study"[42]) there may occur — in addition to chlorophyll — a water-soluble substance, related to the blue-green color of this moss, reminiscent of *Oscillatoria*.[42] As for some of the lichens studied:

"Careful observation of the optical properties of the dissolved colouring substances of *Collema* and *Peltigera* leads me to the assumption that both are mere mixtures in different proportions of two colouring substances . . . . The second of these, if it could be prepared in a pure state, would show some similarity with phycoerythrin as regards both absorption and fluorescence. The first of these hypothetical colouring substances or at least one which totally corresponds to it does occur in the vegetable kingdom, and I have successfully extracted it from Oscillatorineae in substantial quantities."[43]

Cohn's bird's eye view of the predicament of lichens provides a convenient link to the wealth of relevant studies published in 1868, the year from which the new insight into the dual nature of lichens is generally dated.

Surely, the volume which contains more valuable information relevant to our story than any other is the *Botanische Zeitung* for 1868. First, is the article of Dr. Hermann Itzigsohn, listed in the index as I(tzigsohn), Hermann.[109] Then, there is a letter by S. Schwendener "On the relations between algae and lichen-gonidia";[47] W. Fuisting's "On the developmental history of gonidia and the formation of zoospores in the lichens";[110] and E. Stitzenberger's "De lecanora subfusca ejusque formis commentatio".[111]

However, the most interesting item signed "dBy" (for A. de Bary), is a report on Boranetzky's "Contribution to the knowledge of the independent life of the gonidia of lichens":[44]

"The results reported in Boranetzky's study agree in the main (disregarding some details) with those described in Itzigsohn's article and it will be superfluous to point out that both studies were made totally independently one of the other. Boranetzky was published in Petersburg on 10.12.1867; Itzigsohn's article was submitted to the editors (literally: 'Redaction') at the beginning of December of last year. The agreement of these articles consists in that the so-called gonidia, even of the phycochrome-bearing lichen thallus, are able to grow and reproduce by themselves (a process in which) they behave exactly like those plants designated as green algae."[44]

After a brief outline, de Bary concludes:

"The 'lichen' [note the inverted commas — J. L.] would therefore be a hybrid ('Zwitterling'), an alga occupied by a parasite and subservient to the vegetation of the invader . . . . Boranetzky has cultivated thin sections of *Peltigera canina*, and obtained flourishing cultures of gonidial cells arranged in dense clusters. In *Collema pulposum* the gonidial chains likewise developed into numerous globules of *Nostoc*, reaching the dimensions of the head of a needle. In both cases the hyphae died. Just as in Boranetzky's earlier paper, written in collaboration with Famintzin, no thallus with hyphae was obtained."[45]

de Bary then comes back to a question he had raised earlier, to wit "either the gelatinous lichens are the fully developed, fruit-bearing phases of plants the incompletely developed forms of which were until now placed in the algae as Nostocaceae (and) Chroococcaceae of the accepted systems, or they are typical algae which take the form and structure of 'lichens' in that certain ascomycetes penetrate into them, spread their mycelium between them, and frequently adhere to the gonidia. To decide the situation in other algae, it is recommended to spread viable spores of the lichens to be studied on the freely vegetating 'gonidia'. According to an oral communication, the validity of the above supposition for all lichens has recently been suggested from another source which the reviewer is not authorised to mention by name."[46]

Also in the *Botanische Zeitung*, on May 1, 1868, the editors — Hugo von Mohl and Anton de Bary — publish a letter from S. Schwendener, "On the relations between algae and the lichen-gonidia".[47] There, in a footnote, Schwendener is identified as "the author who refused to be mentioned"[47] in de Bary's review, extracted above. This is what Schwendener has to say:

"Together with the last part of my researches on foliacious and gelatinous lichens I feel that I must add some remarks concerning the relations between algae and lichen-gonidia, which have been repeatedly discussed since the submission of my manuscript in the spring of 1867."[47]

"Facts reported in the appendix to my researches, and the observations of Famintzin and Boranetzky and those of Itzigsohn provide, as I believe, decisive proof that the gonidia of the lichens studied and the corresponding unicellular and filamentous algae are identical."[47]

"True, Famintzin and Boranetzky are led by their discovery to strike the algae concerned off the list of independent plants and to consider them henceforth as freely vegetating algal cells (gonidia). On the contrary, I adhere to the view that the lichens, all and sundry, are not independent plants but fungi of the subdivision 'Ascomycetes', to which the respective algae — whose independence I do not doubt — serve as hosts.''[47]

"Already in the winter of 1866/67 I have informed several friends and acquaintances of this view, orally or by letter. However, only at last year's Naturforscherversammlung of Switzerland, held in Rheinfelden (in September 1867) did I express it publicly and without reservation. There, I vindicated my view with several recent and as yet unpublished researches. True, these researches relate only to certain groups of the lichens in which they showed, as seems to me, the algal nature of gonidia without any doubt whatsoever. I place special weight on the following direct observations, which I illustrated in my lecture with the necessary figures:

1. Observations of hyphae which penetrate into colonies of lichens about 150—300 microns in diameter and branch inside them.
2. Observations on colonies of *Gloeocapsa*, which fully conform to the former.
3. Observations of hyphae which have penetrated into the sheaths of Rivulariaceae.[48]

Items 4 and 5 concern certain Racoblenaceae, as well as individuals of *Chroolepus* and *Cystonemus*.

"To explain in more detail the reason why — unlike Famintzin and Boranetzky — I consider not only these but also others, reported in my supplement, and all the observations of others as parasitic proliferation, that would lead us too far. Suffice it to remark here that those considerations which for me have been decisive are in part general principles which, by previous experience, I believe to have general validity in botany, . . . in part the closely related analogies with *Ephebe*, *Ephebella*, *Coenogonium*, etc., i.e., with all those cases in which the parasitism of the hyphae now seems hardly in doubt; in part — lastly — impressions which I have gathered in the course of previous observations on the relations between gonidia and hyphae. I do not doubt that further observation and attempts of cultivation will decide those still hovering questions concerning the interpretation of the gonidia.''[49]

This, then, was the message that all mankind had been waiting for — if one may use the manner of writing in which Linnaeus had written about one of his own books. Intense interest in lichens, from 1850 or so onwards, had much to do with these developments in which contributions from within the lichenologist camp mingled with those by "outsiders" with wider interests. Like the man who thought of something unspeakable and said it, Schwendener had overcome his "childhood error" of 1860 to arrive at a generalization which others — equally suspicious of the traditional view — did not have the moral courage to pronounce. In its birth, the role of personality and character of the pioneers was not less important than that of new observations or what one would like to call facts.

The study of Famintzin and Boranetzky referred to by Schwendener was read to the St. Petersburg Academy of Science on March 13, 1867, and published in its *Mémoires* one year later, on March 13, 1868. It was also printed in the *Botanische Zeitschrift* for 1868 and in the *Annales des Sciences Naturelles* where it is dated 1867.[50] The fact that within one year this study was published in Russia as well as in France and Germany surely greatly aided its reverberations. (I shall quote from the version published in Germany.)

Any section across the lichen-thallus allows the free gonidia to get dispersed in the surrounding water. "In this state, they totally resemble unicellular algae. It is this that led us to the idea to attempt the culture of gonidia outside the thallus. We found but few data on this subject, especially Koerber (1839) who denies the possibility that gonidia ('*gonidia primaria* Koerb.') may develop into new individuals outside a thallus. Speerschneider, working with *Hagenia ciliaris*, holds the opposite view and claims to have observed the development of complete lichens from free 'gonimial cells'. Sections of *H. ciliaris* cultivated in humid air showed that within two months the hyphae decomposed, whereas the gonidia not only remained perfectly healthy, but grew markedly and divided intensively. Soon after, there appeared in the centre of the decomposing section of thallus very small, punctiform light green corpuscules which transformed into several lichen primordia. These observations of Speerschneider were ignored for 14 years. Even A. de Bary's *Morphologie und Physiologie der Pilze, Flechten und Myxomyceten* fails to mention them (though at the end of the chapter on lichens de Bary does quote Speerschneider's study)."[50] Famintzin and Boranetzky feel that this omission is due to the vague manner in which the development of the lichen-thallus from the gonidia is described in Speerschneider's study. In any case, they claim to have confirmed the latter's observations on the manner in which the gonidia are released through the dissolution of the hyphae, and their subsequent growth and division. But "we did not yet have the opportunity to observe the development of the thallus from the gonidia."[51]

The authors claim to have successfully cultured gonidia "from sections, but preferably from thallus macerated under water for several weeks or under a constant stream of water. In these conditions the hyphae softened and disintegrated; the gonidia however remained totally fresh and healthy. Spread on pieces of bark of lime, the free gonidia (as also those in the intact thallus) proved to be totally similar to *Cystococcus* described by Nägeli."[51]

"The most peculiar alteration of the gonidial cells consists in that in most of the cells the cell contents form zoospores. However, other cells disintegrate and variously divide into a mass of cells which slowly round up and eventually separate as spheres . . . . The zoospores lack special features; they are oval-elongated, the pointed colorless tip bearing two cilia is directed forward".[51]

In several zoospores, the authors claim to have observed how after moving for some time through the water they came to rest before growing to twice or three times their initial dimensions. "However, we were unable to clear up their further development."[52]

All considered — I am quoting two of the authors' conclusions —

1. The formation of zoospores characterizes not only algae and fungi, but also lichens. By this, the latter are brought significantly closer to the two former groups.
3. The free-living gonidia have proved to be identical with Nägeli's genus of free-living algae, *Cystococcus*. Therefore, the latter must no longer be conceived as an independent form but as a developmental stage of lichens.[52]

The authors conclude with the hope that other lichens besides *Physcia*, *Cladonia*, and *Evernia* studied by them will yield analogous results, "and we are now engaged in the continuation of our research in this field."[53]

Before proceeding with an outline of other studies published in 1868, let us make a brief visit to Paris in 1869, to the Académie des Sciences. With Delaunay in the chair, the Académie gathered on Monday, June 7, 1869, to decide on the annual distribution of prizes. During that session the Académie decided that the Prix Bréant, open "to whomever will find the means of curing Asian cholera or discover the causes of this terrible disease",[54] should not be awarded. Indeed, Claude Bernard complained bitterly that none of the numerous works submitted over the previous 15 years was worthy of the prize.[54]

The Académie then turned to the Prix Desmazières, founded in his testament, April 14, 1855, by B. H.-J. Desmazière and financed from a 3% interest on an endowment of 35,000 francs. The prize was to be awarded annually "to the author, either French or foreign, of the best or the most useful study published in the preceding year on the whole or a part of 'Cryptogamie'."[55]

The case of the Prix Desmazières was — so it seems — much less complicated than that of the Prix Bréant. The special committee — Tulasne, Trécul, Decaisne, Duchartre, and Brongniart (reporter) — had received the works of but two authors, both more or less concerning "the family of the lichens".

Prof. Famintzin, of St. Petersburg, had sent two "mémoires" published in 1867, one was a study on the development of gonidia and the formation of zoospores in lichens, the other was on the influence of light on *Spirogyra*. The following is translated from the report of the committee for the Desmazière prize, as submitted to the Academy:

"The first of these memoirs presents very singular ('bien singuliers') and very unexpected facts, if they had been confirmed by new observations and if they should not receive an interpretation different from that which Monsieur Famintzin has given them. According to this savant the gonidia of lichens, i.e., the spherical cells filled with chlorophyll which are dispersed in the parenchyma of the lichen thallus — when maintained on the surface of a bark in a state of favorable moisture for several months — will give birth inside them to zoospores, i.e., very uniform corpuscles, endowed with movement by means of vibratory hairs, like the zoospores of algae."[56]

"The results obtained by Famintzin are described with great precision, but up to now attempts to reproduce them have not been successful. In fact, they require great precision and perseverance. Besides, one may ask oneself what is the role of these zoospores and whether, in fact, they do serve the reproduction of lichens, or are products unrelated to the normal functions of these plants."[56]

"In this state of incertitude, the Commission considers it inadequate to take the memoir of M. Famintzin into consideration, though reserving for him the right to present himself for another 'Concours' once the facts which he reports will have had the opportunity to be better evaluated."[56]

"As for the *Spirogyra* memoir, that was less relevant to the study of Cryptogams in general than the other works submitted to the commission, i.e., the studies of Monsieur Nylander. Praising the work of the latter, in general and in particular, the Commission decided to award the prix Desmazières to Nylander!"[56]

For F. Cohn of Breslau, the discussions about the relationship between gonidia and the hyphae of ascospores in lichens provide an important stimulus for the study of algae parasitic on *Lemna*.[57] Indeed, the almost simultaneous discovery of parasitic algae in a variety of higher plants in the early 1870s was conducive to the acceptance of the new theory by botanists with an interest in physiology. At almost the same time, von Janczewski unraveled the parasitic mode of *Nostoc* in some liverworts,[58] following the recognition, by Milde, of the algal nature of those green corpuscles which Hedwig and Schmidel had previously interpreted as male reproductive organs. He even claimed to have observed the penetration of *Nostoc* filaments into *Anthoceros* by way of the stomata (sic!) on the lower side of its thallus. Also, J. Reinke had discovered gonidia-like formations in the dicotyledonous *Gunnera*. He described the luxuriant development of Nostoceae of uncertain affinity in slime produced by the buds of several species of *Gunnera* and their subsequent penetration, by way of the resin-canals, into the parenchymatous cells. Which, as Cohn points out, "makes the relationship of the gonidia in *Gunnera* the opposite of that which Schwendener assumes to exist in the lichens."[59]

"Since — as accepted by most researchers in recent times — the gonidia are not seen as integrated tissue-cells of the lichen-thallus but as foreign algae, capable of reproducing independently, this unusual life-community of heterogeneous thallophytes is generally interpreted as if the algae are enmeshed by the hyphae of an ascomycete. The consortium rests on the principle that the alga is provided by the fungus with raw inorganic nutrients, whereas the fungus obtains from the alga the organic substances required by it. In this manner, the fungus vegetates parasitically on the algae. (Reference 57, p. 92)

ANDANTE — 1872

The missing link in the Schwendenerian interpretation was the synthesis of representative lichen species from their partners. It was the purpose of Dr. Max Reess of Halle, in Germany, to supply this link in a study submitted to the Royal Prussian Academy of Science in Berlin, at its session of October 26, 1871, "on the formation ('Entstehung') of the lichen *Collema glaucescens* Hoffm. through the sowing of its spores on *Nostoc lichenoides*."[60]

de Bary's "either-or" theory (either *Ephebe* and the so-called gelatinous lichens are the perfect stage of plants the imperfect stage of which have up to now been classified as algae of the Nostocaceae and Chrooscoccaceae, or these families are typical algae which adopt the form of *Collemas* or *Ephebes*, etc., as a consequence of their penetration by the mycelia of certain parasitic ascomycetes) tended to be considered as irrelevant for other groups of lichens because of the divergent character of gelatinous lichens, e.g., *Ephebe*. As soon as the gonidia of other groups of lichens were recognized as identical with certain algae, the extension of de Bary's alternatives to those other lichens became self-evident. "Whereas some authors preferred the first alternative, Schwendener cleared the way for a more fertile conception, in that he generalized the second alternative and interpreted all lichens as parasitic ascomycetes with algae of diverse families as hosts."[60]

"Anyone without preconceived ideas must admit that Schwendener's facts do not allow any interpretation different from his. Against it, there is no argument whatsoever, except the two-centuries-old habit of botanists to regard the lichens as organisms on their own. However, the main proof for his view, i.e., the production of a lichen by sowing its spores on the gonidia-forming algae and the subsequent culture of the developing parasitic fungus with the alga, Schwendener has neither provided nor sought-for."[60] (Reference 60, p. 524)

After some critical remarks on Schwendener's articles, Reess proceeds to describe his own study on the synthesis of *Collema glaucescens* Hoffm., the *Nostoc lichenoides* Vauch. gonidia of which were readily obtainable in pure culture over a prolonged period. Anticipating the success of his study, he designated the hyphal part of the lichen briefly as the fungus *Collema glaucescens* (emend.), and the gonidial part as *N. lichenoides* Vauch.

Reess attempted to reproduce natural conditions as far as soil and humidity were concerned. Hyphae from germinating spores grew and branched freely, until the reserve material in the spores became exhausted. But hyphal clusters brought in contact with colonies of *Nostoc*, in one way or another, continued to develop. Soon, the hyphae stopped their elongation, but swelled at their tips and often elsewhere where they attached themselves strongly to the *Nostoc*. Attachment was observed even right at the point of germination.

"At the point of attachment, the hypha produces a branch which penetrates into the gelatinous envelope of *Nostoc*; the intruder mostly branches to form a bundle of hyphae (Reference 60, Figures 2 and 3), a phenomenon which corresponds entirely to what is known about the penetration of parasitic fungi into their hosts . . . ." (Reference 60, p. 527) "A single germination hypha sufficed to gradually proliferate throughout an entire colony of *Nostoc* (Reference 60, Figure 5) . . . . The artificially produced lichens could not be brought to fructification." (Reference 60, p. 528)

Reess points out that "in heteromerous lichens, the parasitic lichen-fungus, unlike all other parasitic plants, must absorb the 'raw nutrition' both for itself and for its 'assimilation alga'. This is the role of the 'root-hairs'. In *Collema*, too, the penetration of 'root-hairs' into colonies of *Nostoc* devoid of hyphae did produce a new *Collema*. Indeed, in such cases the penetration of *Nostoc* colonies by hyphae was much faster (in cultures — 8 to 10 days) than when such colonies were infected with hyphae from germinating spores, owing to the more favorable nutritional status of established lichens. Nor is it necessary to mention that in nature the most prolific reproduction of *Collema* is due to 'prolifications' analogous to the soredia of other lichens." Lastly, Reess mentions that older colonies of *Nostoc*, in the open as well as in culture, are occasionally rapidly penetrated by the mycelium of a fungus with spores and hyphae reminiscent of *Mucor*, which he failed to determine with complete confidence.[60]

When, in the summer of 1872, M. Treub — at Voorschoten near Leiden — began his study of the synthesis of lichens he took Reess for his inspiration. He, too, set out to confirm experimentally the theory of a double nature of lichens, with heteromerous lichens as his target group.[61]

Treub attempted to germinate the spores of *Xanthoria parietina*, *Lecanora subfusca*, and *Ramalina calycaris* under diverse conditions so as to be able to observe for himself the development of young gonidia on the prothallus. He collected the spores at night, and in the morning deposited them in a drop of distilled water on a glass slide, on a slide covered with condensation, and on pieces of bark after several minutes' immersion in boiling water. Some substrates were enriched with ash, others were left for a shorter or longer period in the dark. At first, many cultures were ruined by blight; but neither then nor after he found the means to prevent premature rotting did he find any trace of gonidia.

However, Treub's main concern was with cultures in which the spores of the above lichen species were sown together with the algae which produce their gonidia, either simultaneously or 10 to 14 days apart (the algae being "sown" after the spores). From May to October, Treub observed numerous cultures prepared after he had found a way of obtaining pure *Cystococcus humicola*. The source of these, sown together with the spores of *Lecanora subfusca*, was always a different species, i.e., *Ramalina calycaris*.

By January 1873, Treub had discovered that, contrary to his intuition, humid air could replace water without detrimentally affecting germination. The consequent reduction in fungal infection now made it possible to follow the growth of his cultures for over 3 months, as against 3 to 4 weeks at most for his earlier cultures.

Treub's illustrations help to corroborate his conclusion that "upon coming into contact with an alga of the species to which pertain the gonidia of the particular lichen, the germinating thread of a spore strongly attaches itself to the surface of the individual alga attacked, and continues to grow for a shorter or longer stretch."[61] The first consequence of the attachment is the increased branching of the germination thread. Some of the new branches attach themselves to the surface of the same alga, with renewed branching, so that in the end the algal colony is totally enveloped by hyphae. "It immediately becomes self-evident that such a cluster of hyphae cannot possibly be produced from the reserve materials contained in the spore. Also, a comparison of spores germinated under the same conditions and over the same period, without becoming attached to algae, clearly shows the influence of this attachment (on hyphal growth)."[61]

"Though until now I have not been able to construct from its components a complete heteromerous lichen thallus, I yet believe that I am entirely justified in claiming that the cases observed by me and drawn in Figures A5 and 6 can only find their explanation on the assumption of the double nature of lichens. The defenders of the organic individuality of heteromerous lichens are hereby deprived by empirical means of all their arguments, just as Schwendener has earlier refuted their arguments on anatomical grounds."[61]

In 1873, Schwendener published a 33-page survey entitled "The lichens as parasites of algae".[113] The title implies that lichens exist, at least conceptually, apart from algae — as if Schwendener himself lacked complete confidence in this theory. But from the text we learn that Schwendener was from the first skeptical about de Bary's view that species in which the ascomycete hyphae were observed to envelop the algae should be classed as pseudolichens, all other species being lichens. "I was sceptical," Schwendener says, "but I lacked the support ('Anhaltspunkte') to substantiate a more satisfactory solution, until in the winter of 1866/7 I actually observed how hyaline hyphae penetrated into colonies of *Nostoc* and *Gloeocapsa*, and how as a consequence there developed *Collema* and *Omphalaria* thalli. What was more important, filamentous algae (Scytonemeae, Rivulariaceae) were also penetrated and woven-through and turned into gonidia which I confidently recognized as juvenile stages of a lichen related to *Racoblenna*."[113]

Schwendener concludes with a brief summary of the reception of "my" theory by the botanical public. The lichenologists mostly refused to accept it, whereas the microscopists and physiologists generally regarded it as well founded. Already in *Flora* in 1872,[114] Schwendener had argued with the opponents of the theory, but here he presents an argument not previously put forward by him: "The lichen parasitism as it must be deduced from my theory is indeed unique. However, in recent times adaptations which are in a certain sense analogous have been described. [Here follow the references to *Azolla* (Strasburger), *Gunnera* (Reinke), and *Anthoceros* (Janczewski)] . . . . though here the algae are mostly endophytes. In this manner, my theory of the algal nature of lichen-gonidia may contribute to place a series of remarkable adaptations in plants in the proper light."[113]

Bornet (1873) very decisively grinds his axe in favor of the theory first suggested by de Bary, but first given shape by Schwendener. In Bornet's view, Schwendener had failed to dwell as much as he should have on the nature of the connection between hyphae and gonidia, and on the manner in which this connection is established. Bornet points out that the structure of gonidia which pertain to *Trentepohlia* and *Phyllactinium* is too complicated to elucidate claims about their origin from hyphae. "But where (the gonidia) under consideration consist of isolated cells, like those of *Protococcus* in which multiplication does not occur in a readily discernible direction, one is more easily made to accept *a priori* that they are born from hyphae. Even though the diverse phases of this birth have been described with singular precision, I am in a position to confirm that everything that has been said in this matter is quite unfounded."[62]

"Upon following the hypha in its flexous trajectory among the gonidia, one observes that it successively produces numerous branches. Not one of these, of whatever age, shows any trace of that terminal swelling which becomes filled with chlorophyll and turns into gonidia, as (claimed by) the accepted theory. If, moreover, one observes that frequently one and the same gonidium is attached to several branches of hyphae and even to branchlets which originate from quite distinct hyphae; that the adherence (of the gonidium) is not confined to one point nor to one cell of the hypha, but is often elongated, and comprises numerous cells; that the double contour of the gonidia at the point of contact with the hypha is not interrupted; and that in most cases even the slightest pressure suffices to detach the gonidia without leaving any discernible scar (at the point of detachment) one will convince oneself — merely by considering the anatomical relations — that the gonidia are not, nor can they be, the products of the inflation of a branchlet or of the copulation of different branches. May I add — so as not to return to the matter again — that by putting oneself in good optical conditions, and by not admitting . . . any but precise and clear observations, one will never observe, in any lichen, gonidia produced by hyphae, nor any indication that such could be produced."[62]

Bornet's study, in its age and place the most wide-ranging of its sort, provides comparative anatomical and developmental data on the intimate structure of different types of lichens, with emphasis on the relations between gonidia and hyphae. To do it justice, much more space should be devoted to its numerous new insights. The diligence of its author, as well as his adamant no-nonsense stand on the side of Schwendener, have led to his name being sometimes hyphenated onto the latter's name, as co-author of our theory.

Another study which might serve to illustrate the challenge and the risks of open-minded microscopical observations is due to J. Müller — the author of diverse contributions to lichenology:

"Since, in my studies of lichens, I began using in a routine manner Hartnack's Immersion System, I often see very small corpuscles which revolve horizontally, at great speed, around their vertical axes or otherwise execute rapid and prominent movements. In *Verrucaria viridula* I could clearly see their origin: these corpuscles are produced in the asci, prior to the formation of spores, when they fill the ascus with a great mass of short-vermiform thinglings ('Dingerchen'), about 1 to $1^{1}/_{2}$ μm long. Occasionally they exhibit even inside

the ascus (seen once) a distinct rapid motion, resembling the motion of the so-called 'cor-
pusculi mobilibus' in the extremities of *Closterium* . . . . The entire process resembles that
observed when the antherozoids of moss-antheridia are released. The material used had been
in the herbarium for a year, and though I did not see any cilia, I yet assume that with higher
magnifications the 'thinglings' can be studied in greater detail.''[63]

Müller concludes with a recommendation for the study of these objects by those with
time to spare and better objectives at their disposal.

It was in 1874 that Just began to publish his *Botanische Jahresberichte,* a remarkable
pioneering work of heavy tomes dedicated to abstracts and reports on all manner of botanical
subjects. Volume I, for 1873, was published in the following year; subsequent volumes saw
the light with a delay of 2 years. In Volume I, the lichens were presented by H. Lotka, who
was succeeded by A. Minks in Volumes II for 1874 (published in 1876) and III for 1875
(published in 1877). But Minks' ascendancy did not last long, as we shall see below.

A reviewer worth his mettle will not miss the opportunity for some proselytizing and
Minks was no exception. As a strident opponent of Schwendener's theory he had ample
openings to air his views in the 140 pages of his section of the *Botanischer Jahresbericht*
for 1874.[64] His report on A. Grisebach and J. Reinke's volume, *A. S. Oersted's System of
Fungi, Lichens and Algae,* shows his combative mood at its best:

"Of the 188 pages of this work, 88 deal with fungi, 78 with algae and only 19 pages
treat the lichens (as do only 13 of the 93 woodcuts), as if this group, so important in all
respects, were dealt with merely because it was there, to get it over and done with. Why,
on such occasions, are the lichens not omitted, or described by a lichenologist? Or perhaps
one merely wishes, by means of such handbooks, to feed the traditional underestimation of
this group, particularly among students? Works of this kind do, indeed, achieve the conversion of increasing numbers of disciples to the increasingly modern mycology, because
the beginner deduces the true status of a science from the material presented to him. Minds
trained in this manner later provide the best soil for a crop like that fantastic hypothesis on
the nature of the lichens, which excites mycologists and algologists more than lichenologists.
Therefore, one should not criticize those researchers whose entire thinking is involved with
this class of plants, i.e., the lichenologists, when such works, which treat their subject from
an angle which strains to dictate to them their judgment on the lichens, force upon them a
certain smile. In the present case, a mere report without critical overtones becomes particularly difficult, or rather impossible. Of course, there is in the book nothing new, besides
the Schwendenerian hypothesis. . . . .''[64]

"The lichens constitute a group of thallophytes between the fungi and the algae. As against
the fungi, whose thallus consists exclusively of hyphae, and the algae, which are made up
of cells, the lichens are characterized by a thallus made of hyphae and chlorophyll-bearing
gonidia . . . . In the description of the structure of lichens, almost half is taken up by the
Schwendenerian hypothesis. The relationship which is assumed to prevail between hypha
and gonidium is referred to in a manner surely more commendable to the disciples, as
'consortium'. The remaining pages crowd the barest data on anatomy, physiology, and
morphology. Even Schwendener's research (on the structure of the lichen thallus) is inad-
equately reported.''[64]

In 1874, Weddell, in an introductory discourse to what is essentially a local lichen flora,
pronounces at length on the paramount theoretical battle of the day. While he follows the
system of Nylander, ''which has been adopted by the majority of lichenographs of France
and England'', he remarks that the ''celebrated cryptogamist has too consistently ignored
the works of his contemporaries . . . . Nor do I hesitate to express the hope that in the new
works which one expects from him he will fill a truly regrettable lacuna.''[65]

Weddell confesses an error, in a previous publication: he had wrongly included *Sirosiphon*
among the lichens, like Nylander and others before him, though ''of rights it belongs to the

algae. Even so, *Sirosiphon* plays no mean role in the lichens, as constituent element of some of these singular plants. They are its gonidia, just as *Cystococcus* and *Trentepohlia* are the gonidia of many lichens of higher rank."[66]

"Granted this, two cases may present themselves: either the alga is sufficiently invaded by the hypha (or filamentous tissue of the lichen) to lose some of its normal physiognomy or else the lichenous element merely forms under the sheath of *Sirosiphon* (for example) a layer ('lacis') so transparent that it will be invisible to the unwary observer. Then, the apothecium produced by it may very easily be taken for a fructification of the gonidium proper, and this incomprehensible association may come to be regarded as an intermediate entity between these two classes."[66] From this, there is but a step to the more or less complete "assimilation" of "lichenized" *Sirosiphon* with *Sirosiphon* not attacked by any hypha. The latter will be considered as imperfect lichens ("liberated gonidia") and both will bear the name "*Pseudo-algae*". Weddell goes on:

"I may be permitted to point out here that there is no reason whatsoever to fear lest the adoption of Schwendener's hypothesis might jeopardize the autonomy of lichens: such as it was, so shall it remain. The presence of gonidia will continue to be a most remarkable character of these complex beings. Their rank in the scale of vegetables will not 'en realité' be lowered except in the eyes of those who wanted to elevate them too highly. In a word, such as we know them today, the lichens cannot, by themselves, constitute a class [of their own — J. L.]. Nor can they be made into a family of fungi. Their true place is in the class of Ascophytes or Cryptogames thecasporées, of which — I believe — they must constitute one of its great divisions."[67]

In a footnote,[65] Weddell quotes Nylander (in *Flora*, 1874, p. 58) who insisted on the characters which separate the hyphae of Lichens and Fungi: "Hyphae Fungorum certe nihil structura commune habent cum hyphis Lichenum", a view shared by other authors, too. Indeed, it "appears more rational to consider the alga as an annex of the lichen which performs in relation to it the role of 'instrument' (so as not to say 'organ') of nutrition, than as a feeding plant, in the proper sense of the word. For this, I require no other proof than the luxuriant health of the gonidia in contact with the hyphae."[68]

In *Ephebe*, and in other lichens with filamentous gonidia "the gonidia flourish with the lichen with which they are united, and the two vegetables share a common fate to the end — the true ideal of a consortium. [One would be tempted, adds another footnote, to say 'cooperative society', if that term had not too much fallen into the public domain.]"[68]

In the review of this work, written for the *Botanischer Jahresbericht*, Minks concludes by quoting Weddell himself:

"We must not demand the impossible. Like other groups of organic beings the lichens are totally typical in their center, and often become blurred [literally 'indeterminate'] at their borders. Yet their autonomy is beyond question."[81]

It may be in part because of the wide diffusion of Weddell's views that Nylander decided to swallow the bait, as it were, and go into the fray, a decision which was of more than merely dramatic interest. A possible pretext, if it was needed, may have been provided by two studies of Weddell, of predominantly taxonomical interest. Referring to Weddell's support of Schwendener's theory, Nylander retorts:

"Those who have promoted Schwendener's theory have brought forward nothing confirmatory of it, but only anatomical reasons long ago well known."[69]

Oblivious of the view that the identical derogatory remark was made about the chemical theories of Lavoisier, whose thought dwarfed his experiments, Nylander goes on:

"The absurdity of such an hypothesis is evident from the very consideration that it cannot be the case that an organ (*gonidia*) should at the same time be a parasite of the body of which it exercises vital functions; for with equal propriety it might be contended that the liver or the spleen constitute parasites of the *Mammiferae*. A parasite's existence is autonomous. The parasite lives upon a foreign body, of which nature prohibits its being at the same time an organ. This is an elementary axiom of Physiology. But direct observation teaches that the green matter originally arises within the primary chlorophyll- or phyco-chrome-bearing cellule, and consequently is not introduced from any external quarter, nor arises in any way from any parasitism of any kind. I have already enunciated this in a note (*Flora*, 51, 353, 1868.), and in vain can it be denied. The cellule at first is observed to be empty and then, by the aid of secretion, green matter is gradually produced in its cavity and assumes a definite form. It can, therefore, be very easily and evidently demonstrated that the origin of green matter in Lichens is entirely the same as in other plants. What need is there then of any fuller refutation of the but too notorious thesis of Schwendener?"[69]

Weddell published his study in France in 1873. Nylander ground his axe in Germany, in *Flora*, No. 4, for 1874. The Rev. J. M. Crombie, F.L.S., etc., was asked by Nylander to translate these remarks in *Flora* into English for publication in *Grevillea, A Monthly Record of Cryptogamic Botany and its Literature*, then in its second year (more exactly, in its 22nd month).

Crombie, the loyalist author of the relevant entry in the *Encyclopedia Britannica*, was one of the most esteemed lichenologists of Great Britain. He introduces his translation of Nylander with the following sermon which speaks for itself:

"Since what I have elsewhere written is true, that 'truth itself consists of the continual demolition of error', I have always believed that it tends very much to the interests of science to oppose fanciful or erroneous opinions; nay, it may legitimately be considered one's duty to point out and refute such opinions, for the progress of science depends not a little upon their subversion. Nothing, indeed, as is evident, is more readily received and propagated than erroneous opinions, and, consequently, there is so much greater difficulty in opposing their propagation, though we may not on that account depart from our duty."[70]

It so happened, that a friend of H. A. Weddell, living in Paris, forwarded to him Nylander's angry review which included several misrepresentations. Weddell answered Nylander in some detail in a personal letter, hoping that this would settle the matter. In spite of this letter Nylander proceeded to have the translation of his *Flora* review published in *Grevillea*, Weddell felt it his duty to retaliate:

"Few among those who have passed some part of their life in botanical pursuits, and more specifically in the study of Lichens, can boast of having committed even but a small amount of errors. I for one have to confess many such . . . ."[71]

Weddell goes on: "the truth is, that in the face of many facts lately adduced, specially by Mr. Bornet, it is difficult to deny that many Lichens during the first stage of their life are connected parasitically with some of the inferior Algae. At a later period, however, when the Alga, assuming the form of Gonidia, becomes included within the tissue of the Lichen, the connection, if still kept up, can hardly continue to be considered as parasitical."[71]

Weddell adds, in a footnote, that Nylander seems to misunderstand the Schwendenerists as thinking that the alga lives parasitically on the substance of the lichen, whereas they believe the contrary. He becomes more courageous as he goes on: "as regards Dr. N's special objections to an Algo-Lichen hypothesis, I do not see that they are in any way conclusive, not one of them coming really to the point. They prove undoubtedly the importance of gonidia as instruments (I dare not say organs) of lichen nutrition, but do not, I find, in any manner demonstrate that *true* Gonidia are not Algae . . . ."[71]

The Weddell-Nylander episode really deserves to be quoted in its entirety, as a sample of the acrimonious style which purported to have as its sole objective the salvation of lichenologists from error — the more so as Weddell is not at all an all-out apostle of Schwendenerism. Still, in view of Nylander's importance to lichenology, here are some excerpts from his reply to Weddell to be saved from oblivion: First, a footnote: "they who would admit that science consists of a congeries of *opinions* (to which are referable the notions called in German *Anschauung* and in French *intuition*) are ignorant of what science really is, or confound it with what is fabulous."[72] Nylander controverts Weddell on numerous points. To him, gonidia are exactly that, organs of lichens. Then, he proceeds to tear apart Weddell's just quoted view on the changing relation between gonidia and lichens, with time. "It pleases the author that it should be so. Thus theories are invented (truly it is an easy matter) which may be called 'fancies'. To consult nature and to rely upon observations is superfluous . . . ."[72]

What most angered Nylander was Weddell's statement that of his (Nylander's) objections, not one really comes to the point: "the author evidently strives to attain the Friesian and Lindsayan glory. But I have demonstrated (by *observations*, not opinions) that the filamentose elements of lichens are not the same as the hyphae of fungi. I have demonstrated that the gonidia arise from the first within closed cellules of the thallus, and do not in any way come from any external quater. This is sufficient, and more than sufficient to refute Schwendenerism. Dr. W., not without cause, thinking vague and indeterminate arguments the best, speaks also of 'many facts lately adduced', but what these are he does not mention, nor, in truth, are there any such existing. The first germs of lichens and fungi are seen affixed in corpuscles of some kind or other. Were it not so . . . the gonidia would constitute parasites of the thalli. Another opinion, which would make the thallus and apothecia parasites of the gonidia, is even more absurd."[73]

ADAGIO — 1874

A mere 6 years after the first presentation of Schwendener's theory, it had become the focal event in the life of many devotees of the *scientia amabilis*. In line with this, in May 1874, at the International Botanical Congress in Firenze a special session was held on behalf of the lichens. After the Congress elected — by acclamation — as its President the foremost member from England, Giuseppe Dalton Hooker, President of the Royal Society of London, 22 subjects were proposed for discussion: subject 12 was "On the nature and function of the gonidia in lichens". In due course this was introduced by Weddell and discussed by Famintzin, Caruel, Gibelli, Suringar, Schimper, Kanitz, and Delpino.

This is how Weddell opened the discussion:

"Gentlemen, among the questions included in the program of this Congress, there is probably none that deserves better to fix the attention of the Congress than that which concerns the 'appreciation' of the nature and the role of gonidia in lichens. The interest which attaches to this subject is so vivid that I do not hesitate to bring it before you, convinced as I am that the little I shall have to say will suffice to provoke more interesting and more original contributions from among those members of this assembly who have had the opportunity to direct their work in that direction. On my part, I have no new observations to submit to you. In the middle of often lively, occasionally irritating discussions — which this question has raised — I have tried to keep *au courant* and to form an opinion, taking into consideration all material at my disposal. It is this opinion that I wish to express here and I shall sum it up very briefly by saying that: If to this day the algolichenic theory has not had all the success that it is destined to achieve, that is above all because in speaking of it its authors as well as its propagators have employed an unfortunate word, a word which, to my mind, is not an exact expression of what happens in nature. That word, as you will surely have guessed, is the word 'parasitism'. Let it then disappear from the definition or else have it replaced by another, more true or less precise word, and without a doubt the theory, thus amended, will at last be accepted by all those who do not have a preconceived opinion ('parti pris')."[74]

Weddell presents for consideration two capital questions:

1.   Are the gonidia algae?
2.   Should the relation between the hyphae and the lichen, or ascophyte, be considered as parasitism?

His answer to the first question is in the affirmative, though "at first I, like so many others, experienced a repugnance to admit such a singular fact. Diverse studies on the subject have, step by step, brought conviction to my spirit. That conviction has become total when, quite recently, I saw the series of preparations which M. le docteur Bornet has shown to me. I was able to compare those preparations with the drawings thereof, which he had published, and I can testify to their perfect conformity. I may add that those preparations have also been examined by Messrs. Bentham and Hooker, who have authorized me to make a similar declaration on their behalf."[74]

As for the rest, Weddell agrees that the connections observed between hyphae and gonidia do, undoubtedly, have the appearance of parasitic connections. Later, though, he concludes that what is involved is a bizarre alliance between the lichen and the alga, but surely not parasitism in the ordinary sense of the word.

Famintzin, whose study — in collaboration with Boranetzky — has been referred to above, did not consider the problem of the nature of lichens as resolved. Though he had failed to observe the genesis of gonidia from hyphae, he wished to draw the attention of the congress to a report to the Wiesbaden Congress of Naturalists[115] where A. B. Frank claimed to have seen in *Variolaria communis* how gonidia, colorless at first, and later increasingly green, developed from certain cells of the hyphae.

Caruel then expressed his satisfaction with this report of Frank's work, which corroborated his own studies of *Collema* made 10 years previously, but which were totally ignored except by A. de Bary in his book (1860). He closes his brief report on his findings:

"I must add that these observations were made independently of any opinion on Schwendener's theory, of which I was totally ignorant at the time."[75]

The other contributions to the lichen session came from Suringar, Schimper, Kanitz, Delpino, and Gibbelli. Suringar reported on findings by Treub in Leyden, who followed the first stages in the development of a thallus. Schimper had seen in a certain lichen gonidia of two different colors, and believed that these may belong to two species of algae which otherwise live freely outside the lichen.[76]

Kanitz comes back to Weddell's reference to a "bizarre alliance". He recommends Grisebach's term *consortium*, already applied by Reinke to the relation between *Gunnera* and its alga. Delpino entered at some length into the matter of algae living in and on a variety of higher and lower plants. To him, these show the very same phenomenon as that seen in lichens, i.e., a consortium. He wished to correct the view of Janczewski and of others who consider the *Nostoc* as parasitic, and the enmeshing tissue as host.[77]

The longest contribution, after Weddell, was from Gibbelli who reported on his observations of the gonidia of lichens. Proceeding from his earlier observation (published in the *Nuovo Giornale Botanica Italiano*, Vol. 2, 1870), Gibbelli corroborates some of the findings of Schwendener and others. However, he considers *Collema* to be a dimorphous organism with a complete phase with hyphae and fructifications, and an incomplete phase devoid of hyphae, known under the name *Nostoc*. Gibbelli is satisfied with the proof that one of the partners of the lichen, i.e., the gonidia, is able to reproduce independently. What remains to be shown is that the fungus, too, is able to multiply and reproduce from gonidia alone.[78]

From Firenze, the center of the storm moved briefly in the direction of Paris. There, on November 23, 1874, the assembled members of the Académie des Sciences, with Claude Bernard in the chair, were given a brief report on the situation — again by H.-A. Weddell.

"Botanists who have for long agreed on the nature and origin of gonidia are today of divided opinion . . . . It is above all due to the good observations of Dr. Bornet, checking the works of Schwendener and others, that the (view of the latter) has gained favor in France, but as yet not all spirits are rallied around it."[79]

Weddell deplores the terms *Pseudo-Algae* by which certain authors have united not only algae which resemble lichens but also diverse genera of algae from which those lichens derive their gonidia. "For those authors [not mentioned by name — J. L.] *Stigonema* and *Scytonema* are henceforth no longer algae but incomplete lichens, *liberated gonidia*. Likewise, *Cistococcus, Chroococcus*, or *Trentepohlia* . . . are the *free gonidia* of most foliaceous and crustaceous algae."[80]

"The detailed studies by Janczewski, Thuret and Bornet on the reproduction by spores of *Nostoc* absolutely invalidate the hypothesis which regards these plants as but an initial or deformed state of *Collema*."[80]

Weddell also reports on Gibbelli's contribution to the Firenze Congress, not yet published at the time, concerning the formation of zoospores in the gonidia inside the thallus of *Lecanora subfusca*, an observation "the exactitude of which was confirmed by several witnesses".[80] He fails to mention that Gibbelli is at most a half-hearted supporter of Schwendener. On the other hand, he brings to the attention of the academy that, after reading Bornet's memoir, Gibbelli proceeded to culture gonidia of *Opegrapha varia*. "After some time he had the pleasure of seeing these gonidia growing into magnificent *Chroolepus (Trentepohlia)* which successively yielded both zoosporangia and zoospores."[80]

The year 1874 drew to its close. Yet the opposition — including all the gray and graying eminences of lichenology — remained steadfast in their anti-Schwendenerian stance. The band of the converted, on the other hand, found themselves going in different directions.

Minks, who — as mentioned above — had used his reviews of lichenological papers for Just's *Botanischer Jahresbericht* of 1874 to denounce anyone who supported Schwendener, shows clear signs of tension in his introduction to the lichenological section of the 1875 *Jahresbericht*:

"In general, the report for 1875 has been prepared on the same lines as that for the preceding year. In particular, the reviewer did not feel called upon to deviate from the accepted division. By its very nature any division will involve drawbacks . . . . The reviewer *saw* and *sees* in this division such advantages as have made him retain the previously used division."[81]

In the same volume Minks introduces the review of E. Tuckerman's *Genera Lichenum* with the following words:

"Tuckerman is one of the not numerous lichenologists who — when those 'wonderful' discoveries were made one after another in recent decades — examined them calmly and seriously, but as seriously and decidedly discarded them as sterile and even as entirely fortuitous."[81] Minks was not impressed by microscopical observations.

In the following volume of *Botanischer Jahresbericht*, of 1876, Just reshuffled the board of reviewers: the loyalist Minks was replaced by E. Stahl, an adherent of the new theory. "(B.) Lichens" were reviewed — as hitherto — between "(A.) Algae" and "(C.) Fungi". However, the traditional listing fails to hide the underlying unrest:

"Preliminary Remark: Actually, it had been the intention of the reviewer as well as the publisher of the *Jahresbericht*, from now on, to present the reviews of lichen literature — which until now had a separate status — as an appendix to the reviews of ascomycetes, in accordance with the present state of science. It is for external reasons alone that this change has not been adopted in the current volume. However, in the following volume the planned attachment of the lichens to the ascomycetes will take place . . . ."[82]

"On the basis of all correctly observed facts there can no longer be any doubt that there are no genetic relations between gonidia and hyphae and that the lichens or better the lichen-fungi are nothing else but ascomycetes distinguished from their relatives merely by a peculiar parasitism and the concomitant morphological and physiological features."[82]

In that same year Minks published a lengthy paper on the structure and life of the lichens, a fact which may have been aided by his relief as reviewer. With the aid of two double-plates and 126 pages of text, Minks proves what he had set out to prove, i.e., that the product of every germinating spore is a layer of hyaline hyphae from which, as secondary products, colored secondary hyphae and gonidia-producing organs are developed. The latter — for which Minks coins the terms "Gonandien" and "Gonocystien" — have in common that their final product, the gonidium, is produced inside them, by free cell division.[83]

While the Schwendenerians tended to look down somewhat disdainfully on the loyalists, the latter became the victims of increasing viciousness.

In 1876, the author of the *Nature* review of the English version of E. Haeckel's *Natürlicke Schöpfungsgeschichte* (Reimer, Berlin, 1968.) mentioned that Haeckel had outlined "the truth about the nature of lichens". This casual allusion elicited two letters to the editor of *Nature* by W. Lauder Lindsay. Lindsay, who had long contemplated the publication of *Outlines of Lichenology*, refers to his close interest in all relevant publications, "duly recorded, with abstracts and relative criticism, in my lichenological memorandum book."[84]

"My opinion of the speculations of Schwendener and his followers has all along been, and still is, that so far from 'clearing up' the 'true nature of Lichens' they introduce elements of very decided confusion; and that they are to be regarded merely as illustrations of German transcendentalism, comparable to the fanciful notions of his countryman Bayrhoffer, in 1851, concerning Lichen-Reproduction. The dogmatic assertions of anonymous critics concerning the 'clearing up' of 'the true nature of Lichens' notwithstanding I hold what I have always held, that the Lichens as an Order are quite as natural, important, and distinct as any other Order of the Cryptogamia. And in so saying I do not forget the fact that they overlap the *Algae* and the *Fungi* . . . . I long since proposed the establishment of *intermediate and provisional groups* of *Algo-lichenes* and *Fungo-lichenes*. Such groups would have the advantage of attracting attention to those *passage-forms*, which appear to me to be of the highest interest to the philosophical botanist."[84]

Lindsay cites Berkeley, Thwaites, and Cooke, from England, "than whom we have certainly no botanists better qualified or entitled to form or to offer opinions on such a subject,"[84] as well as Bentham, P.L.S. (i.e., President of the Linnaean Society of London). "But other German botanists not inferior in status or experience to Prof. Schwendener regard the Hypothesis that Lichens are the product of a union of Parasitic Ascomycetes and Algae as far from being proved. . . . If, by artificial cultivation, such a Union could be made to produce a Lichen, the Theory might be held as proven. But this has not yet been effected, and I venture to think and say it never will be . . . ."[84]

After describing several difficulties with which the Schwendenerians have to contend, Lindsay sums up:

"In short, the mantle of Bayrhoffer appears to have fallen on Schwendener; and his Parasitic Theory is merely the most recent instance of German transcendentalism."[84]

By 1877 the dispute seemed to have calmed down with the parties dug in. Now, then, was the time for all good men to get down to the job on hand, the unbiased, thorough study of representatives of the major groups of lichens. One of these, A. B. Frank, soon published an 80-page study of the "Biological relations of the thallus of several crustaceous lichens", proceeding from the unfortunate circumstance that,

"As is well known, we do not possess for any lichen a developmental history without gaps, from the germination of the spore to the complete, typical form of the fertile thallus . . . "[85] even though it is easy to come up with the thought that "in very many instances nature is perhaps not aware of the process of rejuvenation by spores."[85]

In his summary, Frank is burdened by the overlap of the semantic fields of the terms relevant to phenomena lumped together under the heading "parasitism".

"We must bring all those cases in which two organisms live on or in each other ('Auf-

oder Ineinanderwohnen') under one wide concept which does not take into account the role of these organisms, and concerns itself merely with their living together ('Zusammenleben'). For this the term 'Symbiotism' may be recommended."[86] "The several phases of this relationship would then be designated as *pseudoparasitism*, i.e., the accidental coming together, not essential to either of the organisms. Only the next higher, and at least for one partner essential, relationship can fairly be designated as *parasitism*; here the foreign partner merely takes, and gives nothing in return."[86]

"Whereas in all of these instances one organism is foreign to the other, the last and highest stage of symbiotism may be suitably designated as *homobium*, preferable to consortium which only poorly suggests the living together and moreover presupposes a similarity in the roles played by the partners."[87]

"The future will have to decide how closely the two-part organisms are linked together by the bonds of homobium in each case; in other words, whether and how it is possible that one or both parts may live free from each other. As yet it has not been possible to grow the hyphal bodies of typical, gonidia-bearing lichens but as is well known, the gonidia of some of the lichens discussed have been made to grow independently, and have shown reactions analogous to those of related algae. However, there is no evidence against the possibility that in other lichens the homobium has reached such a degree that the gonidia, too, by virtue of nutritional relationships which have become constant and hereditary, are no longer able to vegetate by themselves."[87]

Two more relevant papers were published — with illustrations — in the *Nuovo Giornale Botanico* for 1875,[88.90] the first presumably and the latter expressly a consequence of the 1874 International Botanical Congress in Firenze.

A. Borzi's thoughts and observations on the role of the gonidia in lichens had already been published in *Scienza contemporanea*[116] and are merely supplemented by one plate. Borzi says that he began his study in 1872 with the prime motive to extend Reess' observations to *Collema* and other lichens, and to study the effects of the parasitism of the hyphae on its gonidia. His work is mainly concerned with the cultivation of species of *Parmelia*. On his own admission, it led him away from Krempelhuber and other loyalists, towards Famintzin and Boranetzky.[88]

Borzi's illustrations show details of the germination of *Parmelia stellaris* and others, in conjunction with *Protococcus* and diverse gonidia. His conclusions: the gonidia have no genetic relation to the hyphae. The relations between hyphae and gonidia are throughout identical with those between the histological elements of any fungus and the substrate which nourishes it. In consequence: lichens are ascomycete fungi parasitic on algae, represented by the gonidia.[89]

G. Arcangeli is impressed by the co-existence, in some families of phanerogams, of parasitic and autotrophous genera. "Why should it therefore not be possible to admit, likewise, that lichens should be considered as belonging to the class of Fungi, as fungi with chlorophyll?"[90]

Proceeding to Bornet's observations and those of others (Treub, Borzi), Arcangeli does not wish to double the facts noted by these distinguished observers, "but perhaps these may be susceptible to an interpretation different from that given".[90] For his part, he notes the tendency of hyphae to adhere to and tenaciously envelop whatever they approach — such as the bark of trees, rocks, and walls. Initially, they derive nutrition from any near object, before developing on their own and producing gonidia.

"Is it not possible that the Nostocacee, Protococcacee and Rivulariee, now considered as independent organisms and included in the class algae, could be particular forms of lichens? Can it be said that their vegetative cycle is so well known as to exclude such a possibility?"[91]

Arcangeli is, of course, aware of the proposal to replace "parasitismo" by "consorzio". "Nevertheless, it seems to me that this kind of consortium has little probability in its favor.

In fact, it is hard to understand how two organisms can be associated in such a manner as to exchange their nutrient materials without being greatly disturbed in the exercise of their functions . . . .''[92]

Arcangeli's own observations bear upon the anatomy of *Alectoria*, *Cladonia* and others, with emphasis on their gonidia. Summing up, he has a ''spontaneous'' thought, i.e., that the multiple relations of the lichens which deserve consideration should also include the mosses: ''Just as these begin to develop from a nematoid proembryo, on which grow the plantlets which are organs of higher structure, so in lichens a similar event occurs. Hence, it is very likely that the filaments which grow from the spore are, up to a certain stage of their development, capable of producing organs of a structure more elevated than that of the pseudo-parenchyma. This would make it possible to understand how gonidia are not directly formed from filaments which germinate from a spore, but from more highly developed organs produced from them.''[93]

Two years later, in 1877, Arcangeli returns to the subject, seemingly even more relaxed, but still unconvinced and unconverted. Again, his observations of organogeny in *Dermatocarpon* yielded results contrary to those of Gibelli, a supporter of Schwendener. He had attempted to germinate spores on glass or mica under conditions close to those prevailing in nature, so as to be able to observe their germination. Numerous species were involved. Among others, he germinated spores of *Verrucaria macrostoma* in contact with cells of *Gloeocapsa muralis*, gathered from humid walls.[94] After germinating, the germinal filments did not show any preference in growing towards the algae. Nor did spores of *Collema microphyllum*, upon germination, show any response to a colony of *Nostoc palustre*: they germinated, grew actively at first, then more slowly, and died.[95]

Arcangeli's conclusion is that ''the facts currently at our disposal are more against than in favor of the algolichenic theory.''[96]

In Austria, in 1878, H. Zukal writes about the ''Flechtenfrage'' (lichen-question). In his brief introduction to the algo-lichenic theory he prefers the term ''convivium''. He seems to be more impressed by Stahl's contributions in favor of Schwendener than by those of Minks against.

''However, admitting that the presence of hymenial gonidia favors the development of the spores of *Endocarpon pusillum* as well as those of *Thelidium minutulum*, and even if it is essential to it, this does not imply that that is how all lichens behave, nor that in all lichens the development of the thallus is linked to the pre-existence of gonidia and that attempts to grow lichens from their spores alone will never be successful. That would be a very bad *fallacium fictae universalitas*.''[97]

Zukal ''does not write in order to add to the balance a weight against Schwendener's theory, but merely to accelerate the flow of the discussion on the nature of lichens . . . a problem which appears ripe for its definitive solution.''[98] Still, he provides some observations of his own, obtained by using ''with great success system No. 9 with ocular III of Rudolf Wasserlain'',[98] which he heartily recommends. On the Collemaceae, he says: ''Whoever will study, without preconception, a section of *Physma compacta* at 600 × magnification will note that the hyphae — considered as cylinders — never touch gonidium-like cells by their sides but always with their tips . . . . He who will look sharply will always remark beneath the gonidial cell a small, unique, hyaline support-cell (''Stützzelle''), intercalated between the hypha and the gonidial cell. These large greenish-blue cells at the tip of the hyphae are the mother-cells of the so-called *Nostoc*-strings, i.e., true gonocysts in the sense of A. Minks.''[98]

Two reports published at about the same time in the New World by E. Tuckerman — the foremost lichenologist of the U.S. — also came out whole-heartedly in favor of the Old School. For such was the spirit of loyalty to an old teacher, even if only during a brief stay abroad, that Tuckerman, too, never accepted Schwendenerism.[99]

This would be the proper place to intercalate a review of all available dictionaries and encyclopedias in order to reveal the diffusion of the new truth throughout the Old World and elsewhere. For the good old days of innocence were over. Back in 1771 a *Lichen* was a "liverwort, in botany a genus of cryptogamic algae. The receptacle is roundish, plain, and shining; the farina is dispersed upon the leaves. There are 85 species, all natives of Britain."[100] By now, things had become more complex. To illustrate the fascination of dictionaries and encyclopedias one can do no better than to open the 1875—1889 edition of the *Encyclopedia Britannica*. Here, the lichens — treated at great length — are "briefly defined as cellular perennial plants furnished with a vegetative system containing gonidia and with a reproductive system consisting of female thecasporous fruits and male spermogonous organs. They constitute a distinct class of cellular cryptogams, intermediate between algae and fungi, to both of which in some respects they present certain affinities. By the earlier authors they were regarded as being Aerophyceae or terrestrial algae, while of recent years they have been viewed by some writers as being Ascomycetous fungi. From both of these, however, they are sufficiently distinguished . . . . Their relations to these neighboring classes, and their true systematic place, will be best elucidated on considering their structure and its bearings upon some recent speculations."[6]

The hand is that of the Reverend J. M. Crombie.

A "Publisher's Note" to the 1899 reprint of this edition declares that "The present edition, which comprises all the changes and additions . . . has also been revised with the same care, and will be found to be abreast of the times."[101] In the article on lichens, there is no trace of this.

Crombie freely uses the wide expanses of the *Britannica* to win an easy, Quixotean victory over "Schwendenerism" . . . according to which "a lichen is not an individual plant, but rather a community . . . a master-fungus and colonies of algal slaves, which it has sought out, caught hold of, and retains in perpetual captivity in order to provide it with nourishment. To such a singular theory, which from its plausibility has met with considerable support in certain quarters, various *a priori* objections of great validity may be taken. . . . (1) The parasitism described is of a kind unknown in the vegetable kingdom, inasmuch as the host (the *Algae*), instead of suffering an injury, only flourishes the more vigorously. Moreover, the algal slaves being entirely enclosed in the master-fungus, can evidently supply no nourishment to it whatever. . . . (2) As is well known, lichens shun such habitats as are most frequented by algae and fungi . . . . Either of these arguments is sufficient to throw more than doubt upon Schwendenerism."[6] In fairness to the publisher of the *Britannica* it should be added that in the entry for *Symbiosis* in the same edition, the lichens are presented as the outstanding examples of this relationship!

In M. H. Baillon's *Dictionnaire de Botanique*, the lichens are treated in the third volume in an entry by Ch. Manoury. Nylander's classification, in which the more highly developed lichens grade towards fungi on the one hand and towards algae on the other, is particularly praised. Not a word about Schwendener![102]

Summing up, the eventful and strife-ridden years in which the dual nature of lichens was first surmised, then asserted and eventually proven, reflect — as do other case histories in science — that the assembling of observations and experiments was inadequate to provide enlightenment without the elaboration of the relevant conceptual armory. Whether by cohesion, adhesion, analogy, modeling, or muddling, the latter was obviously and inexorably linked to ideas prevalent in other domains of duality. Matter and spirit, body and soul were old hunting-grounds for such ideas. Through the temporal and spiritual, the executive and legislative, the stock and the scion, one glides readily to parasitism, commensalism, symbiosis, and Schwendenerism.

The relative emphasis on the algal and fungal partners may also reflect social attitudes. In an article written in the 18th century, which I am unable to retrace, mistletoes and other parasitic plants are described as the aristocracy of the plant kingdom, for unlike ordinary plants they only imbibe distilled nutrients fit for the chosen. In the first half of the 19th century, when discussions about the relations between insects and plants reached their peak, the attitudes of the contestants presage the arguments brought forth in the present story. J. G. Kurr, the German author of an 1832 prize-winning study of the function of nectar writes: "... in nature everything has an internal purpose and nothing exists for the sake of another. If this were not so, one might as well claim that pollen is produced to supply the requirements of insects and the seeds of grasses to feed the birds."[103] This was echoed in France by de Candolle: "In organized beings, the various functions have until now been found to be related to their own nature, not to the nature of beings which are foreign to them."[104]

The most enthusiastic opponent of Sprengel's theory on the relations between insects and plants was J. L. G. Meinecke: "... there is nothing that the physiologist needs to avoid more carefully than the theory of usefulness to animals and man ...."[105] "If consistently interpreted, Sprengel's view on the relation between insects and plants makes the butterfly appear but like a flower let loose, and the flower but as a chained butterfly. With only one difference, that this poetic truth is not founded on any prosaic necessity of nature."[105]

POSTLUDE — 1913

One cannot conclude this eventful history without a brief report on a 20th century revival of the old battle cries, spearheaded by F. Elfving, in 1913 and again in 1931.

Elfving admits that in nature green cells may be enmeshed by hyphae, but he seems haunted by gonidia sprouting hyphae and turning green.

"The Schwendenerian theory was, in its time, the necessary, logical expression of known facts. New facts have now been added, and the theory becomes invalid. This is how science develops, and it is illuminating to see that the most erudite researchers employing elaborate methods and sharp logic do not always pronounce the truth but others who intuitively see the truth, long before the proof can be supplied. Krempelhuber, Nylander and those who shared their views were much closer to the truth than their opponents .... Nylander's concise statements, not accompanied by drawings, will not permit to decide in which cases he merely intuited the truth and in which he actually observed it."[106]

Elfving returned to the battle in 1931, in a long and lavishly illustrated study.

"When Schwendener first stated the theory according to which lichens are dual beings consisting of a fungus and an alga, many, probably most (of those who became acquainted with it) considered this a pure absurdity. Over the decades, botanists and biologists in general have become totally intimate with it. Newly discovered facts prove that the old view, of lichens as uniform organisms, must be accepted. This will not happen without considerable resistance. In particular, it will be refuted because it leads to the bizarre consequence that parts of an organism, when removed from it, are able to develop independently to produce new organisms known from nature. Surely this will appear to most biologists as a total absurdity. However, the logic of the facts is irrepressible and indeed, when the time will be ripe for it, the Schwendenerian theory will take its place in history, similar to that occupied by the phlogiston theory, which leading chemists supported in their time."[107]

"I hope that younger botanists will be moved by the present study, and by the false theory exposed in it, to study lichens in nature, not in books. It is merely the salvation of the honor of an old conception. They will find that work in this field is as rewarding as the study of chromosomes, of pH, and other modern questions of popular appeal. Of the older gentlemen, nothing can be expected."[107]

"This is not the place to prophesy on the impact of the new absurdity on general biology."[107] Elfving did not pass unnoticed: F. Tobler devoted a lengthy paper to put him to rest. Tobler finds Elfving's illustrations not clear enough. Neither were Elfving's microscopical preparations, which he had at his disposal, satisfactory. All considered, "I do not see how these studies have shaken Schwendener's theory."[108]

One hundred and twenty years have passed since Schwendener's insight first disturbed the calm waters of lichenology. But many of the intimacies exchanged between algae and fungi under the probing eyes, minds, and instruments of lichenologists still await their deciphering.

ACKNOWLEDGMENT

The author wishes to express his infinite gratitude to Mrs. B. Yedlitzky for her patience and typing. The staff of the library of the Linnaean Society of London has also been extremely helpful throughout this endeavor.

REFERENCES

1. **von Krempelhuber, A.,** *Geschichte und Litteratur der Lichenologie bis 1865,* Vol. 1 and 2, Wolf und Sohn, München, 1867 and 1869.
2. *e.g.,* **Schneider, A.,** *A Guide to the Study of Lichens,* Knight and Millet, Boston, 1904; **des Abbayes, H.,** *Traité de Lichenologie,* Lechevalier, Paris, 1951.
3. **Fries, E. M.,** Fries knew what he was talking about when he said: "Lichenologia est quasi civitas institutionibus obruta cujus reformatio difficilior quam novae constructio", *Lichenographia Europaea Reformata,* Lund, Sweden, 1831, vii.
4. *e.g.,* **Koerber, G. W.,** *Zur Abwehr der Schwendener-Bornetschen Flechtentheorie,* Kern (Müeller), Breslau, 1874, 30.
5. *e.g.,* **Archer, W.,** A further resumé of the gonidia question, *Q. J. Microsc. Soc. N.S.,* 14, 115—139, 1874.
6. *e.g.,* **Crombie, J. M.,** Lichens, *Encyclopedia Britannica,* Vol. 9, 1875—1889 ed., Black, Edinburg, p. 552 ff, reprinted 1899; also, **Lindsay, W. L.,** The true nature of lichens, *Nature (London),* 13, 247, 1876.
7. **von Flotow, J.,** Aus einem Brief an Fries in Upsala, Sept. 1848, *Bot. Z.,* 8, 18 (columns 361—369; 377—382 [column 362]), 1850.
8. Ibid., column 363. The term "gonidium" — intended to reflect the presumed reproductive role of these cells — is due to **Wallroth, G.,** *Naturgeschichte der Flechten,* 2 vols., Williams, Frankfurt, 1825 and 1827.
9. **Thwaites, G. H. K.,** On the gonidia of lichens, *Ann. Mag. Nat. Hist. Ser. 2,* 3, 219—222, (p. 220), 1 pl., 1849.
10. Ibid., p. 222.
11. **Tulasne, L.-R.,** Note sur l'appareil reproducteur dans les lichens et les Champignons, *Ann. Sci. Nat. Bot. Ser. 3,* 15, I-pp. 427—430; II-pp. 470—475 (p. 427), 1851.
12. Ibid., p. 429.
13. Ibid., p. 430.
14. Ibid., p. 475.
15. **Tulasne, L.-R.,** Nouvelles recherches sur l'appareil reproducteur des Champignons, *Ann. Sci. Nat. Bot. Ser. 3,* 20, 129—182, (p. 129), 2 pls., 1853.
16. Ibid., p. 131.
17. **Itzigsohn, H.,** Zur Frage über die Abgrenzung der niederen Gewächsklassen, *Bot. Z.,* 12, columns 77—80 (column 78), 1854.
18. Ibid., columns 78 and 79.
19. Ibid., column 80.
20. **Itzigsohn, H.,** Wie verhält sich *Collema* zu *Nostoc* und den Nostochineen?, *Bot. Z.,* 12, columns 521—527 (column 523), 1854.
21. Ibid., column 525.
22. Ibid., column 526.
23. **de Bary, A.,** Die neuesten Arbeiten über Entstehung und Vegetation der niederen Pilze; insbesondere Pasteur's Untersuchungen, *Flora,* 45, 355—365 (p. 356), 1862.

24. **Abbé Hué,** Abbé Hué sur W. Nylander, *Bull. Soc. Bot. Fr.,* 46, 153—160, 1899.

25. **Nylander, W.,** Om den systematiska skillnaden emellan svampar och lafvar, 10.1.1855, *Ofver. Kongl. Vet. Akad. Forh.,* 12, 7—11, 1855.

26. Ibid., p. 10.

27. **Speerschneider, J.,** Mikroskopisch-anatomische Untersuchungen über *Ramalina calicaris* Fr., *Bot. Z.,* 13, columns 345—354; 361—369; 377—385 (column 369), 1855.

28. **Bornet, E.,** Description de trois lichens nouvaux, *Mem. Soc. Imp. Sci. Nat. Cherbourg,* 4, 225—234, 4 pls., 1856; and *Ann. Sci. Nat. Bot. Ser. 3,* 18, 231, 1856.

29. **Bayrhoffer, J. D. W.,** Entwicklungs- und Befruchtungsweise von *Thrombium Nostoc* Wallr., *Bot. Z.,* 15, columns 137—145 (column 137), 1 pl., 1857.

30. Ibid., column 144.

31. **Thuret, G.,** Observations sur la reproduction de quelques Nostocinees, *Mem. Soc. Imp. Sci. Nat. Cherbourg,* 5, 19—32 (p. 20), 3 pls., 1857.

32. **Schwendener, S.,** Untersuchungen über den Flechtenthallus. *Die Strauchartigen Flechten,* Vol. 1, Engelmann, Leipzig, 1860, 78, 7 pls. Reprinted from *Beitr. Wiss. Bot.,* 2, 109—186, 7 pls.; 3, 127—198; 4, 161—202; 1860. Summary published in *Bot. Z.,* 18, column 258—259, 1860; and in On the nature of the gonidia of lichens, *Q. J. Microsc. Soc.,* 13, 235—251, 1863.

33. **de Bary, A.,** *Morphologie und Physiologie der Pilze, Flechten und Myxomyceten, Handbuch der Physiologischen Botanik,* Vol. II, Hofmeister, W., Ed., Engelmann, Leipzig, 1860, 316 (p. vii), 1 pl. de Bary published what may be regarded as a second edition in 1884, available in English as *Comparative Morphology and Biology of Fungi, Mycetoxoa and Bacteria,* Clarendon, Oxford, 1887, 525 pp. Further information on de Bary in **Large, E. C.,** *The Advance of the Fungi,* Jonathan Cape, London, 1940.

34. Ibid., p. 241.

35. Ibid., p. 242.

36. Ibid., p. 258.

37. Ibid., p. 259.

38. Ibid., p. 260.

39. Ibid., p. 290.

40. Ibid., p. 291.

41. **Cohn, F.,** Über Oscillarien und Florideen, *Bot. Z.,* 25, columns 38—39, 1867.

42. **Askenasy, E.,** Beiträge zur Kenntnis des Chlorophylls und einiger dasselbe begleitenden Farbstoffe, *Bot. Z.,* 25, columns 233—239 (column 236), 1 pl., 1867.

43. Ibid., column 235.

44. **dBy (de Bary, A.),** (Review of) **Boranetzky, J.,** Beiträge zur Kenntnis des selbstständigen Lebens der Flechtengonidien, *Bot. Z.,* 26, columns 196—198, March 20, 1868; (Report on) **Boranetzky, J.,** Contribution to the knowledge of the independent life of the gonidia of lichens, *Mel. Biol. Bull. Phys. Math. Acad. Imp. Soc. St. Petersburg,* 7, 473—493, 8 pls., 1868; and *Jahrb. Wiss. Bot.,* 7, 1—16, 1869.

45. Ibid., column 197.

46. Ibid., column 198.

47. **Schwendener, S.,** Über die Beziehungen zwischen Algen und Flechtengonidien, *Bot. Z.,* 26, columns 289—292 (column 289), January 5, 1868. Also published in Verh. Schw. Naturf. Ges. Rheinfelden, 88—90, 289—290, 1867. See also **Schwendener, S.,** *Die Algentypen der Flechtengonidien,* Programm für die Rectoratsfeier der Universität Basel, Schulze, Basel, 1869, 12 pp., 3 pls.

48. Ibid., column 290.

49. Ibid., column 292.

50. **Famintzin, A. and Boranetzky, J.,** Zur Entwickelungsgeschichte der Gonidien und Zoosporenbildung der Flechten, *Bot. Z.,* 26, columns 169—178 (column 170), 1 pl., March 13, 1868; also in *Mem. Acad. Imp. Soc. St. Petersburg Ser. 7,* 9, 13, 1868; and in *Ann. Sci. Nat. Bot. Ser. 5,* 5, 1867.

51. Ibid., column 171.

52. Ibid., column 173.

53. Ibid., column 176.

54. **Anon.,** *C. R. Acad. Sci.,* 68, 1379, 1869.

55. Ibid., p. 1417.

56. Ibid., p. 1398. In the *Dictionary of Scientific Biography,* Nylander is wrongly awarded the Prix Des Mazieres(!). Currently, the prize is worth 4,000 NF, and is available every third year to French and foreign scientists. Applications are to be addressed to the Permanent Secretaries of the Academie des Sciences.

57. **Cohn, F.,** Über parasitische Algen, *Beitr. Biol. Pfl.,* 2, 86—108, 1 pl., 1872.

58. **von Janczewski, E.,** Zur parasitischen Lebensweise des *Nostoc lichenoides, Bot. Z.,* February 2, 1872; article enlarged, with some additional illustrations in Le parasitisme du *Nostoc lichenoides, Ann. Sci. Nat. Bot. Ser. 5,* 16, 306—316, 1 pl., 1872.

59. **Cohn, F.,** op. cit., p. 93, 1872.

60. **Reess, M.,** Über die Entstehung der Flechte *Collema glaucescens* Hoffm. durch Aussaat der Sporen derselben auf *Nostoc lichenoides, Monatsber. K. Preuss. Akad. Wiss. Berlin,* 16, 523—532, 1 pl., 1872.

61. **Treub, M.,** Lichenencultur, *Bot. Z.,* 31, columns 721—727 (column 726), 1 pl., November 14, 1873. Abstracted in Archer (see Reference 5); many of the statements there attributed to Treub, such as his view about Müller, do not occur in the original article!

62. **Bornet, E.,** Recherches sur les gonidies des lichens, *Ann. Sci. Nat. Bot. Ser. 5,* 17, 45—110, (p. 70), 11 pls., 1873; abstracted in **Phillips, W.,** *Grevillea,* 2(13), 36—40, 1874.

63. **Müller, J.,** Lichenologische Beiträge. I, *Flora,* 57, 185—192, 1874.

64. **Minks, A.,** in *Botanischer Jahresbericht 1874,* 1876, 61.

65. **Weddell, H. A.,** Florule lichenique des lave d'Agdé, *Bull. Soc. Bot. Fr.,* 21, 330—351, 1874; extracted as Quelques mots sur la théorie algolichenique, *C. R. Acad. Sci.,* 79, 1172—1175, 1874; and in C. R. Hebd. Scéances Acad. Sci., 79, p. 333, November 23, 1874.

66. Ibid., p. 334.

67. Ibid., p. 335.

68. Ibid., p. 336.

69. **Nylander, W.,** Nylander on the algo-lichen hypothesis, etc., *Grevillea,* 2(22), 145—152, 1874; translation, by J. M. Crombie, of Nylander's article in *Flora,* 57, 394—399, 1874 (in Latin).

70. Ibid., p. 145.

71. **Weddell, H. A.,** Remarks on a paper published (Jan. 1874) by Dr. W. Nylander in the *Flora* and lately reissued in *Grevillea, Grevillea,* 2(22), 182—185, (p. 182), 1874.

72. **Nylander, W.,** On Dr. H. A. Weddell's Remarks in *Grevillea, Grevillea,* 3(25), 17—22, (p. 17), 1874.

73. Ibid., p. 18, footnote.

74. *Atti del Congresso Internationale Botanico, Firenze, Mayo 1874,* Firenze 1876, 65.

75. Ibid., p. 68.

76. Ibid., p. 70.

77. Ibid., p. 72.

78. Ibid., p. 69.

79. **Weddell, H. A.,** Quelques mots sur la théorie algo-lichenique, *C. R. Acad. Sci.,* 79, 1172—1175 (p. 1172), 1874. In 1860, Weddell had solicited support for the "honour of being included in the list of presentations for the nomination to the place which has fallen vacant in the Académie des Sciences on the so unforeseen death of Mr. Payer." Just in case, he included in his letter his "acte de naturalisation". Not successful at the time, he may have hoped to attract the attention of the Académie on this occasion, 14 years later (Museum d'Histoire Naturelle, MS, II, No. 440, 1860).

80. Ibid., p. 1175.

81. **Minks, A.,** in *Botanischer Jahresbericht 1875,* 1877, 53.

82. **Stahl, E.,** in *Botanischer Jahresbericht 1876,* 1878, 70.

83. **Minks, A.,** Beiträge zur Kenntniss des Baues und Lebens der Flechten. II, *Verh. Zool. Bot. Ges. Wien,* 42, 377—508, 1876.

84. **Lindsay, W. L.,** The true nature of lichens: letter to the editor, *Nature (London),* 13, 247—248 (p. 248), January 27, 1876.

85. **Frank, A. B.,** Über die biologischen Verhältnisse des Thallus einiger Krustenflechten, *Beitr. Biol. Pfl.,* 2(2), 123—201, (p. 123), 1 pl., 1877.

86. Ibid., p. 195.

87. Ibid., p. 197.

88. **Borzi, A.,** Intorno agli officii dei gonidii de 'Licheni', *Nuovo G. Bot. Ital.,* 7, 193—204 (p. 195), 1 pl., 1875.

89. Ibid., p. 203.

90. **Arcangeli, G.,** Sulla questione dei gonidi, *Nuovo G. Bot. Ital.,* 7, 271—292 (p. 273), 3 pls., 1875.

91. Ibid., p. 274.

92. Ibid., p. 289.

93. Ibid., p. 290.

94. **Arcangeli, G.,** Di nuovo sulla questione dei gonidi, *Nuovo G. Bot. Ital.,* 9, 231, 1877.

95. Ibid., p. 232.

96. Ibid., p. 234.

97. **Zukal, H.,** Zur Flechtenfrage, *Oesterr. Bot. Z.,* 28, 226—229 (p. 227), 1878.

98. Ibid., p. 229.

99. **Tuckerman, E.,** The question of the gonidia of lichens, *Am. J. Sci. Arts,* 17, 254—256, 1879; **Tuckerman, E.,** Lichens or fungi?, *Bull. Torr. Bot. Cl.,* 7, 66, 1881.

100. *Encyclopedia Britannica,* Vol. 2, Bell & MacFarquhar, Edinburgh, 1771, 973.

101. Ibid., Vol. 9, 1875—1889 ed., 552; reprinted in 1890, 1892, 1895, 1896, and 1899.

102. **Baillon, M. H.,** *Dictionnaire de Botanique,* Hachette, Paris, 1891.

103. **Kurr, J. G.,** *Untersuchungen über die Bedeutung der Nektarien,* Henne, Stuttgart, 1833, 138.

104. **de Candolle, A.,** *Physiologie Végétale,* Vol. 2, Béchet, Paris, 1832, 588.

105. **Meinecke, J. L. G.,** Über die Bedeutung des Nectarium, in *Neue Schriften der Naturforschenden Gesellschaft zu Halle,* 1(1), 21—32 (p. 29), 1809. For that fascinating story in full, see **Lorch, J.,** The discovery of nectar and nectaries and its relation to views on flowers and insects, *Isis,* 69(249), 514—533, 1978.

106. **Elfving, F.,** Untersuchungen über die Flechtengonidien, *Acta Soc. Sci. Fenn. Ser. B,* 44, 1—71 (p. 58), 8 pls. 1913.

107. **Elfving, F.,** Weitere Untersuchungen über die Flechtengonidien, *Acta Soc. Sci. Fenn. Ser. B,* 1, 1—26, (p. 26), 13 pls., 1931.

108. **Tobler, F.,** Elfving's Untersuchungen über Flechtengonidien, *Hedwigia,* 72, 68—74, (p. 74), 1932.

109. **Itzigsohn, H.,** Kultur der Glaucoconidien von *Peltigera canina, Bot. Z.,* 27, 185—196, 1 pl., 1869.

110. **Fuisting, W.,** Beitrage zur Entwickelungsgeschichte der Lichenes, *Bot. Z.,* 26, 641—7, 657—65, 673—84, 1868.

111. **Stitzenberger, E.,** De lecanora subfusca ejusque formis commentatio, *Bot. Z.,* 26, 1868.

112. **Schwendener, S.,** On the relations between algae and the lichen-gonidia, *Bot. Z.,* May 1, 1868.

113. **Schwendener, S.,** *Die Flechten als Parasiten der Algen,* Schweighauser, Basel, 1873. (Reprinted from *Verh. Naturf. Ges. Basel,* 1873.

114. **Schwendener, S.,** *Erörterungen zur Gonidienfrage,* Neubauer, Regensburg, 1872. (Reprinted from *Flora,* 55, 161—166, 177—183, 193—202, 225—234, 1872.)

115. **Frank, A. B.,** Report to the Wiesbaden Congress of Naturalists, *Bot. Z.,* 32, p. 242, 1874.

116. **Borzi, A.,** *Scienza Contemporanea,* Messina, 2, 5, 1874.

Section II: Lichen Components

Chapter II.A

THE FUNGAL PARTNER

David L. Hawksworth

I. INTRODUCTION

Lichenization is one of the most successful methods by which fungi fulfill their requirement for carbohydrates. Of the 64,200 species of fungi accepted in the latest edition of the *Dictionary of the Fungi*[1] some 13,500 are lichenized, that is about 20% of all fungi. However, the proportion of the species within the different divisions of the kingdom Fungi which are lichenized varies considerably (Table 1), and lichen-like associations can also arise with some filamentous bacteria (Actinomycetes).

This chapter reviews the fungal partners, or "mycobionts",[2] to be found in lichen associations. In most cases, the mycobiont is the exhabitant[3] while the photosynthetic partner, or "photobiont"[4] (see Preface), is the inhabitant.[3] It is the exhabitant which normally retains the capacity to reproduce sexually in mutualistic symbioses,[3] and, as it is these structures which form the basis of the classification of fungi, most of the mycobionts in lichen associations can be referred to particular classes or orders of Fungi. It should be noted that the names given to lichens are treated as referring to the mycobiont; photobionts can therefore have separate scientific names (see Chapter II.B).

II. ASCOMYCOTINA (ASCOMYCETES)

Lichenization reaches its zenith in the Ascomycotina, with almost half the group forming lichens (Table 1). The development of a satisfactory classification of this, the largest division of the Fungi, has proved particularly difficult due to a variety of factors,[5,6] but significant progress has been made in the last few years in a downward construction of a system. Only ranks of order and below are currently being accepted, the latest outline recognizing 46 orders and 239 families.[7] Of these orders, 16 include lichen-forming species: Arthoniales, Caliciales, Dothideales, Graphidales (includes Gomphillales), Gyalectales, Helotiales, Lecanorales, Opegraphales, Ostropales, Patellariales, Peltigerales, Pertusariales, Pezizales, Pyrenulales, Teloschistales, and Verrucariales. Synopses of the diagnostic characters of these orders and lists of the lichenized families included within them are to be found in Chapter X of this handbook. However, it is of interest that just five orders, the Graphidales, Gyalectales, Peltigerales, Pertusariales, and Teloschistales, comprise only lichenized species.

It is not only orders which include both lichen-forming fungi and species with different biologies, but also families and genera. Examples of the 20 or so genera known to include both lichen-forming *and* plant parasitic, saprobic, or lichenicolous species are *Arthonia*, *Arthopyrenia*, *Arthothelium*, *Chaetothecopsis*, *Mycomicrothelia*, *Opegrapha*, *Orbilia*, *Sphinctrina*, *Thelocarpon*, and *Xylographa*.[8,9] In some genera, different species can be found which have up to five distinct nutritional methods of which one option is the formation of a lichen association.[9]

This situation appears to rise in two ways. First, anciently lichenized groups which had lost the ability to produce lichenized thalli adopted lichenicolous or saprobic biologies, as in *Arthonia* (Arthoniales), *Agyrium*, *Dactylospora* and *Sarea* (Lecanorales), and *Opegrapha* (Opegraphales). Second, rather more recent, predominantly plant parasitic or saprobic groups are found which have species apparently experimenting or starting to adopt the lichen method of nutrition, as is the case in *Baeomyces* and species of *Cudoniella*, *Orbilia*, and *Pezizella*

Table 1
NUMBERS OF LICHEN-FORMING SPECIES IN THE VARIOUS SUBDIVISIONS OF THE FUNGI

Subdivision	Number of species	Number of lichenized species	Lichenized species (%)
Ascomycotina	28,650	13,250	46.25
Basidiomycotina	16,000	50	0.31
Deuteromycotina	17,000	200	1.18
Mastigomycotina	1,170	1	0.09
Myxomycota	625	?2	?0.32
Zygomycotina	765	0	0
Total (rounded)	64,200	13,500	21

(all Helotiales), *Arthopyrenia* (Dothideales), *Schaereria* (? Pezizales), and *Thelopsis* (Ostropales).

Evidence is accumulating that many of the exclusively lichenized orders and families are of great antiquity, especially from a comparison of present distribution patterns with what is now known of continental movement over the last 400 million years[10-12] and due to the exceptionally large numbers of lichenicolous fungi only found on species in certain orders.[13] This has previously been hinted at on the basis of anatomical studies of ascomata and, especially, ascus types,[5,13-16] and it now seems that it is among the lichenized orders that the ascomycetes with the most ancestral traits are to be expected. Certain families of the Lecanorales (especially Rhizocarpaceae), Peltigerales, and Pertusariales seem to be of particular antiquity. When the orders of the Ascomycotina are analyzed biologically, it is seen that a high proportion of the nonlichenized orders occur on substratas which have only arisen in the last 200 million years[17] — after certain extant lichen species appear to have fixed not only species limits but also the chemotypes, as in *Dimeleana*.[10]

In comparison to the Ascomycotina, lichenization is relatively infrequent in the other groups of Fungi treated below. In these, it occurs sporadically in a manner suggestive of the nutritional experimentation to be seen in predominantly plant parasitic or saprobic orders of the Ascomycotina.

III. BASIDIOMYCOTINA (BASIDIOMYCETES)

In the Basidiomycotina, lichenization appears to be confined to the Hymenomycetes.[18,19] The most familiar genus, including basidiolichens in the Northern Hemisphere, is *Omphalina* (Agaricales, Tricholomataceae) which also comprises saprobic, lichenicolous, and bryophilous species.[9] The basidiomata in *Omphalina* are typically mushroom-like, and the photobiont cells are included in either granules near the base of the stipe (''*Botrydina*'') or in distinct squamules (''*Coriscium*'').[19,20] The possibility that mycobionts in such cases also obtain nutrients from decaying plant materials in the ground cannot, however, be discounted in the absence of physiological studies on these associations.

Some bracket-forming fungi (Aphyllophorales) often have algal cells growing on their surfaces (e.g., *Pseudotrametes gibbosa*), but such occurrences may be largely fortuitous. The mainly tropical genus *Dictyonema* (Thelephoraceae),[21] in contrast, is exclusively lichen-forming with the photobiont cells localized in a distinct layer within the basidiomata in the larger species. Some species of *Athelia* (Corticiaceae) have also been regarded as lichenized.[22] A further genus of the Aphyllophorales which includes a few lichenized species is *Multi-*

clavula (Clavariaceae); in this case, the basidiomata arise from a slimy photobiont film on rotting wood. In the exclusively lichenized *Lepidostroma calocerum* (Clavariaceae) scale-like thalli are produced.[19]

IV. DEUTEROMYCOTINA (DEUTEROMYCETES)

The Deuteromycotina, formerly referred to as the Fungi Imperfecti, mainly comprise fungi producing asexual mitospores termed conidia. Many are anamorphs (i.e., conidial states) of known species of Ascomycotina, but in others no teleomorph (i.e., sexual state) is known or may have been lost in the course of evolution.

Numerous lichen-forming ascomycetes produce conidia which have a sexual role as spermatia,[23,24] and, in a few, the conidia act as diaspores, dispersing the mycobiont. In some species, teleomorphs are produced rather infrequently and the anamorphs have a key role in dispersal (e.g., *Cliostomum griffithii*, *Lecanactis abietina*, *Opegrapha vermicellifera*).[25] It is not surprising, therefore, to find that some lichens only produce conidia and never a sexual state (teleomorph).

Forty-seven exclusively conidial genera of lichen-forming fungi have been described,[26,27] but many more remain to be formally recognized. Some of these are referrable to the Coelomycetes, with pycnidial (flask-shaped) or acervular (cupuliform) conidiomata, while others belong to the Hyphomycetes, with conidia produced on hyphae or aggregations of hyphae.

In addition to fungi forming conidia, completely sterile mycelial fungi are often also referred to the Agonomycetes in the Deuteromycotina; mostly genera of filamentous lichens such as *Byssophytum*, *Cystocoleus*, and *Racodium*.

The conidial lichen-forming fungi are considered in further detail in Chapter V.B.

V. MASTIGOMYCOTINA (MASTIGOMYCETES)

The single lichenized species referred to this group is the enigmatic terricolous *Geosiphon pyriforme* in which cyanobacteria are consistently contained in special elongated vesicles.[28,29] While the precise nature of this association awaits critical study, the association appears to be mutualistic and to merit the recognition as a "lichen" since a definite specialized structure is produced.

VI. MYXOMYCOTA

Plasmodia of *Fuligo cinerea* and *Physarum didermoides* have been found to engulf and form associations with three *Chlorella* species in culture. The symbiotic plasmodia grow more luxuriantly,[30] but such "myxolichens" are perhaps best viewed as facultatively lichenized. I can find no evidence of constant mutualistic relationships between myxomycetes and algae occurring in nature.

VII. OTHER LICHEN-LIKE ASSOCIATIONS

In addition to the wide variety of fungus-alga associations,[8] certain Actinomycetes can form lichen-like associations ("actinolichens"). For example, *Chlorella xanthella* and *Streptomyces* form composite-structured thalli on potato slices and in culture,[31] and gelatinous thalli formed between *Chlamydomonas reisiglii* and an unidentified actinomycete have been found in nature.[32] Such associations of *Chlamydomonas* species are probably not as rare as generally assumed, and, in these cases, the phycobiont cells loose their ability to form or maintain their flagella,[33] a type of suppression characteristic of inhabitants in mutualistic symbioses.[3]

REFERENCES

1. **Hawksworth, D. L., Sutton, B. C., and Ainsworth, G. C.,** *Ainsworth and Bisby's Dictionary of the Fungi*, 7th ed., Commonwealth Mycological Institute, Kew, England, 1983.
2. **Scott, G. D.,** Lichen terminology, *Nature (London)*, 179, 486, 1957.
3. **Law, R. and Lewis, D. H.,** Biotic environments and the maintenance of sex: some evidence from mutualistic symbioses, *Biol. J. Linn. Soc.*, 20, 249, 1983.
4. **Ahmadjian, V.,** Holobionts have more parts, *Int. Lich. Newsl.*, 15(2), 19, 1982.
5. **Eriksson, O.,** The families of bitunicate ascomycetes, *Opera Bot.*, 60, 1, 1981.
6. **Hawksworth, D. L.,** Problems and prospects in the systematics of the Ascomycotina, *Proc. Indian Acad. Sci. (Plant Sci.)*, 94, 319, 1985.
7. **Eriksson, O. and Hawksworth, D. L.,** Outline of the ascomycetes — 1986, *Syst. Ascomycetum*, 4, 1, 1986.
8. **Hawksworth, D. L.,** The taxonomy of the lichen-forming fungi: reflections on some fundamental problems, in *Essays in Plant Taxonomy*, Street, H. E., Ed., Academic Press, New York, 1978, 211.
9. **Hawksworth, D. L. and Hill, D. J.,** *The Lichen-Forming Fungi*, Methuen, New York, 1984.
10. **Sheard, J. W.,** Palaeogeography, chemistry and taxonomy of the lichenized ascomycetes *Dimeleana* and *Thamnolia*, *Bryologist*, 80, 100, 1977.
11. **Sipman, H. J. M.,** A monograph of the lichen family Megalosporaceae, *Bibl. Lich. Vaduz.*, 18, 1, 1983.
12. **Tehler, A.,** The genera *Dirina* and *Roccellina* (Roccellaceae), *Opera Bot.*, 70, 1, 1983.
13. **Hawksworth, D. L.,** Co-evolution and the detection of ancestry in lichens, *J. Hattori Bot. Lab.*, 52, 323, 1982.
14. **Chadefaud, M., Letrouit-Galinou, M. A., and Janex-Favre, M.-C.,** Sur l'origine phylogenetique et l'evolution des Ascomycètes des lichens, *Mem. Soc. Bot. Fr.*, 1968 *(Coll. Lich.)*, 79, 1970.
15. **Letrouit-Galinou, M. A.,** Les asques des lichens et la type archaeasce, *Bryologist*, 76, 30, 1973.
16. **Honegger, R.,** Ascus structure and function, ascospore delimitation and phycobiont cell wall types associated with the Lecanorales (lichenized ascomycetes), *J. Hattori Bot. Lab.*, 52, 417, 1982.
17. **Dick, M. W. and Hawksworth, D. L.,** A synopsis of the biology of the Ascomycotina, *Bot. J. Linn. Soc.*, 91, 175, 1985.
18. **Oberwinkler, F.,** Die Gattungen der Basidiolichenen, *Votr. Bot. Ges., Dtsch. Bot. Ges.*, n.f., 4, 139, 1970.
19. **Oberwinkler, F.,** Fungus-alga interactions in basidiolichens, *Beih. Nova Hedwigia*, 79, 739, 1984.
20. **Watling, R.,** Lichenicolous agarics, *Bull. Br. Lich. Soc.*, 49, 28, 1981.
21. **Parmasto, E.,** The genus *Dictyonema* ('Thelephorolichens'), *Nova Hedwigia*, 24, 99, 1978.
22. **Jülich, W.,** A new lichenized *Athelia* from Florida, *Persoonia*, 10, 149, 1978.
23. **Jahns, H. M.,** The trichogynes of *Pilophorus strumaticus*, *Bryologist*, 76, 414, 1973.
24. **Honneger, R.,** Scanning electron microscopy of the contact site of conidia and trichogynes in *Cladonia furcata*, *Lichenologist*, 16, 11, 1984.
25. **Vobis, G.,** Studies on the germination of lichen conidia, *Lichenologist*, 9, 131, 1977.
26. **Vobis, G. and Hawksworth, D. L.,** Conidial lichen-forming fungi, in *Biology of Conidial Fungi*, Vol. 1, Cole, G. T. and Kendrick, B., Eds., Academic Press, New York, 1981, 245.
27. **Hawksworth, D. L. and Poelt, J.,** Five additional genera of conidial lichen-forming fungi from Europe, *Plant Syst. Evol.*, 154, 195—211, 1986.
28. **Mollenhauer, D.,** Botanische Notizen Nr. 1: Beobachtungen an der Blaualge *Geosiphon pyriforme*, *Nat. Mus.*, 100, 213, 1970.
29. **Jahns, H. M.,** *Farne-Moose-Flechten, Mittel-, Nord- und Westeuropas*, BLV Verlagsgesellschaft, München, 1980.
30. **Lazo, W. R.,** Growth of green algae with *Myxomycete plasmodia*, *Am. Midl. Nat.*, 65, 381, 1961.
31. **Lazo, W. R. and Klein, R. M.,** Some physical factors involved in actinolichen formation, *Mycologia*, 57, 804, 1965.
32. **Hawksworth, D. L.,** unpublished data, 1979: photobiont determined by Dr. H. Ettl (Czechoslovakia).
33. **Ettl, H.,** personal communication, 1979.

Chapter II.B

THE ALGAL PARTNER

Elisabeth Tschermak-Woess

I. INTRODUCTION

Since the recognition of the dualistic nature of lichens more than a hundred years ago, only a limited number of lichenologists have become interested in the algal partner of this intimate association between fungi and algae. Until about the middle of this century the phycobiont was called gonidium. This term had its origin in an early and eventually disproved view that the gonidia were apical segments of the fungal hyphae (see Chapter I). In 1957, Scott[1] introduced the term phycobiont for the algal partner and mycobiont for the fungal partner of lichens and other consortia.

The taxonomic study of lichen phycobionts lags far behind that of the mycobionts. This resulted in part from the fact that lichenologists interested in the taxonomy of lichens mostly study herbarized material which in many cases did not permit the indentification of the algal partner even with respect to genus. In addition, phycobionts in the lichenized state often are modified to such an extent that algae must be cultivated free from the fungal partner in order to determine their systematic position. In recent years, standardization of cultural techniques has been suggested as a basis for the description of new phycobiont species.[2] However, it is probably more important to use each of the following methods: study *in situ*, study in free-living state in nature, and exact investigation of characters and developmental cycle free-living in culture. Use of all three methods will have to take into account the observation that certain characters may be lost in one or the other state.

II. SYSTEMATIC POSITION

In some lichens, the systematic position of the phycobiont *in situ* can be determined to the genus level; species determinations can rarely be made. In other cases only the algal class can be determined. Thus, culturing of phycobionts is, in most cases, a prerequisite to proper identification.[3-6] Details for the isolation, purification, and growth of phycobionts may be found in Chapter XIII.

The majority of algae occurring as phycobionts belong to the Chlorophyceae (estimated to comprise about 90% of all lichens), and to a lesser extent to the Cyanophyceae (about 10%);[3,7] in two or three cases, phycobionts belong to the Xanthophyceae and, in one not yet sufficiently documented case, to the Phaeophyceae. Among the Chlorophyceae, the Chlorococcales, which include trebouxioid algae, are most represented (see Table 1 and comment 55). Ahmadjian[2] estimates that 50 to 70% of the lichens contain trebouxioid phycobionts. The second most common order among the Chlorophyceae is the Trentepohliales, with *Trentepohlia* as a very widespread genus. In the Cyanophyceae *Nostoc* (*Nostocales*) is the most common phycobiont.

It is not possible to state the exact number of algal genera recorded from lichens. It fluctuates around 40, depending on different views on the justification of certain algal genera, on the nature of certain consortia, and on the accuracy of the determination (Tables 1 and 2). The number of algal species presently identified is roughly 100; however, this is probably a small percentage of the number of algal species actually present in lichens.

An intriguing and repeatedly discussed question concerns the specificity of fungal partners toward certain algae. Are mycobionts always associated with the same algal species or may

Table 1
ALGAL GENERA AND SPECIES IDENTIFIED AS LICHEN PHYCOBIONTS[a]

Phycobiont	Lichen[b]	Type of observation[c]	Ref.
	Cyanophyceae [1][d]		
Chroococcales			
Chroococcaceae	T. *Lichinella* (?) *applanata* Henssen[e]	s	40, p. 74
	T. *L. robusta* Henssen	s	40, p. 73f
	T. *L. stipatula* Nyl.	s, c	40, 154
	T. *Phyllisciella aotearoa* Henssen et Bartlett	s	155
	T. *P. polymorpha* Henssen	s	155
	T. *Pterygiopsis atra* Vain.	s	40
	T. *P.* (?) *foliacea* Henssen	s	40
"Chroococcalean"	T. *P. submersa* Büdel, Henssen et Wessels	s	269
Anacystis sp.	T. *Peltula* sp.	s	157
	C. *Stereocaulon austroindicum* Lamb [2]	s, *	105
A. *montana* (Lightf.) Dr. et Daily [3]	T. *Peltula polyspora* (Tuck.) Wet.	s, c	32
	T. *P. richardsii* (Herre) Wet.	s, c	32
	T. *P.* spp. (various)	s, c	32, compare 33
Aphanocapsa sp. [4]	C. *Stereocaulon haumanianum* Ouvign.	s	158, p. 102ff
Chroococcus sp. (?)	T. *Jenmania osorioi* Henssen	s	159
Chroococcus (?) or *Gloeocapsa* (?) *sanguinea* (Ag.) Ktz. emend. Jaag [5]	T. *Phylliscum endocarpoides* Nyl. [6]	s, f, ff	34, p. 395f
Gloeocapsa sp.	C. *Amygdalaria consentiens* (Nyl.) auct. (= *Huilia consentiens* (Nyl.) Hertel = *Lecidea consentiens* Nyl.)	s, *	160, 161
	C. *A. pelobotryon* (Wahlenb. in Ach.) Norm. (= *Lecidea pelobotryon* (Wahlenb.) Leight.)	s, *	37
	T. *Edwardella mirabilis* Henssen	s, f	270
	T. *Anema nummularium* (Mont.) Nyl. (= *Omphalaria notarisii* Mass.)	s, *	56
	T. *Gonohymenia* spp.	s	35
	T. *G. sinaica* Galun	s	162
	T. *Heppia* sp.	s	37
	T. *Jenmania goebelii* Wächter	s	37
	T. *Phylliscidium monophyllum* (Kremplh.) Forss.	s	163, p. 38
	T. *Pyrenopsis conferta* Nyl. (= *Synalissa conferta* Bornet)	s	56, p. 93
	T. *P. phaeococca* Tuck.	s	164
	T. *Rechingeria cribellifera* Serv. (= *Gonohymenia mesopotamica* Stnr.)	s, 2	37, 162
	C. *Stereocaulon alpinum* Laur. [7]	s, f, *	56
	C. *S. ramulosum* (Sw.) Räusch f. *elegans* Th. Fr.	s, *	18, p. 543
	C. *S. ramulosum* (Sw.) Rausch f. *ramulosum* [8]	s, *	18, p. 543

Table 1 (continued)
ALGAL GENERA AND SPECIES IDENTIFIED AS LICHEN PHYCOBIONTS[a]

Phycobiont	Lichen[b]	Type of observation[c]	Ref.
"*Gloeocapsa*-artig"	T. *Heppia turgida* (Ach.) Nyl. (= *Gloeoheppia turgida* (Ach.) Gyel.)	s	165
G. kuetzingiana Näg. emend. Jaag [9]	T. *Anema moedlingense* Zahlbr.	s, ff	166
	T. *Peccania coralloides* Mass.	s, f, *	147
	T. *Psorotichia murorum* Mass.	s, f, several	34, p. 399
	T. *P. pelodes* Krb.	s, f	34, p.401
	T. *P. recondita* Arn.	s, f	34, p. 401
	T. *P. rehmica* Arn. [10]	s, f	34, p. 401
G. muralis Kütz (?)	T. *Psorotichia schaereri* (Mass.) Arn.	s, f, ff	147
	T. *Thyrea pulvinata* (Schaer.) Mass.	s, f	147
G. sanguinea (Ag.) Kütz, emend. Jaag [5, 11]	C. *Huilia panaeola* (Ach.) Hertel (= *Lecidea panaeola* (Ach.) Ach. [12]	s, *	167, 168
	T. *Pyrenopsis foederata* Nyl. [13]	s, f	34, p. 392
	T. *P. fuliginoides* Rehm [13]	s, f	34, p. 391f
	T. *P. fuliginosa* Nyl. [possibly misspelled and corrected to *P. fuliginea* Nyl. (= *Psorotichia lignyota* Forss.)]	s, f, 2	34, p. 388f
	T. *P. granatina* (Sommerf.) Nyl.	s, f, 3	34, p. 391
	T. *P. grumulifera* Nyl. Th. Fr.	s, f	34, p. 391
	T. *P. haematopsis* (Sommerf.) Th. Fr.	s, f	34, p. 391
	T. *P. pulvinata* (Schaer.) Th. Fr. (= *P. haemelea* (Sommerf.) Norrl.)	s, f	34, p. 391
	T. *P. sanguinea* Anzi	s, f, 2	34, p. 388f
	T. *Synalissa acharii* Fries [14]	s, f, 2	34, p. 393
	T. *S. phylliscina* (Tuck.) (= *Pyrenopsis phylliscina* (Tuck.) Tuck.?)	s, f	34, p. 393
	T. *S. symphorea* (Ach.) Nyl. (= *S. ramulosa* (Hoffm.) Fr., including *S. violacea* Geitler) [15]	s, f, ff, > 200, *	34, 44, 56, 146, 147, 169, 170
Entophysalidaceae	T. Lichinaceae	s	171 [16]
	T. *Peccania* sp.	?	172 according to 171
Hormathonema sp. (?)	T. *Phylliciella marionensis* Henssen	s, c, f?	155, 173
Entophysalidaceae-type	T. *Gonohymenia nigritella* (Lett.) Henssen	s, c	173
Pleurocapsales Pleurocapsaceae			
Chroococcidiopsis (4 strains) [11]	T. *Anema nummularium* (Mont.) Nyl.	s, c, *	41, 173
	T. *Gonohymenia* sp.	s, c	173
	T. *Peccania cerebriformis* Henssen et Büdel	s, c	41, 171, 173, also see 30
	T. *Psorotichia columnaris* Henssen et Büdel	s, c	41, 171, 173, also see 30

Table 1 (continued)
ALGAL GENERA AND SPECIES IDENTIFIED AS LICHEN PHYCOBIONTS[a]

Phycobiont	Lichen[b]	Type of observation[c]	Ref.
Chroococcidiopsis-	T. Lichinaceae		171
Myxosarcina-group	T. Lichinaceae and Peltulaceae		172 according to 171
	T. *Lichinella intermdia* Henssen	s, c	173
	T. *Peccania coralloides* (Mass.) Mass.	s, c, *	173
pleurocapsalean alga	T. *Thelochroa montini* Mass.	s	171
Scopulonemataceae			
Hyella caespitosa Bornet et Flah.	T. *Arthopyrenia halodytes* (Nyl.) Arn. emend. Swinscow	s, c, f, ff	174, also see 175, 176
Stigomenatales			
Stigonemataceae			
Stigonema sp.	C. *Amygdalaria* spp.	s	160
	C. *A. consentiens* (Nyl.) auct. (= *Huilia consentiens* (Nyl.) Hertel = *Lecidea consentiens* Nyl.)	s, *	160
	C. *A. pelobotryon* (Wahlenb. in Ach.) Norm. (= *Lecidea pelobotryon* (Wahlenb.) Leight.)	s, *	37, 161
	C. *Argopsis friesiana* Müll. Arg. [17]	s	45
	T. *Ephebe lanata* (L.) Vain. (= *E. pubescens* Fr.) [18]	s, f, ff, several	40, 46, 146
	T. *E.* spp. (11 further spp.)	s	40
	C. *Huilia panaeola* (Ach.) Hertel (= *Lecidea panaeola* (Ach.) Ach.)	s, *	168
	C. *Pilophorus fibula* (Tuck.) Th. Fr.	s	177
	C. *P. strumaticus* Nyl. ex Cromb.	s, f, ff	125, 177
	T. *Spilonema paradoxum* Born.	s, f	40, 178
	T. *S. revertens* Nyl.	s	40
	C. *Stereocaulon alpinum* Laur. (= *S. paschale* var. *alpinum* (Laur.) Du Rietz)	s, *	18
	C. *S. arcticum* Lynge	s, *	105, p. 250
	C. *S. austroindicum* Lamb	s, *	105, p. 205
	C. *S. botryosum* Ach. emend. Frey	s, *	179
	C. *S. brassii* Lamb	s, *	105, p. 271
	C. *S. caespitosum* Redgr.	s	18, p. 538
	C. *S. condensatum* Hoffm.	s	179
	C. *S. cornutum*, Müll. Arg.	s	180, p. 216
	C. *S. crambidiocephalum* Lamb	s	105, p. 290
	C. *S. curtatum* Nyl.	s	18, p. 543
	C. *S. dactylophyllum* Flörke	s	179
	C. *S. delisei* Bory	s	179
	C. *S. denudatum* Flörke	s	158
	C. *S. didymicum* Lamb	s, *	105, p. 291
	C. *S. dusenii* Lamb	s, *	105, p. 293
	C. *S. evolutum* Graewe	s	179
	C. *S. executum* Nyl.	s	18, p. 543
	C. *S. fribrillosum* Lamb	s	105, p. 251
	C. *S. furfuraceum* Duvign.	s	158

Table 1 (continued)
ALGAL GENERA AND SPECIES IDENTIFIED AS LICHEN PHYCOBIONTS[a]

Phycobiont	Lichen[b]	Type of observation[c]	Ref.
	C. S. graminosum Schaer.	s	180
	C. S. grande (Magn.) Magn.	s, *	105
	C. S. intermedium (Sav.) Magn.	s, *	105, p. 223
	C. S. japonicum Th. Fr.	s	18, p. 543
	C. S. karisimbiens Duvign.	s	158
	C. S. loricatum Lamb	s	181
	C. S. mamillosum Duvign.	s	158
	C. S. microthuja Duvign.	s, f	158
	C. S. nanodes Tuck.	s	179
	C. S. nigromaculatum Duvign.	s	158
	C. S. octomerellum Müll. Arg.	s, *	105, p. 265
	C. S. paschale (L.) Hoffm.	s, *	18, 180, p. 220
	C. S. penicillium Duvign.	s	158
	C. S. philippinense Räs.	s	105, p. 299
	C. S. pileatum Ach.	s	179
	C. S. pityricans Nyl.	s	180, p. 215
	C. S. ramulosum (Sw.) f. ramulosum (= S. furceatum Fr.)	s, f, *	18, 56, 105
	C. S. ramulosum f. elegans Th. Fr.	s, *	18, p. 543; 180, p. 246
	C. S. ramulosum f. simplicus Lamb	s, *	105, p. 284
	C. S. rubiginosum Pers.	s, *	105, p. 300
	C. S. ruwenzoriense Duvign.	s	158
	C. S. tomentosum Fr. var. tomentosum	s, *	18
	C. S. verruciferum Nyl.	s, *	180, p. 216
	C. S. vesuvianum Pers.	s	179, 180, p. 216
"stigonemoid"	C. Placopsis baculigera Lamb	s	192, p. 220ff
	C. P. cribellans (Nyl.) Räs.	s	192, p. 227
	C. P. papillosa Vain.	s	192, p. 230ff
Capsosiraceae			
Hyphomorpha antillarum Borzi [19]	T. Spilonema dendroides Henssen	s, 2	40, 182
H. perrieri Frémy [19]	T. S. schmidtii (Vain.) Henssen	s, 2	40, 182
Nostocales			
Rivulariaceae	T. Lichina minutissima Henssen	s	159
	T. L. polycarpa Henssen	s	159
Rivulariaceae (?)	T. Steinera radiata James et Henssen	s	191
Calothrix sp.	T. Calotrichopsis filiformis Henssen	s	40
	T. C. granulosa Henssen	s	40
	T. C. insignis Vain.	s	40
	C. Coccotrema maritimum Brodo	s, c	78
	C. C. pocillarium (Cummings) Brodo	s, c	78
	T. Hertella chilensis Henssen	s	183
	C. Lepolichen granulatus (Hook.) Müll. Arg.	s	8
	T. Porocyphus coccodes (Flot.) Körb.	s, c	40
	T. Porocyphus (6 further species)	s	40

Table 1 (continued)
ALGAL GENERA AND SPECIES IDENTIFIED AS LICHEN PHYCOBIONTS[a]

Phycobiont	Lichen[b]	Type of observation[c]	Ref.
Calothrix sp. (probably)	T. *Lichina rosulans* Henssen	s	184
C. crustacea Thuret [20]	T. *Lichina confinis* (Müll.) Ag.	s, c, ff, *	185
C. parietina (Näg.) Thuret [21]	C. *Stereocaulon octomerellum* Müll. Arg.	s, *	105
C. pulvinata Kg. [20]	T. *Lichina confinis* (Müll.) Ag.	s, c, ff, 2, *	185
C. scopulorum (W. et M.) Ag. [20]	T. *L. confinis* (Müll.) Ag.	s, ff, *	56
	T. *L. pygmaea* (Lightf.) Ag.	s	56
Dichothrix sp.	T. *L. tasmanica* Henssen	s	184
Dichothrix sp. (?)	T. *L. macrospora* Henssen, Büdel et Wessels	s, f	156
D. baueriana (Grun.) Born. et Flah.	T. *L. willeyi* (Tuck.) Henssen	s, c, ff	184
D. orsiniana Born. et Flah. (very probably)	T. *Placynthium nigrum* (Huds.) Gray (= *Pannaria triptophylla* var. *nigra* Nyl.)	s, c, f, ff several	103
Nostocaceae			
Anabaena sp. [22]	C. *Stereocaulon mikenoense* Duvign.	s	158
	C. *S. tumbertii* Duvign.	s	158
Anabaena sp. (?)	C. *S. pomiferum* Duvign.	s	158
Nostoc sp.	T. Arctomiaceae (all)	s	37, 271
	T. *Biatora numida* Kullhem	s	186
	C. *Chaenotheca trichialis* (Ach.) Th. Fr.	s, *	8
	T. Collemataceae (all)	s	37, 53
	T. *"Dendriscocaulon"*	s	Compare 8
	T. Heppiaceae [23]		37
	T. *Lepidocollema* spp.	s	37
	C. *Nephroma arcticum* (L.) Torss. [24]	s	187
	C. *N. expallidum* (Nyl.) Nyl.	s	187
	T. C. Pannariaceae (part of)	s	37, 188
	T. C. Peltigeraceae (part of)	s	37
	C. *Pilophorus acicularis* (Ach.) Th. Fr.	s	177
	C. *Pilophorus* (3 further spp.)	s	177
	T. *Polychidium* spp. (3)	s	40
	C. *Psoroma durietzii* James et Henssen	s	189
	C. *P. hypnorum* (Vahl.) Gray	s, f, ff	190, see 37
	T. *Steinera* spp. (3)	s	191
	T. C. Stictaceae (part of)	s	37
Nostoc sp. resp. ''nostocoid''	C. *Stereocaulon alpinum* Laur. (= *S. paschale* var. *alpinum* (Laur.) Du Rietz)	s, *	18
	C. *S. alpinum* Laur. var. *erectum* Frey.	s	180
	C. *S. arcticum* Lynge	s, *	180
	C. *S. argus* Hook. f. et Tayl.	s	180
	C. *S. botryosum* Ach. emend. Frey	s, *	37, 179
	C. *S. brassii* Lamb	s, *	105
	C. *S. claviceps* Th. Fr.	s	18
	C. *S. coniophyllum* Lamb	s	179

Table 1 (continued)
ALGAL GENERA AND SPECIES IDENTIFIED AS LICHEN PHYCOBIONTS[a]

Phycobiont	Lichen[b]	Type of observation[c]	Ref.
	C. S. *corticulatum* Nyl.	s	179
	C. S. *farinaceum* Magn.	s	179
	C. S. *glareosum* (Sav.) Magn.	s	179
	C. S. *grande* (Magn.) Magn.	s, *	105
	C. S. *incrustatum* Flörke	s	179
	C. S. *intermedium* (Sav.) Magn.	s, *	105
	C. S. *macrocephalum* Müll. Arg.	s	18
	C. S. *paradoxum* Lamb	s	105
	C. S. *paschale* (L.) Hoffm.	s, *	18
	C. S. *ramulosum* (Sw.) Räusch. f. *ramulosum*	s, *	105
	C. S. *ramulosum* f. *elegans* Th. Fr.	s, *	180
	C. S. *rivulorum* Magn.	s	179
	C. S. *strictum* Th. Fr.	s	18
	C. S. *tomentosum* Fr. var. *orizabae* (Th. Fr.) Lamb	s	18
	C. S. *tomentosum* Fr. var. *tomentosum*	s, *	18, 179
	C. S. *verruciferum* Nyl.	s, *	180
	T. C. Stictaceae	s	37
"nostocoid"	C. *Argopsis* (?) *cymosoides* Lamb	s	45
	C. A. *megalospora* Th. Fr.	s	45
	C. *Placopsis albida* Lamb	s	192
	C. *Placopsis* (11 further spp.)	s	192
N. *commune* Vaucher	T. *Collema flaccidum* (Ach.) Ach.	s, c, ff, 2	53
	T. C. *furfuraceum* (Arn.) Du Rietz	s, c, 2	53
	T. C. *occultatum* Balg. var. *populinum* (Th. Fr.) Degel.	s, c	53
	T. C. *subnigrescens* Degel.	s, c, 3	53
	T. C. *tuniforme* (Ach.) Ach.	s, c, 9	53
	C. *Lobaria* cf. *erosa* (Eschw.) Nyl.	s, c	193
N. *commune* Vaucher (?)	T. *Collema multipartitum* Sm.	s, c, 4	53
N. *muscorum* Ag.	T. C. *bachmanianum* (Fink) Degel. var. *bachmanianum*	s, c, ff	53
	C. *Lobaria hallii* (Tuck.) Zahlbr.	s, c	193
	C. L. *pulmonaria* (L.) Hoffm.	s, c	193
N. *punctiforme* (Kütz.) Hariot [25]	C. *Peltigera aphthosa* (L.) Willd.	s, c	194
	T. P. *canina* (L.) Willd.	s, c, ff, several	21, 194
	T. P. *collina* (Ach.) Schrad. (= P. *scutata* (Dicks.) Duby = P. *limbata* Del.)	s, c	194
	T. P. *horizontalis* (Huds.) Baumg.	s, c	194
	T. P. *malacea* (Ach.) Funck	s, c	194
	T. P. *polydactyla* (Neck.) Hoffm.	s, c, several	21, 194, 263
	T. P. *pruinosa* (Gyeln.) Inum.	s, c	263
	T. P. *rufescens* (Weis.) Humb. [26]	s, c, 2	44, 194
	T. P. *rufescens* var. *virescens* Stnr. (= P. *virescens* (Stnr.) Gyeln.)	s, c	263

Table 1 (continued)
ALGAL GENERA AND SPECIES IDENTIFIED AS LICHEN PHYCOBIONTS[a]

Phycobiont	Lichen[b]	Type of observation[c]	Ref.
	T. *P. spuria* (Ach.) DC (= *P. errumpens* (Tayl.) Vain.)	s, c	194
	C. *P. venosa* (L.) Hoffm.	s, c	194
N. sphaericum Vauch.	T. *Collema crispum* (Huds.) Web. in Wigg. var. *crispum*	s, c, ff, 3	53
	T. *C. cristatum* (L.) Web. in Wigg. var. *cristatum*	s, c, 4	53
	T. *C. limosum* (Ach.) Ach.	s, c, 2	53
	T. *C. polycarpon* Hoffm. var. *polycarpon*	s, c, 2	53
	T. *C. tenax* (Sw.) Ach. emend. Degel.	s, c, 6	53
	T. *Hydrothyria venosa* Russ.	s, c	264 according to 3
	T. *Leptogium issatschenki* Elenk.	s, c	51
	T. *Pannaria pezizoides (Web.)* Trevis.	s, c, f	44
	T. *P. rubiginosa* (Ach.) Bory. var. *lanuginosa* (Hoffm.) Zahlbr.	s, c	44
Scytonemataceae *Scytonema* sp.	T. *Compsocladium archiboldianum* Lamb [27]	s	195
	T. *Coccocarpia adnata* Arvidss.	s	54
	T. *Coccocarpia* further spp. [28]	s	7, 54, 56, 59
	T. *Degelia* spp. (3)	s	196
	T. *Dictyonema* sp.	s, 2	197
	T. *Dictyonema* spp.	s	35, 55, 57
	T. *D. irpicinum* Mont.	s, f, ff	198
	T. *D. ligulatum* (Kremp.) Zahlbr.	s, 4	57
	T. *D. moorei* (Nyl.) Henssen	s, 5	40, 57
	T. *D. pavonia* (Sw.) Parm. (= *Cora pavonia* (Sw.) Fr.) [29]	s, several	199—201
	T. *D. sericeum* (Sw.) Berk.	s, ff, several	197, 200, 201
	T. *Erioderma* spp.	s	38
	T. *E. hypomelaenum* Hue (= *Pannaria hypomelaena* Nyl.)	s	56
	T. *E. unguingerum* Nyl.	s	56
	T. *Heppia* spp.	s	38
	T. *H. adglutinata* Mass.	s, f	178
	T. *H. echinulata* Marton et Galun [30]	s, c	157
	T. *H. lutosa* (Ach.) Nyl. [31]	s, c	3
	T. *Hertella subantarctica* Henssen	s	183
	T. *Lichinodium ahlneri* Henssen	s	40
	T. *L. sirosiphoideum* Nyl.	s	40
	T. *Pannaria* [32]	s?	3
	T. *Petractis clausa* (Hoffm.) Krempelh.	s	37
	T. *Placynthium arachnoideum* Henssen	s, 17	202
	T. *Polychidium contortum* Henssen	s	40
	T. *P. dendriscum* Henssen	s	40
	T. *Pyrenothrix nigra* Riddle (= *Lichenothrix riddlei* Henssen)	s, 11	203—205, see also 124

Table 1 (continued)
ALGAL GENERA AND SPECIES IDENTIFIED AS LICHEN PHYCOBIONTS[a]

Phycobiont	Lichen[b]	Type of observation[c]	Ref.
	C. *Stereocaulon brassii* Lamb	s, *	105
	C. *S. coniophyllum* Lamb	s,*	37, 179
	C. *S. didymicum* Lamb	s, *	105
	C. *S. dusenii* Lamb	s, *	105
	C. *S. myriocarpum* Th. Fr.	s	2
	C. *S. ramulosum* (Sw.) Räusch. f. *ramulosum*	s, *	18, 105, 180
	C. *S. ramulosum* f. elegans Th. Fr.	s, *	18, 180
	C. *S. ramulosum* var. *perpumilum* Lamb	s	105
	C. *S. ramulosum* var. *simplicus* Lamb	s, *	105
	C. *S. rubiginosum* Pers.	s, *	105
	C. *S. togashii* Lamb	s	105
	C. *S. trachyphloeum* Lamb	s	181
	T. *Thermutis velutina* (Ach.) Flot.	s, f, several	40, compare 58
	T. *Zahlbrucknerella calcarea* (Herre) Herre	s	40
	T. *Z. maritima* Henssen	s	40
	T. *Zahlbrucknerella* (5 further spp.)	s	206
S. hoffmannii Ag.	T. *Heppia* North American spp.	s, c	32
Scytonema sp. probably or "Scytonemaceae" scytonemoid or scytonemiform	C. *Lasioloma* spp.	s	7
	C. *Lopadium* spp.	s	7
	T. *Peltula* spp.	s	38
	C. *Placopsis asahinae* Lamb	s	192
	C. *Placopsis* (16 further spp.)	s	192
	T. *Placynthium rosulans* (Th. Fr.) Gyeln.	s, c	272
	C. *Solorinaria* spp.	s	38
Tolypothrix sp.	T. *Hertella chilensis* Hensen	s	183

Tribophyceae (= Xanthophyceae) [33]

Tribonematales			
Heteropediaceae			
Heterococcus caespitosus Vischer	T. *Verrucaria funckii* (Spreng. in Funck.) Zahlbr. (= *V. elaeome-laena* auct.)	s, c, ff many [34]	61, 62
	T. *V. laevata* Ach.	s, c	62
	T. *V. maura* Wahlenb. in Ach. [35]	s?, c, *	207

Fucophycee (= Phaeophyceae) [33]

Ectocarpales			
Lithodermataceae			
Petroderma maculiforme (Wollng.) Kuck.	T. *Verrucaria* sp. [36]	s, c, ff	148

Table 1 (continued)
ALGAL GENERA AND SPECIES IDENTIFIED AS LICHEN PHYCOBIONTS[a]

Phycobiont	Lichen[b]	Type of observation[c]	Ref.
	Chlorophyceae [37]		
Volvocales			
Chlamydomonadaceae			
Chlamydomonas augustae Skuja	T. *Pyromonas laetissimum* Schroeder (facultatively lichenized)	s, f, mycobiont, ff	208
Chlorococcales			
Coccomyxaceae			
Coccomyxa sp. [39]	T. *Baeomyces phycophyllus* Ach.	s, c	209 according to 210
	T. *Nephroma arcticum* (L.) Torss.	s	187
	T. *N. expallidum* (Nyl.) Nyl.	s	187
Coccomyxa sp. (?) [38]	T. *Multiclavula mucida* (Fr.) Petersen (= *Clavaria mucida* (Fr.) Corner)	s, c, f, several	211, 212
C. ellipsoidea Zehnder	T. *Solorina saccata* (L.) Ach.	s, c, *	102
C. glaronensis Jaag.	T. *S. saccata* (L.) Ach	s, c, ff, 2, *	68, 128
C. icmadophilae Jaag.	T. *Baeomyces roseus* Pers. [40]	s, c	128, compare 68
	T. *Icmadophila ericetorum* (L.) Zahlbr.	s, c, 4	68, 102
	T. *Omphalina hudsoniana* (Jenn.) Bigelow (= *Coriscium viride* (Ach.) Vain.)	s, several	200, 213
C. mucigena Jaag.	T. *Peltigera aphthosa* (L.) Willd.	s, c, *	68
C. ovalis Jaag. [41]	T. *Solorina saccala* (L.) Ach.	s, c, *	68
C. peltigerae Warén	T. *Peltigera aphthosa* (L.) Willd.	s, c, 5, *	68, 87
	T. *P. venosa* (L.) Hoffm.	s, c, *	68
C. peltigerae variolosae Jaag.	T. *P. aphthosa* (L.) Willd.	s, c, *	102
	T. *P. leucophlebia* (Nyl.) Gyeln. (= *P. variolosa* (Mass.) Sch.)	s, c	68
C. peltigerae venosae Jaag.	T. *P. venosa* (L.) Hoffm.	s, c, *	68
C. pringsheimii (*botrydinae*) Jaag.	T. *Omphalina* sp. (= *Botrydina vulgaris* Bréb.)	s, c	68, also compare 214—217
C. solorinae bisporae Jaag.	T. *Solorina bispora* Nyl.	s, c	68
	T. *S. octospora* Arn.	s, c	68
C. solorinae croceae Jaag	T. *S. crocea* (L.) Ach.	s, c, 2, *	68, 104
C. solorinae saccatae Chodat	T. *S. saccata* (L.) Ach.	s, c, 2, *	68, 104
C. subellipsoidea Acton emend. Jaag	T. *Omphalina* sp. (= *Botrydina vulgaris* Bréb.)	s, c, several	68, compare also 214—216
C. tiroliensis Jaag	T. *Solorina crocea* (L.) Ach.	s, c, 2, *	68, 128
"*Coccomyxa*-ähnlich"	T. *Multiclavula corynoides* (Peck) Peterson (= *Clavulinopsis septentrionalis* Corner)	s, 4	218
Glyoeocystis sp. [42, 43]	T. *Bryophagus gloeocapsa* Nitschke ex Arn. (= *Gyalecta gloeocapsa* (Arn.) Zahlbr. = *Gloeolecta gloeocapsa* (Arn.) Lettau = *G. bryophaga* (Krb. ex Arn.) Vezda)	s	22, 219
G. botryoides Näg. (?)	T. *Lecidea fuliginosa* Ach. var. *chthonoblastes* Erichs. (= *Stereonema chthonoblastes* Braun in Kütz. emend. Kupffer) [44]	s	72

Table 1 (continued)
ALGAL GENERA AND SPECIES IDENTIFIED AS LICHEN PHYCOBIONTS[a]

Phycobiont	Lichen[b]	Type of observation[c]	Ref.
G. polydermatica Kütz.) Hind. (= *Coccomyxa epigloeae* Jaag. et Thomas) [45]	T. *Epigloea bactrospora* Zukal	s, c, f	73
Chlorococcaceae			
Asterochloris phycobiontica [46]	T. *Anzina carneonivea* (Anzi) Scheidegger (= *Varicellaria carneonivea* (Anzi) Erichs.)	s, c, 7	75, 150, 220
Dictyochloropsis Geitler emend. Tsch.-Woess. sp.	T. *Lobaria* spp. (7)	s	77
	T. *L. adscripturiens* (Nyl.) Hue [47]	s, c	77
	T. *L. spathulata* (Inum.) Yoshim.	s, c	77
D. reticulata (Tsch.-Woess) Tsch.-Woess (= *Myrmecia reticulata*) Tsch.-Woess = *Disctyochloris reticulata* (Tsch.-Woess) Reisigl) [48]	T. *Bacidia nanipara* Lett.	s, c, ff [49]	62
	T. *Catillaria chalybeia* (Börr.) Mass.	s, c	221
	T. *Lobaria amplissima* (Scop.) Forss.	s, c, 2	80
	T. *L. laetevirens* (Lightf.) Zahlbr.	s, c, 4	80
	T. *L. pulmonaria* (L.) Hoffm. var. *pulmonaria*	s, c, 33	80, 222
	T. *L. pulmonaria* var. *meridionalis* (Vain.) Zahlbr. [50]	s, c, 5	80
	T. *Phlyctis argena* (Spreng.) Flot. [50a]	s, c,	80, 223
	T. *Sarcogyne regularis* Körb. (= *S. pruinosa* Körb. var. *pruinosa*) [51]	s, c, *	80
D. splendida Geitler emend. Tsch.-Woess var. *splendida* Tsch.-Woess	T. *Chaenotheca brunneola* (Ach.) Müll. Arg. [52]	s, c, ff, *	12
D. splendida Geitler var. *gelatinosa* Tsch.-Woess	T. *Catinaria grossa* (Pers. ex Nyl.) Vain.	s, c	92
D. symbiontica Tsch.-Woess var. *ellipsoidea* Tsch.-Woess	T. *Chaenothecopsis consociata* (Nádv.) A. Schmidt	s, c, *	17
D. symbiontica var. *pauciautosporica* Tsch.-Woess	T. *Megalospora atrorubicans* (Nyl.) Zahlbr. subsp. *australis* Sipm.	s, c	92
	T. *M. gompholoma* (Müll. Arg.) Sipm. *gompholoma*	s, c	92
	T. *Pseudocyphellaria aurata* (Ach.) Vain.	s, c	92
D. symbiontica var. *symbiontica* Tsch.-Woess	T. *Chaenothecopsis consociata* (Nádv.) A. Schmidt	s, c, several, *	17
	T. *Pseudocyphellaria aurata* (Ach.) Vain.	s, c, 2	92
Myrmecia biatorellae (Tsch.-Woess et Plessl) Petersen [48, 53]	T. *Catapyrenium rufescens* (Ach.) Breuss (= *C. lachneum* (Ach.) Sant. = *Dermatocarpon rufescens* (Ach.) Zahlbr.	s, c, ff, many	62, 128, 224
	T. *Dermatocarpon hepaticum* (Ach.) Th. Fr.	s, c, several	81, 225

Table 1 (continued)
ALGAL GENERA AND SPECIES IDENTIFIED AS LICHEN PHYCOBIONTS[a]

Phycobiont	Lichen[b]	Type of observation[c]	Ref.
	T. *D. tuckermani* Zahlbr.	s, c	22, 128
	T. *D. velebiticum* Zahlbr.	s, c	62
	T. *Lecidea berengeriana* (Mass.) Th. Fr.	s, c	129
	T. *Lobaria linita* (Ach.) Rabenh.	s, c, 2	128, 134
	T. *Polysporina simplex* (Dav.) Vězda (*Sarcogyne simplex* (Dav.) Nyl. = *Biatorella simplex* (Dav.) Branth et Rostr.	s, c	226
	T. *Psora decipiens* (Hedw.) Hoffm. (= *Lecidea decipiens* (Hedw.) Ach.)	s	227 [54]
	T. *P. globifera* (Ach.) Mass.	s, c	228
	T. *P. hypnorum* (Vahl.) Gray (= *Pannaria hypnorum* (Vahl.) Körb.)	s, c	229
	T. *Sarcogyne privigna* (Ach.) Mass. var. *calcicola* Magn.	s, c	80
	T. *Verrucaria submersella* Serv. (= *V. submersa* Schaer.)	s, c, 2	62
Pseudotrebouxia spp. [55]	T. very widespread, e.g., in many Lecanoraceae, Parmeliaceae, Physciaceae, Teloschistaceae, Umbilicariaceae	s, c, ff	e.g., 11, 87, 231
P. aggregata Arch. [55a]	T. *Lecidea fuscoatra* (L.) Ach.	s, c	97
	T. *Xanthoria parietina* (L.) Th. Fr. [56]	s, c, *	89, 231
P. decolorans (Ahm.) Arch. (= *Trebouxia albulescens* Guid. deNic. et diBen.)	T. *Buellia punctata* (Hoffm.) Mass.	s, c	88
	T. *Xanthoria parietina* (L.) Th. Fr. [56]	s, c, several, *	88, 232, comp. 89
P. impressa (Ahm.) Arch.	T. *Physcia stellaris* (L.) Nyl.	s, c	88, comp. 89
P. incrustata (Ahm.) Arch.	T. *Lecanora dispersa* (Pers.) Sommerf.	s, c	2, 89
P. simplex (Tsch.-Woess) Tsch.-Woess	T. *Chaenotheca chrysocephala* (Turn. ex Arch.) Th. Fr. [57]	s, c, many, *	12, 17
	T. *C. subroscida* (Eitn.) Zahlbr.	s, c	17
Trebouxia sp. [55]	T. All tested *Cladonia* spp., *Huilia. Pilophorus, Stereocaulon,* part of *Lecidea, Parmelia, Physcia,* and others	s, some c	11, 87, 104, 230, 233
T. crenulata Arch.	T. *Parmelia acetabulum* (Neck.) Duby	c	89
	T. *P. caperata* (L.) Ach.	s, c, *	21 according to 76
	T. *Xanthoria aureola* (Ach.) Erichs.	c	89
T. erici Ahm.	T. *Cladonia cristatella* Tuck.	s, c	88
	T. *Cladonia* further spp.	s, c	21
T. gelatinosa (Ahmad.) Arch.	T. *Parmelia caperata* (L.) Ach. [58]	s, c, several, *	234, compare 21, 89, 101, 149
T. glomerata (Warn.) Ahm.	T. *Cladonia* spp.	s, c	21

Table 1 (continued)
ALGAL GENERA AND SPECIES IDENTIFIED AS LICHEN PHYCOBIONTS[a]

Phycobiont	Lichen[b]	Type of observation[c]	Ref.
	T. *C. boryi* Tuck.	s, c	97
	T. *C. coccifera* (L.) Willd.	s, c	87
	T. *C. rangiferina* (L.) Web. (*Cladonia* spp., 5 others) [59]	s, c	87
	T. *Huilia albocaerulescens* (Wulfen) Hertel	s, c, 2	97
	T. *Stereocaulon pileatum* Ach.	s, c, *	88, compare 89
	T. *S. saxatile* Magn. (= *S. evolutoides* (Magn.) Frey)	s, c	88, 97, compare 89
T. italiana Arch.	T. *Xanthoria parietina* (L.) Th. Fr. [56]	c, *	89
T. magna Arch.	T. *Pilophorus acicularis* (Ach.) Nyl.	s, c	2, compare 89
T. pyriformis Arch.	T. *Cladonia squamosa* (Scop.) Hoffm.	c	89
	T. *Stereocaulon dactyllophyllum* var. *occidentale* (Magn.) Grumm.	c	89
	T. *S. pileatum* Ach.	s, c, *	97
Oocystaceae			
Chlorella sp.	T. *Lecidella elaeochroma* (Ach.) Choisy (= *Lecidea parasema* auct. [60]	c, *	102
Chlorella sp. (?)	T. *Placynthiella* Gyelnik	s	186
''*C. lichina* Chod.'' [61]	T. *Chaenothecopsis exsertum* (Nyl.) Tibell (= *Calicium exsertum* Nyl.)	s(?), c, ff	82
	T. *Chrysothrix chlorina* (Ach.) Laundon (= *Lepraria chlorina* (Ach.) Ach.) [62]	s, c	82
C. saccharophila (Krüger) Migula var. *ellipsoidea* (Gerneck) Fott et Nováková (= *C. ellipsoidea* Gerneck	T. *Trapelia coarctata* (Sm.) Choisy in Werner (= *Lecidea coarctata* (Sm.) Nyl.) [63]	s, c, ff, 2, *	6, 236
	T. *Woessia fusarioides* Hawksw., Poelt et Tsch.-Woess	s, c	235
C. sphaerica Tsch.-Woess	T. *Pseudocyphellaria* (several spp.)	s, c	235
Elliptochloris sp.	T. *Micarea prasina* Fr.	s, c	128
E. bilobata Tsch.-Woess	T. *Baeomyces rufus* (Huds.) Rebent	s, c, 3	133
	T. *Catolechia wahlenbergii* (Ach.) Körb	s, c	132
	T. *Protothelenella corrosa* (Loerb.) Mayhr. et Poelt	s, c	133
	T. *P. sphinctrinoides* (Nyl.) Mayhr. et Poelt	s	133
Pseudochlorella sp.	T. *Stereocaulon microcarpum* Th. Fr.	s, c (?)	2
	T. *S. strictum* Th. Fr. [64]	s, c, *	2, 128
P. pyrenoidosa (Zeitler) Lund (= *Chlorellopsis pyrenoidosa* Zeitler)	T. *Micarea assimilata* (Nyl.) Coppins (= *Lecidea assimilata* Nyl.)	s, c	62
	T. *Trapeliopsis granulosa* (Hoffm.) Lumbsch (= *Lecidea granulosa* (Hoffm.) Ach.)	s, c, 2	62, 128

Table 1 (continued)
ALGAL GENERA AND SPECIES IDENTIFIED AS LICHEN PHYCOBIONTS[a]

Phycobiont	Lichen[b]	Type of observation[c]	Ref.
Chlorosarcinales			
Chlorosarcinaceae			
Chlorosarcinopsis minor (Gerneck) Herndom (= *Chlorosarcina minor* Gerneck)	T. *Lecidea lapicida* (Ach.) Ach. T. *L. plana* (Lahm in Körb) Nyl.	s, c, ff, [65] s, c	213 213
Nannochloris normandinae Tsch.-Woess [66]	T. *Normandina pulchella* (Borr.) Nyl.	s, c, several	134
Chaetophorales			
Chaetophoraceae			
Coccobotrys verrucariae Chod. emend. Vischer [67]	T. *Verrucaria nigrescens* Pers.	s, c	104, compare 108
Dilabifilum arthopyreniae (Vischer et Klement) Tsch.-Woess (= *Pseudopleurococcus arthopyreniae* Vischer et Klement)	T. *V. adriatica* (Zahlbr.) Zahlbr. (= *V. amphibia* Clem. (?))	s, c, ff	176
D. incrustans (Vischer) Tsch.-Woess (= *Ps. incrustans* Vischer)	T. *V. aquatilis* Mudd T. *V.* cf. *rheithrophila* Zschacke	s, c, ff s, c	237 238
Gongrosira sp. (?)	T. *Catillaria minuta* (Mass.) Lettau (= *C. arnoldii* (Krempelh) Th. Fr.	s	213
Leptosira obovata Vischer (= *Pleurastrum obovatum* (Vischer) Tupa)	T. *Vezdaea aestivalis* (Ohl.) Tsch.-Woess et Poelt	s, c, ff, several	137
L. thrombii Tsch.-Woess [67a]	T. *Thrombium epigaeum* (Pers.) Wallr.	s, c, 3	151, 239, 240
Protococcus dermatocarponis miniati (= *Hyalococcus dermatocarponis* Warén (?)) [68]	T. *Dermatocarpon miniatum* (L.) Mann.	s, c, 2	62, 87
	T. *D. weberi* (Ach.) Mann (= *D. fluviatile* (Web.) Th. Fr.)	s, c	234, according to 3, 22
	T. *Endocarpon adscendens* (Anzi) Müll. Arg. (= *E. pallidum* auct.)	s, c	62, compare 241
	T. *E. pusillum* Hedw.	s, c, *	109, according to 62
	T. *Thelidium perexiguum* (Müll. Arg.) Zahlbr.	s, c	
	T. *Verrucaria tristis* (Mass.) Krempelh.	s, c	62
P. staurothelis	T. *Staurothele catalepta sensu* Malme et auct. al.	s, c	62
	T. *S. fissa* (Tayl.) Zw.	s, c	62
P. verrucariae acrothelloides	T. *Thelidium absconditum* (Hepp.) Rabenh.	s, c	62
	T. *T. auruntii* (Mass.) Krempelh.	s, c	62
	T. *T. decipiens* (Nyl.) Krempelh. (= *T. immersum* (Leight.) Mudd)	s, c	62
	T. *T. parvulum* Arn.	s, c	62
	T. *Trapelia coarctata* (Sm.) Choisy (= *Lecidea coarctata* (Sm.) Nyl.) [63]	s, c, *	62

Table 1 (continued)
ALGAL GENERA AND SPECIES IDENTIFIED AS LICHEN PHYCOBIONTS[a]

Phycobiont	Lichen[b]	Type of observation[c]	Ref.
P. cf. *verrucariae* *acrotelloides*	T. *Thelidium* cf. *antonellianum* Bagl. et Car.	s, c	81
	T. *T.* cf. *minutulum* Körb. (= *T. acrotellum* Arn.) [69]	s, c	81
	T. *Verrucaria* cf. *acrotella* Ach.	s, c	81
Trochiscia sp.	T. *Polyblastia hyperborea* Th. Fr.	?	22
"*T. granulata* (Reinsch.) Hansg" [70]	T. *P. amota* Arn.	s, c, f, ff	61, 242
Klebsormidiales			
Klebsormidiaceae			
Stichococcus sp.	T. *Chaenotheca carthusiae* (Harm.) Lettau [71]	s, 195, *	13, 83
	T. *C. chrysocephala* (Turn. ex Ach.) Th. Fr.	s, several, *	15, 16
	T. *C. cinerea* (Pers.) Tibell	s, 39	13
	T. *C. servitii* Nádv.	s	13
	T. *C. xyloxena* Nádv.	s, 154	13
	T. *Coniocybe gracillima* Vain.	s	13, 14
	T. *Staurothele rugulosa* (Mass.) Arn. (= *Polyblastia rugulosa* Mass.)	s, c	109
Stichococcus sp. (?)	T. *S. succedens* (Rehm.) Arn.	s, c	243
S. bacillaris Näg. [72]	T. *Calicium* sp.	s, c, ff	82
	T. *Chaenotheca brunneola* (Ach.) Müll. Arg. [73]	s(?), c, *	82
	T. *C. ferruginea* (Turn. ex Sm.) Migula (= *C. melanophaea* (Arch.) Zw.) [73]	s, c, many	82
	T. *C. furfuracea* (L.) Tibell (= *Coniocybe furfuracea* (L.) Ach.)	s, c	13, 61, 82
	T. *C. stemonea* (Ach. Müll. Arg.	s, c	82
	T. *C. trichialis* (Ach.) Th. Fr.	s, c	82
	T. *Cybebe gracilenta* (Ach.) Tibell (= *Coniocybe gracilenta* Ach.)	s, c	82
	T. *Lepraria* sp.	s, c	82
S. chloranthus Raths [74]	T. *Chaenotheca ferruginea* (Turn. ex Sm.) Migula	s, c, *	82
S. diplosphaera (Bialosuknia) Chod.	T. *Endocarpon pusillum* Hedw. [75]	s, c, several, *	113, 244
S. mirabilis Lagerh.	T. *Staurothele clopima* (Wahlenb.) Th. Fr.	s, c, several	113
S. pallescens Chod. var. *lucida* Raths [74]	T. *Chaenotheca furfuracea* "var. *fulva* Tr."	s, c, *	82
Prasiolales			
Prasiolaceae			
Prasiola sp.	T. *Turgidosculum complicatulum* (Nyl.) Kohlm. et Kohlm. (= *Mastodia tesselata* (Hooker f. et Harvey) Hooker et Harvey ex Hooker = *Guignardia alaskana* Reed = *G. prasiolae* (Winter) Lemmerman [76]	s, ff according to 115, several	65

Table 1 (continued)
ALGAL GENERA AND SPECIES IDENTIFIED AS LICHEN PHYCOBIONTS[a]

Phycobiont	Lichen[b]	Type of observation[c]	Ref.
Ulvales			
Monostromataceae			
Blidingia minima var. *vextata* (Setch. et Gardn.) Norris [77]	T. *T. ulvae* (Reed) Kohlm. et Kohlm. [76]	s, ff(?)	65
Cladophorales			
Cladophoraceae			
Cladophora fuliginosa Kütz.	T. *Blodgettia confervoides* Harv.	s, many	245, compare 65
Trentepohliales			
Trentepohliaceae			
Cephaleuros sp.	T. *Raciborskiella*	s	7
	T. *Strigula* spp. (12)	s	7
C. virescens Kunze	T. *S. complanata* (Fée) Mont.	s, ff	7
	T. *S. elegans* (Fée) Müll. Arg.	s	246, compare 247
Phycopeltis Millardet (= *Phyllactidium* Kütz sensu Bornet) sp.	T. *Microtheliopsis uleana* Müll. Arg.	s, ff	7
	T. *Opegrapha dibbenii* Sérus.	s	248
	T. *O. lambinonii* Sérus.	s, ff, 5	248
	T. *O. puiggarii* Müll. Arg. (= *O. filicina* Mont.)	s, f, ff	56, compare 7 p. 102f
	T. *O. santessonii* Sérus.	s, ff, 7	248
	T. *Phyllophiale alba* Sant.	s	7
	T. *Porina pseudofulvella* Sérus	s	249
	T. 78 spp. of obligate foliicolous lichens of the Arthoniaceae, Gyalectaceae (compare 7, p. 411 on *Semigyalecta*), Opegraphaceae, Porinaceae, Pyrenulaceae, Strigulaceae, Thelotremataceae	s	7
Trentopohlia sp.	T. *Blarneya hibernica* Hawksw., Coppins et James	s	250
	T. *Byssocaulon niveum* Mont.	s, several	56, 251
	T. *Chiodecton nigrocinctum* Mont.	s	56
	T. *C. sanguineum* (Sw.) Vain.	s, several	252—254
	T. *Cystocoleus ebeneus* (Dillw.) Thwaites (= *Coenogonium nigrum* auct.) [78]	s, c, several, *	118
	T. *Dimerella lutea* (Dicks.) Trevis (= *Lecidea lutea* Schoer.)	s	56
	T. *Enterographa sorediata* Coppins et James	s	255
	T. *Glyphis cicatricosa* Ach. var. *lepida* Zahlbr. (= *G. lepida* Krempelh.)	s, c	256
	T. *Lecidea microsperma* Nyl.	s	56
	T. *Opegrapha mougeotii* Mass.	s	61
	T. *O. saxicola* Ach.	s	61
	T. *Phaeographina fulgurata* Müll. Arg.	s, c	256
	T. *Sagenidium patagonicum* Henssen	s	257

Table 1 (continued)
ALGAL GENERA AND SPECIES IDENTIFIED AS LICHEN PHYCOBIONTS[a]

Phycobiont	Lichen[b]	Type of observation[c]	Ref.
	T. *Topelia* spp. (4)	s	258
	T. *Trypethelium eluteriae* Spreng.	s	259
	T. present in all or some members of the Arthoniaceae, Arthopyreniaceae, Caliciaceae, Graphidaceae, Gyalectaceae, Lecanactidaceae, Lecanoraceae (Jonaspis), Microglaenaceae, Mycoporaceae, Opegraphaceae, Porinaceae, Pyrenulaceae, Roccellaceae, Thelotremaceae, *Sclerophora*, Caliciales (8 genera not assigned to families).	s	Compiled from 37, 38, see also 14, 15, 83, 260
Trentepohlia abietina (Flot.) Hansg.	T. *Coenogonium interplexum* Nyl.	s, c, ff, *	117
T. *annulata* Brand	T. *Graphis scripta* (L.) Ach.	s, c, ff, *	261
T. *arborum* (Ag.) Hariot	T. *Coenogonium interplexum* Nyl.	s, c, ff, 3, *	117
	T. *C. interpositum* Nyl.	s, c, 2	117
	T. *C. linkii* Ehrenb.	s, c, 3, *	117
T. *aurea* (L.) Martius	T. *C. interplexum*	s, c, ff, *	117
	T. *C. linkii* Ehrenb.	s, c, 5, *	117
	T. *Gyalecta jenensis* (Batsch.) Zahlbr. (= *G. cupularis* (Hedw.) Schär.)	s, f, ff, many [79]	61, 120
	T. *Jonaspis suaveolens* (Schär.) Th. Fr.	s	61
T. cf. *aurea*	T. *Racodium rupestre* Pers.	s, c, several	262
T. *elongata* (Zeller) De Toni	T. *Coenogonium interplexum* Nyl.	s, c, ff, *	117
	T. *C. linkii* Ehrenb.	s, c, 3, *	117
T. monilia de Wild. (= *Physolinum monilia* (de Wild.) Printz)	T. *C. moniliforme* Tuck.	s, ff, several	7
T. *odorata* (Wigg.) Wittrock	T. *C. leprieurii* (Mont.) Nyl.	s, c, ff, several, *	117
T. cf. *odorata*	T. *Opegrapha rufescens* Pers. (= *O. herpetica* (Ach.) Ach.)	s, many	61
T. *umbrina* (Kütz) Born.	T. *Arthonia asteroidea* Ach. (= *A. radiata* Ach. f. *astroidea* Ach.)	s, c	121 [80]
	T. *Chaenotheca hispidula* (Ach.) Zahlbr. (= *C. phaeocephala* var. *hispidula* (Ach.) Keissler	s, c	82
	T. *Coenogonium leprieurii* (Mont.) Nyl.	s, c, *	117
	T. *Graphis scripta* (L.) Ach.	s, c, *	121
	T. *Opegrapha atra* Pers.	s, c	121
T. cf. *umbrina*	T. *O. varia* Pers.	s, ff	56
	T. *Pyrenula nitida* (Weig.) Ach. (= *Verrucaria nitida* Schrad.)	s, c	56, 256
	T. *Roccella fucoides* (Dicks.) Vain. (= *R. phycopsis* Ach.)	s	56

Table 1 (continued)
ALGAL GENERA AND SPECIES IDENTIFIED AS LICHEN PHYCOBIONTS[a]

[a] Many statements and/or identifications in older literature are questionable; these have not been critically reappraised for this table. For critical analysis, the reader is referred to the cited literature. In some cases, older incorrect information has been corrected or modified.

[b] Nomenclature of lichens mainly according to Santesson.[265] In cases where authors were not given, names were taken from References 266 and 267 (also used for synonymy if newer monographs were not available).

[c] s = Observed and/or identified *in situ* (sometimes not directly mentioned, but assumed); c = identification from cultured isolates; f = free-living in nature, growing out or near vicinity of thallus; ff = free-living in nature, distant from thallus or in general; 1, 2, 3 . . . = number of thalli checked (if mentioned); * = lichens to which more than one phycobiont genus or species have been assigned.

[d] Numbers in brackets refer to comments in Section IV of this chapter.

[e] Letter in front of species name signifies: C = in cephalodia; T = in thallus.

Table 2
ALPHABETICAL LIST OF
ALGAL GENERA REPORTED
AS PHYCOBIONTS

Anabaena	*Heterococcus*
Anacystis	*Hormathonema*
Aphanocapsa	*Hyella*
Asterochloris	*Hyphomorpha*
Blidingia	*Leptosira*
Calothrix	*Myrmecia*
Cephaleuros	*Myxosarcina*
Chlamydomonas	*Nannochloris*
Chlorella	*Nostoc*
Chlorosarcinopsis	*Petroderma*
Chroococcidiopsis	*Phycopeltis*
Chroococcus	*Prasiola*
Cladophora	*Protococcus*
Coccobotrys	*Pseudochlorella*
Coccomyxa	*Pseudotrebouxia*
Dichothrix	*Scytonema*
Dictyochloropsis	*Stichococcus*
Dilabifilum	*Stigonema*
Elliptochloris	*Tolypothrix*
Gloeocapsa	*Trebouxia*
Gloeocystis	*Trentepohlia*
Gongrosira	*Trochiscia*

the latter change within wide or narrow limits? In lichens producing cephalodia, it has long been recognized that one fungus may associate with a green as well as a blue-green algal species. In recent years it has become more generally accepted that one fungal partner may form completely different, separated, or joint thalli (morphotypes, phycosymbiodemes, chimeroid associations) when lichenized with either phycobiont type.[8-10] In the author's opinion, the question of specificity in cephalodiate lichens, as well as noncephalodiate ones, deserves intense investigation. From the available data given in detail in Table 1, one can conclude that a certain degree of specificity prevails. In some older examples in which one mycobiont is said to be lichenized with several species of one algal genus, the algal species concept is generally too narrow, so that the putative species should be reduced to strains (e.g., three *Coccomyxa* species in *Peltigera aphthosa*, see comment 39 and Table 1). In other cases of species or genus variation, reinvestigation on a broad comparative basis seems necessary (e.g., in *Xanthoria parietina* and *Parmelia sulcata* which are said to be lichenized with

Trebouxia and *Pseudotrebouxia*[11]). There are still other cases in which reported variations in algal partners are probably accurate, such is the case in three *Chaenotheca* species. *Chaenotheca brunneola* in Austria is lichenized with *Dictyochloropsis splendida*;[12] this phycobiont is probably also present in specimens collected from other sites in Europe.[12,13] On the other hand, in Costa Rica, New Zealand, and Australia, *Chaenotheca brunneola* is associated with a trebouxioid alga (based on *in situ* examination).[14-16] In Costa Rica, *C. carthusiae* was found to be lichenized with a trebouxioid alga[14] rather than *Stichococcus*, with which it is associated in more northernly areas of the Northern Hemisphere.[13] In the case of *C. chrysocephala*, thalli from South Tyrol (Italy) and different parts of Austria contained *Pseudotrebouxia simplex*; it was formerly thought to be generally lichenized with a chlorococcalean alga.[12,13,17] Recently in New Zealand, thalli with "Trebouxia" (or "Pseudotrebouxia") were found growing in close vicinity of thalli with *Stichococcus*, and "in some cases even different parts of what would seem to be the same thallus contain different phycobionts".[15,16]

In some genera (e.g., *Stereocaulon*), a number of lichen species are reported to form cephalodia (see Chapter III.C) with species from more than one blue-green algal genera, such as *Stigonema*, *Scytonema*, *Nostoc*, *Anacystis*, or *Gloeocapsa* (see Table 1); they are sometimes reported to occur on the same plant and even in the same cephalodium.[18] Some of these statements may be well founded, but often good illustrations showing decisive characters of the different algae are lacking in the literature.

With respect to many other lichens thought to have vicarious phycobionts, documentation is very meager if not altogether lacking. In addition to morphological and developmental criteria, physiological and specifically immunological parameters may be used to characterize taxa and strains of phycobionts (compare References 19 and 20). Such studies may demonstrate that there is a greater variation in the subspecies level of algal symbionts than has been hitherto recognized.

The systematic position of the algal partner may or may not be correlated with that of the fungal partner. In certain cases, related fungal partners can be lichenized with related algal partners. In other cases related fungi can associate with quite different algae. The first case is demonstrated by the genus *Cladonia* in which all species so far investigated (about 28) are lichenized with *Trebouxia* Pulmaly emend. Archibald.[11,21] Other examples include *Peltigera* and *Solorina*; the green algal primary symbionts belong to *Coccomyxa*, the secondary symbionts are *Nostoc*. As an example of a fungal species of one genus associating with various unrelated algae, *Verrucaria* may be mentioned. Until now, six algal genera (*Heterococcus*, *Petroderma*, *Myrmecia*, *Coccobotrys*, *Dilabifilum*, and "*Protococcus*") of quite different systematic position (Tribo-, Fuco-, Chlorophyceae) have been reported as symbionts. One might assume that a diversity of this sort would be characteristic of certain primitive crustose lichens. However, observations disprove this view. In the genus *Lobaria*, with highly developed foliose thalli, two genera of primary phycobionts, *Dictyochloropsis* and *Myrmecia*, are known; in *Pseudocyphellaria*, three, *Trebouxia* (or *Pseudotrebouxia*), *Dictyochloropsis*, and *Chlorella*, can be found (see Table 1). The phycobionts show no apparent specificity. In many cases, the same algal species can be lichenized with fungi of quite different systematic position.

A key to the algal genera found in lichens has been presented by Ahmadjian.[22] Detailed and richly illustrated surveys on all phycobiont genera and species known until 1967 were published by Ahmadjian[3] and Letrouit-Galinou.[23] The reader is referred to these treatises, as such a broad description and illustration of all known phycobionts will not be presented here. In this survey, only some of the most common phycobionts known until 1967 are described and illustrated, and a number of species newly described or first observed as phycobionts since 1967 are treated in some detail (Section V). As far as possible, a complete compilation of all known phycobionts and their fungal partners together with some other data are presented in Table 1 and Section III. Contradictions in the literature and further questions necessitated the addition of Section IV entitled "Comments to List of Phyco- bionts". The reader is requested to give this section careful consideration, as it contains much relevant information.

III. LIST OF PHYCOBIONTS AND LICHENS IN WHICH THEY OCCUR

As already mentioned, algal species and genera known to occur as lichen partners until 1967 have been extensively treated.[3,22,23] Since then, statements on the systematic position of phycobionts appeared scattered in a wide range of scientific journals which may or may not be readily accessible. Thus, an attempt was made to collect old and new respective data into a single list. In Table 1 the algae are ordered according to recent classifications. Dependent on available data, algal species or genera are advanced. In Table 2, an alphabetized list of all treated phycobiont genera is presented.

In order to facilitate establishment of future supplementary lists and perhaps a new key, investigators are requested to send offprints of relevant publications with a mark "algal taxonomy" to the Library of the British Lichen Society in Bristol.

IV. COMMENTS TO LIST OF PHYCOBIONTS

1. On general principles, the author prefers the designations Cyanophyceae (and blue- green algae) to Cyanobacteria (compare References 24 to 28 and also discussion and literature in Reference 29). The system is presented here according to Friedmann.[30]

2. In this case, the phycobiont in the cephalodia was identified by Drouet[31] (probably *in situ*). *Anacystis* is nowadays not acknowledged as a distinct genus. It was revised by Drouet and Daily.[31] The revision of the taxonomy of coccoid Cyanophyceae by these authors is considered inadequate (compare References 29 and 30) and was rejected.

3. *Anacystis montana* was determined by Drouet from cultures of Wetmore.[32] According to Drouet and Daily[31] it comprises among others *Gloeocapsa alpina* Ng., *G. rupestris* Ktz., and *G. sanguinea* Ktz. With regard to the phycobionts of *Peltula* spp., compare also Ref- erence 33.

4. The determination of Duvigneaud[158] appears questionable. In his Figure 30E, he depicted the phycobiont of the cephalodium in chain-like aggregates which is not consistent with characteristics of *Aphanocapsa*. In his general account of the genus *Stereocaulon*, he also mentioned that *Chroococcus* occurred as a phycobiont in cephalodia. This also does not seem to be well founded.

5. According to Jaag,[34] *Gloeocapsa sanguinea* should include *G. alpina* Näg., *G. magma* (Brb.) Ktz., and *G. ralfsiana* (Harv.) Kütz.

6. Ahmadjian[3] quoted Jaag[34] as having found *Chroococcus* as the symbiont of *Pyre- nopsidium* and *Phylliscum*. However, Jaag suggested that the phycobiont of *Phylliscum endocarpoides* might be either *Chroococcus* or, more probably, *Gloeocapsa sanguinea*. He did not investigate *Pyrenopsidium*; he only stated (Reference 34, pp. 388, 395) that Zahlbruckner[35], whom he quoted only indirectly, thought the phycobionts of both lichen genera belonged to *Chroococcus*. Nowadays, it is generally accepted that Zahlbruckner's statements with respect to phycobionts are only provisional since he relied on questionable 19th century information. This should be kept in mind in those cases where Zahlbruckner is given as a reference. According to Letrouit-Galinou,[23] Jaag[34] isolated three different *Gloeocapsa* spp. from *Phylliscum endocarpoides*. The present author could not find any such comment in Jaag's publication.

Henssen[36] stated that the phycobionts of all *Phylliscum* spp. belonged to the Chroococ- caceae and have red gelatinous sheaths, suggesting that they are *G. sanguinea*. Later,[37] the phycobiont of *Phylliscum macrosporum* Henssen was identified as *Chroococcus* sp., a genus normally with colorless or yellow-brown sheaths.

7. According to Poelt[38] the most commonly encountered phycobiont in the cephalodia of *Stereocaulon alpinum* is *Nostoc*; Wirth[39] also stated it to be *Nostoc*. Lamb (Reference 18, p. 542) stressed that different algal types may occur in the cephalodia of the same species of *Stereocaulon*.

8. Since the main phycobionts in the cephalodia of *Stereocaulon ramulosum* are *Stigonema* and *Scytonema* and since *Stigonema* has a *Gloeocapsa*-like chroococcalean state, the presence of *Gloeocapsa* should be reexamined.

9. According to Jaag,[34] *Gloeocapsa kuetzingiana* Näg. emend. Jaag includes *G. rupestris* Kütz. and *G. pleurocapsoides* Něk. *pro parte*.

10. It is not clear where this taxon belongs. According to Henssen,[40] *Psorotichia rehmica* Mass. = *Porocyphus rhemicus* (Mass.) Zahlbr. which is lichenized with a *Calothrix* sp.

11. Büdel and Henssen[41] doubt some or all of the classical identifications of coccoid blue-green algal phycobionts found in the "Pyrenopsidaceae" (now included in the Lichinaceae). Many of these identifications were based on comparative investigations of algal colonies which became free of the fungal partner at the periphery of the thalli, and of free-living algae on the substrate surrounding lichen specimens. In the present author's opinion, at least some of the identifications were reliable. It is also proposed that free-living *Gloeocapsa* spp. be thoroughly investigated in culture for the presence of endospore-like products, and compared to strains isolated by Büdel and Henssen[41] from three Lichinaceae (see *Chroococcidiopsis*). Friedmann[283] thinks the determinations of Büdel and Henssen to be correct.

Bubrick and Galun[33] also question *in situ* identifications of unicellular blue-green algal partners of the Lichinaceae and Heppiaceae. In unialgal cultures they observed a greater diversity of types (not identified as to species) than formerly accepted. The situation was complicated by the fact that *in situ* the algae were modified by the fungus so that some essential characters were modified or lost. Examinations of algae both *in situ* and in culture is, therefore, recommended.

12. According to Hertel[42] *Huilia panaeola* = *Amygdalaria panaeola*.

13. In the two mentioned specimens, Jaag (Reference 34, p. 391 f) besides the two quoted *Pyrenopsis* spp. with *Gloeocapsa sanguinea* as a phycobiont, found what he thought to be *Psorotichia* with *G. kuetzingiana*. Apart from the algal species, he could not find any other differences. This case may parallel that discussed under 15. It seems premature to follow Letrouit-Galinou[23] in interpreting *Pyrenopsis fuliginoides* as lichenization between one fungal and two algal species.

14. This lichen, investigated by Jaag,[34] was derived from specimens belonging to ZT. According to Poelt,[43] many of the old determinations (e.g., *Pyrenopsis* and *Psorotichia* spp.) were probably wrong. *Synalissa acharii* Fries is possibly synonymous to *S. symphorea*.

15. Jaag[34] (p. 395), in his lichen specimen No. 19 between thallus parts with all characters of *Synalissa ramulosa*, found thalli which contained *Gloeocapsa kuetzingiana* with yellow sheaths. He thought they belonged to a different lichen which, apart from the phycobiont, shows identical anatomical details with thallus parts containing *G. sanguinea*. From this and other obsolete statements, later authors[37] concluded that *S. symphorea* may be lichenized with three different algal species (*G. alpina*, *G. sanguinea*, *G. kuetzingiana*). However, Jaag demonstrated that *G. sanguinea* and *G. alpina*, and the respective lichens, were conspecific. This was later confirmed by Schiman.[44]

The parts of Jaag's specimen harboring *G. kuetzingiana* do not belong to *S. symphorea*, but to *Peccania coralloides* Mass. This was established by the present author[169] in the course of a reinvestigation of the specimen from Zürich which contains several small thalli of *S. symphorea*, one of which was intermingled with *P. coralloides*.

16. From cultural experiments of Bubrick and Galun,[33] one can gather that the phycobionts of a number of Lichinaceae belong to the Chroococcales.

17. Lamb[45] describes the alga in the cephalodia in the following way: " . . . nests of blue-green algae of gloeocapsoid appearance (apparently compacted *Stigonema*) . . . ''.

18. From a discussion on *Ephebe pubescens* (= *E. lanata*) by De Bary[46] one can indirectly conclude that the phycobiont was *Stigonema mamillosum* Ag. ex. Born. et Flah. De Bary compared *E. pubescens* with *Stigonema atrovirens* Ag. which, according to De Toni,[47] is really *Stereocaulon mamillosum* and is lichenized.

19. It is possible that Borzi's and Frémy's type materials of *Hyphomorpha antillarum* and *H. perrieri* were not free-living algae, but lichens whose fungal partner had been overlookcd, and which caused a *Stigonema*-like alga to change its growth form. The type material of *H. perrieri* was lost during World War II; if that of *H. antillarum* still exists, it should be reexamined.

20. Future investigations should determine whether *Calothrix crustacea*, *C. pulvinata*, and *C. scopulorum* are separate species. The plasticity of cultured *Calothrix* depending upon environmental factors has been documented.[48]

21. Identification by Drouet.[31]

22. All statements referring to *Anabaena* in cephalodia of *Stereocaulon* should be reexamined by means of cultures, since the more distinct gelatinous sheath characteristic of *Nostoc* might be modified *in situ*.

23. Letrouit-Galinou[23] also mentioned that according to Zahlbruckner[35] most of the Heppiaceae are lichenized with *Nostoc*. The present author could not find such a statement by Zahlbruckner. References from other authors suggest *Scytonema* as a common phycobiont in the Heppiaceae.

24. According to Jordan and Rickson[49] *Nephroma arcticum* may have two distinct morphological forms of blue-green algae in the same thallus and occasionally in the same cephalodium. The authors point out the possibility that the two forms may represent a single taxonomic species (see also Reference 50, page 26).

25. Danilov,[51] Geitler,[52] and Degelius[53] doubt *Nostoc punctiforme* to be a good species. Therefore, new cultural experiments with the *Nostoc* phycobionts of the quoted *Peltigera* spp. should be carried out.

26. The determination of *Peltigera rufescens* by Schiman[44] has been confirmed by the present author[278] and by Poelt.[279]

27. Lamb described *Scytonema* and a chlorococcalean phycobiont as morphologically balanced partners of *Compsocladium*. James and Henssen[8] assume this lichen possesses cephalodia.

28. Santesson[7] states that the phycobiont of *Coccocarpia* resembled a species of *Scytonema* with occasionally separated cells. According to Arvidsson,[54] the phycobionts of various *Coccocarpia* species look very similar and "microscopic studies do not indicate the presence of more than one species." He could not find the false branching typical of *Scytonema* (nor any branching), but thought that the alga belonged to this genus and was very much modified by lichenization.

29. Parmasto[55] transferred *Cora pavonia* to the genus *Dictyonema* and said it was lichenized with "*Chroococcus* (A species of Rivulariaceae)". Both of these latter statements clearly are erroneous. The *Chroococcus* identification relies on obsolete 19th century information. Bornet[56] who is famous for his good observations for his time, was also incorrect in thinking that the phycobiont of *Cora pavonia* belonged to the Chroococcaceae. He also erroneously thought that the fungal partner of *Dictyonema sericeum* grew within the sheaths of the *Scytonema* filaments. Later investigations on several *Dictyonemas* spp. showed it to form a tight sheath outside of the thin algal sheath, and a haustorial system within the algal trichomes.[57] In other cases (*Thermutis velutina* and *Coccocarpia cronia*), the fungal hyphae grow within the sheaths of the *Scytonema* filaments.[58,59]

30. Algal filaments growing out of fragments of thalli of *Heppia echinulata* are notably

more narrow at the apex than at the base. This may result from the fact that algal cells are enlarged as a consequence of lichenization and only gradually regain their normal dimensions. On the other hand, the reduction in breadth could be an inherent character of the alga; in this case it could not belong to *Scytonema*. Whether or not the false branching typical of *Scytonema* occurs has not been reported.

31. According to Wetmore,[32] Drouet identified the phycobiont of *Heppia lutosa* (from cultures established by Wetmore) as *Scytonema hoffmannii* Ag.

32. Ahmadjian[3] also mentioned *Pannaria* to be lichenized with *Scytonema*. This probably goes back to Bornet,[56] who investigated *Pannaria triptophylla* var. *nigra* Nyl., an early synonym for *Placynthium nigrum* (Huds.) Gray. This species is lichenized with an alga belonging to the Rivulariaceae, very probably *Dichothrix orsiniana*. Letrouit-Galinou[23] cited Zahlbruckner[35] as a reference for lichenization of *Pannaria* with *Scytonema*, but Zahlbruckner ascribed "*Nostoc*-Gonidien" to *Pannaria*.

33. Systematics according to Christenssen.[60]

34. Several thalli checked by Tschermak[61] and Zeitler[62] and many more used for teaching purposes by the present author [collected in Gschnitzbach and smaller rivers nearby in North Tyrol (Austria) and a small river near Jaufenpass in South Tyrol (Italy)].

35. Based on *in situ* and culture studies, Tschermak-Woess[63] found the phycobiont of *Verrucaria maura* from Brean Down, North Somerset, England, to belong to the Chaetophorales. Further studies on the identity of the phycobiont of *V. maura* are needed.

36. This marine *Verrucaria* sp. was identified by Swinscow[280] and is still insufficiently known. For other rather "loose", submarine lichenoid associations and their algal partners see References 64 and 65.

37. Systematics according to Silva.[66]

38. The phycobiont of *Multiclavula mucida* possesses a pyrenoid. Its systematic position is dubious, since the species included in *Coccomyxa* have no pyrenoid.

39. The genus *Coccomyxa* is in need of a new monographic treatment. Probably several of the species given in Table 1 should be reduced to intraspecific taxa or strains. Komárek and Fott[67] think that some species should be placed in different genera (e.g., *Gloeocystis*, *Pseudococcomyxa*, *Choricystis*). In their opinion those lichenized taxa which do not produce gelatinous sheaths should be excluded from *Coccomyxa*. The present author found that in certain chlorophycean phycobionts (e.g., *Elliptochloris bilobata*, *Dictyochloropsis* spp., *Coccomyxa icmadophilae*) production of a sheath was variable and could be suppressed by lichenization.

From certain of Jaag's[68] illustrations, Komárek and Fott[67] concluded that some species have polar cushions of jelly by which they adfixed to a substrate, implying a connection with *Pseudococcomyxa*. The present author thinks this interpretation to be erroneous.

40. Formerly *Baeomyces rufus* was thought to be lichenized with *Coccomyxa*, but recently was shown to contain *Elliptochloris bilobata*.

41. Letrouit-Galinou[23] (citing Jaag[68]) also lists *Coccomyxa simplex* as one of the phycobionts of *Solorina saccata*. However, Jaag only mentions that there are some similarities between this free-living species and *C. ovalis*.

42. *Gloeocystis* is a problematical taxon which has not been uniformly defined in phycology. Lichenologists often use the designation *Gloeocystis* indiscriminately for green phycobionts embedded in a gelatinous matrix. If accepted in the sense of Hindák,[69] reexamination in some cases is necessary.

43. Hedlund[70] reported on phycobionts which he thought were *Gloeocystis*-like, but his observations and conclusions do not seem reliable. The same holds true for statements of Hue,[71] who claimed to have found *Gloeocystis* alone or together with *Scytonema* in the cephalodia of a lichen.

44. The statement of Kupffer[72] that *Stereonema chthonoblastes* is lichenized with *Gloeocystis*, *Cystococcus*, and *Chlorococcum* is highly questionable.

45. Jaag and Thomas[73] described the phycobiont of *Epigloea bactrospra* as *Coccomyxa epigloeae*. It was said to be very similar to *C. dispar* Schmidle, but to differ from it by the possession of a pyrenoid. With good reason Komárek and Fott[67] placed the alga in question in the genus *Gloeocystis* as redefined by Hindák[69] (compare Reference 74).

46. With an improving knowledge of the genus *Trebouxia* De Pulmaly emend. Arch., the genus *Asterochloris* will possibly have to be included in it (compare Scheidegger[75] and Gärtner[76]).

47. From Plate 3a in Yoshimura,[77] it appears that the alga in question is not *Dictyochloropsis reticulata*, but a different *Dictyochloropsis* species, perhaps *D. symbiontica*.

48. In the literature, the phycobionts of certain lichens are indicated as "?*Myrmecia*" or "*Myrmecia* sp." (e.g., Brodo[78] with respect to two species of *Coccotrema*, the phycobionts of which were cultured and determined by Ahmadjian). Since *Myrmecia reticulata* had to be transferred into the genus *Dictyochloropsis*, the position of the respective algae (*Dictyochloropsis* or *Myrmecia biatorellae?*) is now uncertain.

49. On free-living *Dictyochloropsis reticulata* compare References 79 and 80.

50. In the quoted publication, *Lobaria pulmonaria* var. *pulmonaria* and var. *meridionalis* were not distinguished. The thalli from Teneriffa belonged to var. *meridionalis*. The material investigated since 1978[80] comprised one thallus from Sardinia, Italy (pass 5 km northwest Bolotana, leg. Brunnbauer, April 4, 1983), cultures of the phycobiont of a thallus collected in the state of Maine and established by Ahmadjian, and one culture established by Brunner from a thallus collected near Klöntalersee, Kanton Glarus, Switzerland, October 19, 1981.

50a. The identification of the phycobiont of *Phlyctis argena* probably was erroneous. On the basis of new unpublished data it does not belong to *Dictyochloropsis reticulata* but to *D. splendida*.

51. Geitler[81] found *Sarcogyne pruinosa* to be lichenized with *Myrmecia biatorellae* (compare discussion of this case by Tschermak-Woess, Reference 80, p. 76f).

52. Raths[82] isolated a variety of *Stichococcus bacillaris* from *Chaenotheca brunneola*. It seems possible that her lichen material was in part incorrectly identified (compare Tibell — Reference 15, p. 639). Tibell[14-16,83] (by *in situ* investigation, probably of herbarized material) found *C. brunneola* from Costa Rica, New Zealand, and Australia lichenized with *Trebouxia* (or *Pseudotrebouxia*) and considered this to be a case of vicarious phycobionts.

53. *Myrmecia biatorellae* was found probably free-living in soil by Trenkwalder,[84] Vinatzer,[85] and Watanabe.[86] Andreeva et al.[273] isolated a strain of *Myrmecia* from soil in the Ukraine which they identified as *M. biatorellae*. Since its zoospores, in contrast to those of *M. biatorellae*, have stigmata, this alga might belong to *M. bisecta* Reisigl (compare Reisigl[274]).

54. The authors[227] state *Lecidea decipiens* to be lichenized with *Myrmecia*. From their Figure 9 it is assumed to be *M. biatorellae*.

55. Considerable uncertainties exist with respect to the taxonomy of trebouxioid algae. A number of authors acknowledge that two groups of species exist in *Trebouxia* de Pulmaly (*Cystococcus* auct., non Nägeli). One group (No. I) reproduces only by zoospores and aplanospores (in the sense of arrested zoospores) which occur in relative high numbers (32, 64, 128). Zoospores, aplanospores, and autospores are typical of the second group (No. II). Autospores are formed in low numbers (4 or 8, sometimes 16) and are at first flattened against each other. Warén,[87] using a different terminology, assembled the species of group I in the subgenus *Eleuterococcus* of *Cystococcus*, and those of group II in the subgenus *Eucystococcus* (compare also Reference 88). Archibald[89] left only the species of group I in the genus *Trebouxia* de Pulmaly emend. Archibald and introduced the genus *Pseudotrebouxia* Archibald for group II; *Trebouxia* was assigned to the Chlorococcales, while *Pseudotrebouxia* was incorporated into the Chlorosarcinales. Tschermak-Woess[90] protested against this wide separation, as it separated apparently closely allied species groups. Beyond that, the Chlorosarcinales should comprise only species with bipartition and not those with multiple division

as seen in *Pseudotrebouxia*. Mattox and Stewart[91] place *Trebouxia* and *Pseudotrebouxia* in the order Pleurastrales of the Pleurastrophyceae.

Another question refers to the maintenance of *Pseudotrebouxia*. Formerly, the present author[90] pleaded for its reinclusion in *Trebouxia* by referring to the analogous situation in *Dictyochloropsis*. In this genus of the Chlorococcales, autospore-producing species and species not producing autospores clearly belong together on account of several common characters. In addition, certain taxa produce autospores only at a very low rate and this might be considered as showing an intermediate behavior (compare also References 17 and 92). In the case of trebouxioid algae, however, the situation is somewhat different because the members of the two groups (group I = *Trebouxia*, group II = *Pseudotrebouxia*) have become more diversified in several characters and no taxa are known which might be considered intermediate in their autospore formation abilities. Thus, retention of the genus *Pseudotrebouxia* (but in the Chlorococcales) seems advisable.

Gärtner[93] (compare also Reference 94) has stressed that the division processes in all *Trebouxia* (including *Pseudotrebouxia*) species is that of "cytogonie" (multiple division) and does not acknowledge the differences between autospores and aplanospores. According to him, *Pseudotrebouxia* should not be maintained. Recently, König and Peveling[95,96] showed differences in cell wall properties between some *Trebouxia* and *Pseudotrebouxia* species and argued for the retention of *Pseudotrebouxia*. Clearly, a well-founded monographic treatment of *Trebouxia* and *Pseudotrebouxia* is urgently needed. The statements in Table 1, therefore, are provisional and do not go into much detail. Species designations of phycobionts and mycobionts are given only in a small number of examples. Lists of *Trebouxia* and *Pseudotrebouxia* species isolated from lichens are given by Ahmadjian[2,11] (see also Reference 97). While finishing the present contribution, proofs of a monographic treatise of the genus *Trebouxia* (including *Pseudotrebouxia*)[76] came into this author's hands. This monograph convinced the present author that by not keeping *Trebouxia* and *Pseudotrebouxia* apart, certain important characteristics may be obscured.

55a. Archibald[89] described *Pseudotrebouxia aggregata* on the basis of cultures established by Quispel:[231] "Quispel from *Xanthoria* sp. as *Cystococcus humicola*". Quispel (Reference 231, p. 442) isolated *Cystococcus* from *Xanthoria parietina* (and a *Parmelia* and *Physcia* sp.), but did not identify it as to species.

56. The phycobiont of *Xanthoria parietina* has been cultured and investigated by several authors and from different locations (e.g., Warén[87], Werner[98]). In general, it belongs to the genus *Pseudotrebouxia* (excepting perhaps *Trebouxia italiana*).

57. Raths[82] found *Chaenotheca chrysocephala* to be lichenized with *Cystococcus cladoniae furcatae* Chod. and *C. cladoniae endiviaefoliae* Chod.; both probably belong to the genus *Trebouxia* de Pulmaly emend. Arch. She gave no cytological details and her lichen material was not preserved. She isolated "*C. cladoniae endiviaefoliae*" (No. 54) from a thallus from Lunz (Niederösterreich). In several thalli from Lunz, Tschermak-Woess[17] found *Chaenotheca chrysocephala* to be lichenized with *Pseudotrebouxia simplex*.

58. Jaag[99] described four taxa of *Cystococcus* isolated from *Parmelia caperata* which should probably be assigned to *Trebouxia* de Pulmaly emend. Arch. From the few cytological characters given, it is questionable whether they can be maintained as separate species. In two specimens from different sites in Austria, Tschermak-Woess[100] also found *Trebouxia*. Wang-Yang and Ahmadjian[101] investigated 32 specimens of *Parmelia caperata* from six countries and found certain differences between their isolates. One can assume indirectly that, in the view of the authors, some of them should be given species rank. Meisch[21] isolated *Trebouxia gelatinosa* from one thallus from Tyrol, but found a *Pseudotrebouxia sp.* in a specimen from Sweden.

59. According to Wang-Yang and Ahmadjian[101] isolates of *Trebouxia* from 34 specimens of *Cladonia rangiferina* from 13 countries fell into 5 groups. These apparently were thought to represent different species.

60. Zehnder[102] was not sure whether the alga isolated from *Lecidea parasema* was its phycobiont. From Geitler's[103] description and drawings, it can be concluded that the phycobiont of this lichen belongs to *Pseudotrebouxia*.

61. Komárek and Fott (Reference 67, p. 602) think that the characters of *Chlorella lichina* as well as of *C. cladoniae*, as described by Chodat,[104] do not harmonize with those of the genus *Chlorella*. Chodat found both algae as epiphytes on *Cladonia* spp. (Reference 104, pp. 92, 108) and not as their phycobionts as presumed by some authors (compare Raths — Reference 82, page 341f).

62. Tibell[15] thinks that the identification of *Chlorella lichina* by Raths[82] most probably was based on an isolation from a misidentified *Lepraria*.

63. Zeitler,[62] in a lichen identified as *Lecidea coarctata*, found *Protococcus verrucariae acrothelloides*. The lichen in question may have been a different species.

64. With respect to the primary phycobiont of *Stereocaulon strictum*, statements are conflicting. Lamb[105] reported it to belong to *Trebouxia*; Ahmadjian[2] mentioned it to be *Pseudochlorella*. The latter author gives no author's name of *St. strictum*, so it remains somewhat uncertain whether he investigated the same lichen as Lamb.

65. Recently, Trenkwalder[84] found *Chlorosarcinopsis minor* probably free-living in soil of a *Pinus silvestris-Quercus pubescens* wood in South Tyrol (Italy).

66. According to light microscopical analysis, this alga divides by bipartition and not by autosporulation. Brown and Elfman[106] showed autosporulation in some other putative *Nannochloris* spp. Therefore, the systematic position of the phycobiont of *Normandina* remains uncertain.

67. This and some other chlorophycean algae with rudimentary filaments are not mentioned in the systematic treatise of Silva.[66] They are here classified according to Bourrelly.[107] Warén[87] also describes *Coccobotrys lecideae* from *Lecidea uliginosa* (Schrad.) Ach. (= *L. fuliginea* Ach.). From his description and figures, it appears questionable whether this phycobiont belongs to the genus *Coccobotrys*.

67a. *Leptosira thrombii*, in the opinion of Tupa,[277] should also be included in *Pleurastrum*.

68. This and the following algae provisionally named *Protococcus* were investigated in detail by Zeitler.[62] She intentionally did not formally describe them as new species. In the cultures of *P. dermatocarponis miniati*, Zeitler observed short filaments; the other ''species'' showed cell packets, irregular assemblages, and isolated cells. Vischer[108] thinks that Zeitler's *Protococcus* belongs to *Diplosphaera* Bial.

69. According to Stahl,[109] the phycobiont of *Thelidium minutulum* is identical with that of *Endocarpon pusillum*.

70. The genus *Trochiscia* is not well founded. Species assumed to belong to it are rather heterogenous. The alga in question was studied *in situ* and in enrichment cultures. Because it forms short filaments, it cannot belong to the Chlorococcales as formerly held (see also Letrouit-Galinou[23]).

71. In Costa Rican specimens (seven investigated), Tibell (Reference 14, p. 229) found *Chaenotheca carthusiae* lichenized with *Trebouxia* (or *Pseudotrebouxia*). He thought this case to parallel that of *C. brunneola* and *C. chrysocephala* and to represent one of vicarious phycobionts.

72. On the basis of biometrical and cultural data, Raths[82] differentiated between eight varieties of *Stichococcus bacillaris*; five of them were found in *Chaenotheca furfuracea*. Their significance as varieties may be doubted (compare Vinatzer[85]); possibly they can be kept as strains. In the table, Rath's varieties are not treated separately. *S. coniocybes*, described by Letellier,[110] probably belongs to *S. bacillaris*.

73. Tschermak-Woess[12] stated that *Chaenotheca brunneola* was lichenized with *Dictyochloropsis splendida*, and *C. melanophaea* with *Pseudotrebouxia* (referred to in the respective publication as *Trebouxia*). Tibell[111,112] also described the phycobionts of these lichens as

belonging to the Chlorococcales (with respect to *C. brunneola*, compare comment 52). Tibell[14] mentions *Trebouxia* (probably including *Pseudotrebouxia*) in five specimens of Costa Rican *C. ferruginea*.

74. It appears doubtful whether this species can be separated from *Stichococcus bacillaris*.

75. The assumption that the phycobiont of *Endocarpon pusillum* is *Stichococcus* is problematical as it has a pyrenoid, while *Stichococcus* as generally conceived has no pyrenoid. In the hymenium it forms packets which is also not characteristic of *Stichococcus* (compare Plate 2c in Ahmadjian and Heikkilä[113]). The taxonomical history of *Diplosphaera chodati* Bial. (= *S. diplosphaera* (Bial.) Chod.) is controversial as discussed by Letrouit-Galinou (Reference 23, p. 66) (also see table under *Protococcus dermatocarponis miniati*).

76. Some authors doubt the two *Turgidosculum* consortia to be true lichens. Printz[114] assumed that the algal partner of *Mastodia tesselata* belonged to *Prasiola fluviatilis* (Sommerf.) Aresch. According to Reed,[115] *Guignardia alaskana* Reed is associated with *Prasiola borealis* Reed which also occurs free from the fungus (compare the literature quoted by Kohlmeyer and Kohlmeyer[65] and Henssen and Jahns[37]).

77. Kohlmeyer and Kohlmeyer[65] raise the question whether *Blidingia minima* var. *vextata* could be *Blidingia minima* var. *minima* modified by the influence of the fungal partner. Free-living var. *minima* occurs in the same habitats as the lichen or lichen-like association (compare also Norris[116]).

78. Uyenco[117] erroneously thought that Skuja and Ore[118] identified the phycobiont of *Coenogonium nigrum* as *Trentepohlia umbrina*. These authors carefully compared material of *T. umbrina* from nature and from cultures with their alga and did not find them identical. Vainio (Reference 119, p. 158) and Puymaly[268] considered the phycobiont of this consortium to belong to *T. aurea*.

79. Reinke[120] depicted filaments and tufts of filaments of *Trentepohlia aurea* growing out of the thallus of *Gyalecta jenensis*. The present author also often saw *T. aurea* apparently growing out of the thallus of this lichen. Henssen and Jahns[37] considered them to represent epiphytes. New, accurate anatomical investigations are needed.

80. From the short communication and very schematically drawn figures of Hérisset,[121] it seems dubious whether his statements are correct.

V. EXAMPLES OF PHYCOBIONTS

Depending on available data and figures, algal genera and/or species are described. As far as possible, each taxon is characterized and depicted in the free-living (in culture and/or in nature) and in the lichenized state.

A. Common Phycobionts

Gloeocapsa sanguinea **free-living (from nature)** — Cells are spherical (diameter is 4 to 6 (8) μm), in colonies of 2 to 8 or often more cells, surrounded by stratified or homogenous gelatinous sheaths, joined to each other or sometimes totally homogeneous (*Aphanocapsa* state); sheaths are red, blue, or sometimes nearly black (pigment gloeocapsin) and in the form of grains or homogeneously distributed in light-exposed parts of the sheaths. Cell division is by binary fission; at high frequency division rates, nannocytes; these may appear similar to endospores (baeocytes); in general storage of cyanophycin. Akinetes ellipsoidal or spherical with narrow sheaths. Growth form in nature is rather variable (comp. Geitler — Reference 122, Abb. 71);[275,276] the variability of *Gloeocapsa* spp. in culture should be checked, as forms identified as *Chroococcidiopsis* possibly may belong to *Gloeocapsa*. Found on naked, often extremely sun-exposed rocks; is often the main component of black crustaceous vegetation, in tufts of mosses, etc. (Figure 1a to 1d).

Gloeocapsa sanguinea **lichenized** — As a phycobiont of *Synalissa symphorea*, cells are

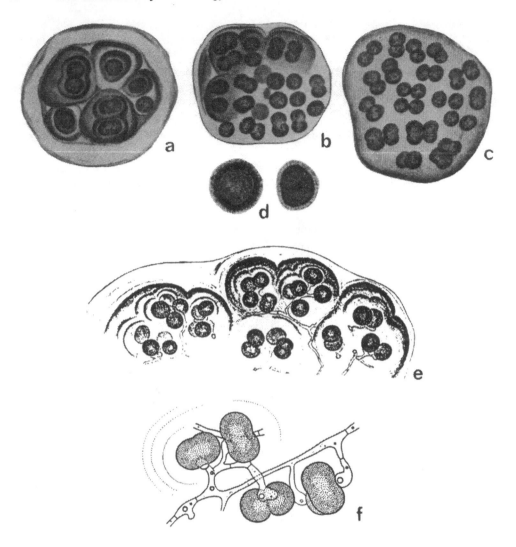

FIGURE 1.   *Gloeocapsa sanguinea.* a to d = free-living (very probably from nature; a and b = with typical sheaths; c = Aphanocapsa state; d = akinetes; e and f = lichenized in *Synalissa symphorea*. (Figures a to d redrawn from Novacek, F., *Sb. Klubu Prirdovedeckeho Brne*, 13, 1, 1930, and Geitler, L., in *Encyclopedia of Plant Anatomy*, Zimmermann, W. and Ozenda, P., Eds., Borntraeger, Berlin, 1960, 86. Figures e and f from Geitler, L., *Arch. Protistenkd.*, 80, 378, 1933. With permission.)

relatively large and without cyanophycin; distinct stratification of sheaths is found only in peripheral parts of thallus (Figure 1e and 1f); akinetes and nannocytes were not observed.

Nostoc **spp. free-living (in nature and in culture)** — Gelatinous thalli, in certain species, are up to several centimeters large, consisting of more or less long filaments in which trichomes* in the form of strings of beads are surrounded by mostly thick sheaths. Sheaths appear individualized and are often stratified, are brown at the periphery of thalli, and merge to an uncolored jelly in their interior. Cells are spherical, ellipsoidal, in general of equal breadth; heterocysts intercalary; akinetes are found often in high numbers in succession. Cell division is perpendicular to the long axis of the trichome; there is no branching (Figure 2a). Occurs in fresh water, seldom in brackish or marine; aerophytical on periodically wettened earth, between mosses, etc.

---

*   Trichomes are filaments without sheath.

Nostoc **spp. lichenized** — Habit is little changed in *Collema*, *Leptogium*, and some species of *Lempholemma* (Figure 2b and 2c); in other *Lempholemma* spp., trichomes are partitioned into small groups of, for example, 5 cells vs. 18 cells in parts where algae have outgrown the fungus.[44] In these latter parts, trichomes are much more narrow than in older intensely lichenized parts (Figure 2d).[123] In other genera (e.g., *Peltigera*), habit is strongly changed; width of sheath is reduced, filaments are more convoluted and shorter than in the free-living state; vegetative cells are enlarged; akinetes are usually not observed.

Scytonema **spp. free-living (in nature)** — Filaments are isolated (exception reported in Reference 281), in turfs, tufts, or crustaceous; trichomes appear singly surrounded by relatively compact, colorless, brownish-yellow to brown sheaths. False branchings are generally one-sided and formed in twos, and are initiated as a loop or by outgrowths beneath and above dead cells or heterocysts. Hormogonia originate from ends of trichomes (Figure 3a). Occurs in fresh water, on earth, in tufts of mosses, etc.

Scytonema **sp. lichenized** — In *Pyrenothrix* and *Dictyonema*, not much modified (Figures 3b to 3d), only trichomes broadened (compare Figures 3d and 3c);[57,124] in other lichens more spectacular varied (compare Henssen and Jahns — Reference 37, pp. 21—23). Filaments may be reduced to apparently unicellular forms (e.g., Reference 157).

Stigonema **spp. free-living (in nature)** — Thallus consists of free filaments in turfs or crusts; trichomes, at least in old parts, are double or multirowed; true branching by longitudinal division of cells. Sheaths are colorless, yellow to brown; heterocysts intercalary or lateral; hormogonia originate from apical regions of young branches. In older parts, *Gloeocapsa*-like stages can be observed (Figures 4a and 4b). Occurs on rocks, bark, earth, in peat-bogs, etc.

Stigonema **spp. lichenized** — Not drastically changed in *Ephebe* spp. (Figure 4c), more altered in *Spilonema* (Reference 37, p. 23) and in fully grown cephalodia of *Pilophorus strumaticus*[125] (Figures 4d to 4f).

Coccomyxa **spp. free-living (in nature and in culture)** — Cells are isolated or in gelatinous colonies, with or without a gelatinous, structureless, or stratified matrix; cells are ellipsoidal, ovoid, or seldom spherical. Parietal chloroplast is situated mostly along the longitudinal wall and without a pyrenoid. Propagation is by autospores (two, four, and sometimes eight), the first (or single) plane of division is mostly oblique; no motile stages (Figures 5a, 6a, and 6b). Occurs acrophytic on trunks of trees, in tufts of mosses, on damp wood, earth, rocks, but is seldom aquatic. Skuja[126] discriminated between the sections *Eucoccomyxa* with gelatinous colonies and *Choricystis* with solitary cells devoid of mucilage. Fott[127] thinks that the two sections should be elevated to the rank of genera. However, it appears that mucilage production in *C. icmadophilae* (and also in *Elliptochloris* and *Dictyochloropsis*) may depend to some extent on growth conditions.

Coccomyxa **spp. lichenized** — In most lichens, the shape of cells is broader and more near spherical than free-living;[68,128] cells apparently enlarged, in most cases are not reported to be surrounded by a gelatinous sheath (Figure 5c). In young superficial thallus parts of *Icmatophila ericetorum* sheaths are not observed; in older thallus parts, sheaths present (Figure 6c).[61]

Dictyochloropsis reticulata **free-living (in nature and in culture)** — Cells are spherical (diameter 4 to 20 μm), solitary, with one peripheral, reticulate chloroplast in the shape of a cup (young cells) or a hollow sphere (more or less fully grown cells) (Figures 7a to 7c). In vegetative cells, the reticulum is usually in one layer, seldom one or two trabecular parts reaching inwards towards the centrally located nucleus; chloroplast without pyrenoid. Between the nucleus and the chloroplast a group of dictyosomes can sometimes be observed by light microscopy (Figures 7f and 7g). Propagation is by 4, 8 (16) autospores (Figure 7d), by zoospores, or aplanospores (in the sense of arrested zoospores); both of the latter originate in high numbers (32, 64) from relatively large mother cells in which the chloroplast lattice

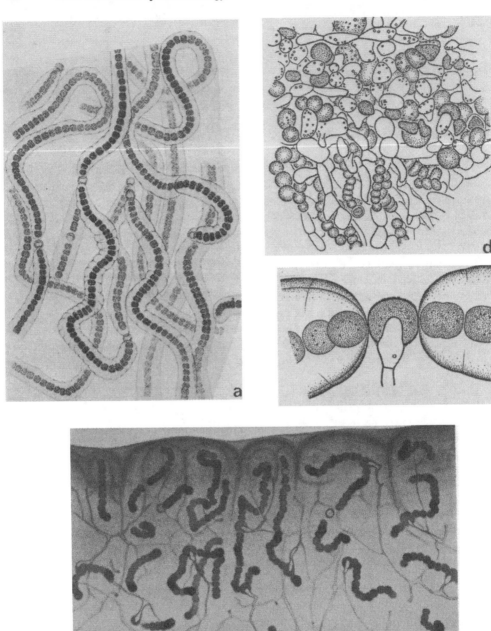

FIGURE 2.   *Nostoc*. a = *N. commune* free-living in nature; b and c = *Nostoc* sp. in *Leptogium chalazanum* (Ach.) De Lesd.; d = *Nostoc* sp. in *Lempholemma* sp., in the lower part growing out of the lichen association and with smaller cells and longer trichomes. (Figure a from Geitler, L., in *Encyclopedia of Plant Anatomy*, Zimmermann, W. and Ozenda, P., Eds., Borntraeger, Berlin, 1960, 34. Figures b and c from Geitler, L., *Arch. Protistenkd.*, 80, 378, 1933. Figure d from Schiman, H., *Osterr. Bot. Z.*, 104, 409, 1958. With permission.)

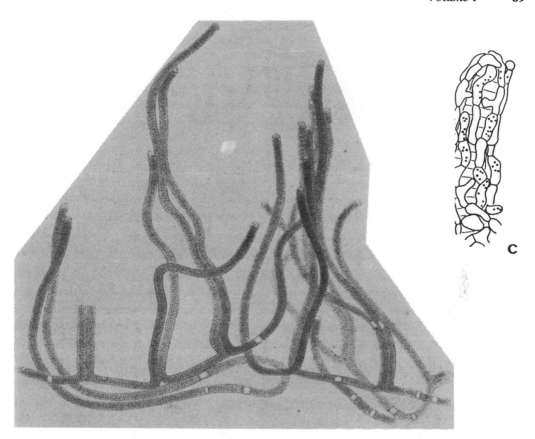

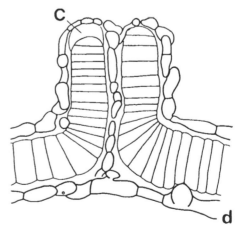

a

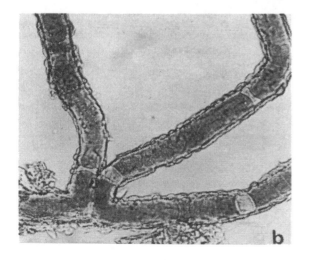

FIGURE 3. *Scytonema.* a = *S. hoffmanni* free-living in nature; b to d = *Scytonema* sp. in *Pyrenothrix nigra* (note difference in trichome breadth between young part in c and main trichome of an older part in d, in d gelatinous cap (C) at tip of one branch). (Figure a from Geitler, L., in *Encyclopedia of Plant Anatomy*, Zimmermann, W. and Ozenda, P., Eds., Borntraeger, Berlin, 1960. Figure b from Henssen, A. and Jahns, H. M., *Lichenes*, Georg Thieme Verlag, Stuttgart, 1974. Figures c and d from Tschermak-Woess, E. et al., *Plant Syst. Evol.*, 143, 293, 1983. With permission.)

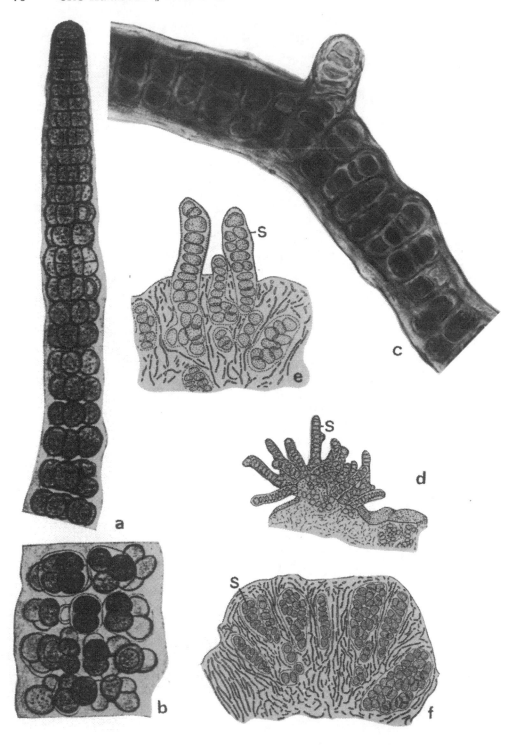

FIGURE 4.   *Stigonema.* a and b = *S. mamillosum* free-living in nature, a young part; b = old gloeocapsoid part; c = *Stigonema* sp. lichenized in *Ephebe orthogonia* Henssen (In the publication of Henssen and Jahns,[37] erroneously listed as *E. tasmanica.*); d to f = successive stages of the incorporation of *Stigonema* sp. in cephalodia of *Pilophorus strumaticus;* S = filaments of *Stigonema.* (Figures a and b from Geitler, L., in *Encyclopedia of Plant Anatomy,* Zimmermann, W. and Ozenda, P., Eds., Borntraeger, Berlin, 1960. Figure c from Henssen, A. and Jahns, H. M., *Lichenes,* Georg Thieme Verlag, Stuttgart, 1974. Figures d to f from Jahns, H. M., *Ber. Dtsch. Bot. Ges.,* 85, 615, 1972. With permission.)

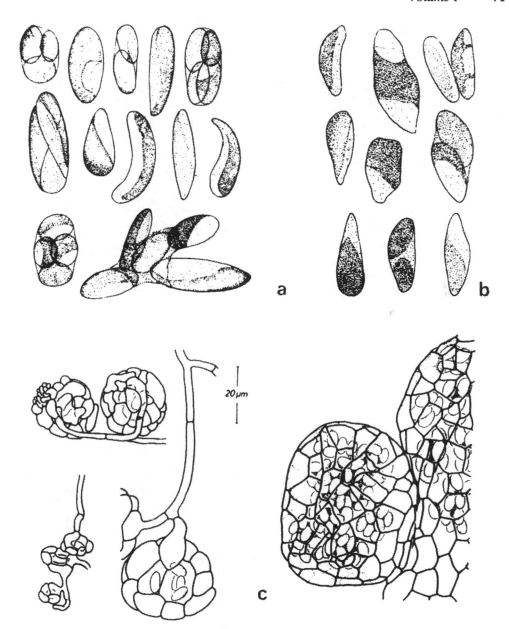

FIGURE 5. *Coccomyxa.* a = *C. pringsheimii* free-living in culture; b = *C. subellipsoidea* directly isolated from the thallus of *Omphalina* sp. (*Botrydina vulgaris*); c = *Coccomyxa* sp. lichenized in *Omphalina* sp. (*Botrydina vulgaris*). (Figures a and b from Jaag, O., *Ber. Schweiz. Bot. Ges.* 42, 169, 1933, Figure c from Oberwinkler, F., *Nova Hedwigia,* 79, 739, 1984. With permission.)

regularly reaches from the periphery into the interior of the cell (Figures 7h to 7j). Zoospores with two flagella separately inserted; one pulsating vacuole, without stigma; during rounding up of zoospores, points of insertion glide backwards (typical for the whole genus). In nature, cell walls are found occasionally with scales which probably originate from gelatinous sheaths (Figure 7e). Occurs aerophytically on bark and rocks, apparently needs a certain degree of humidity.[80,92]

***Dictyochloropsis reticulata* lichenized** — Cells are not as large as incipient zoo- and aplanosporangia and do not accumulate as much starch; propagation is solely by autospores.

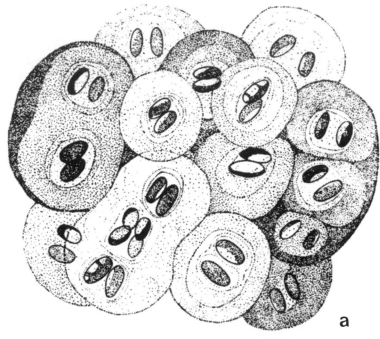

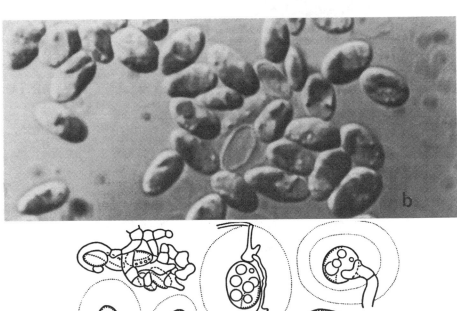

FIGURE 6.   *Coccomyxa*. a = *C. confluens* (Kütz.) Fott free-living from nature; b = *C. icmadophilae* free-living in culture; c = *C.* cf *icmadophilae* lichenized with *Icmadophila ericetorum*. (Figure a from Jaag, O., *Beitr. Kryptogamenflora Schweiz*, 8(1), 1, 1933. Figure c from Tschermak, E., *Osterr. Bot. Z.*, 90, 233, 1941. With permission.)

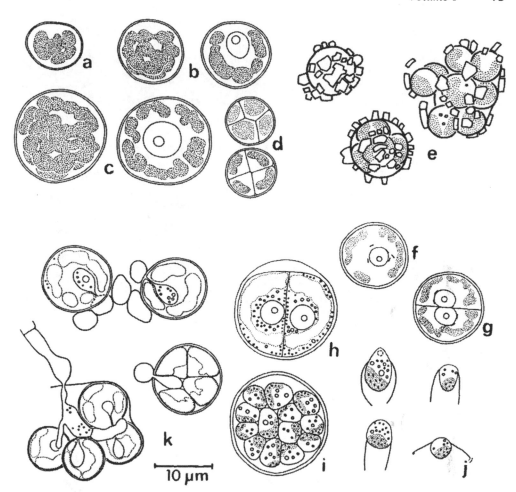

FIGURE 7. *Dictyochloropsis reticulata.* a to j = free-living, with the exception of e, in culture; a and b = young cells with cup-shaped reticulate chloroplast; c = fully grown cell; d = autosporangium (b to d face frontal views and optical sections); e = cells with scales from nature; f and g = with dictyosomes; h and i = young and fully developed zoosporangium; j = zoospores showing the dislocation of flagella; k = lichenized with *Catillaria chalybeia* (f to j are in a different, larger scale). (Figures from Tschermak-Woess, E., *Lichenologist*, 10, 69, 1978; Tschermak-Woess, E., *Plant Syst. Evol.*, 147, 299, 1984; Tschermak-Woess, E., *Osterr. Bot. Z.*, 98, 412, 1951. With permission.)

No scales or sheaths appear on cell walls; dictyosomes are probably relatively small and not visible in the light microscope (Figure 7k).

Myrmecia biatorellae **free-living (in culture, not directly observed in nature)** — Cells are spherical, ellipsoidal, or pear shaped; diameter is 2.5 μm in spherical young cells to above 30 μm in ellipsoidal cells. Cells are single or in 4- or 8-celled colonies, held together by the persisting mother cell wall. Cell walls of young cells are thin, in older cells are often partly thickened. Chloroplast is deeply cup-shaped, with two deep incisions, without pyrenoid; the nucleus is more or less centrally located. Reproduction is by 4 or 8 autospores per mother cell or by zoo- or aplanospores (64, 128); division processes in auto- and zoo(aplano-)sporangia differing. Zoosporangia possess a peg-like extension of inner wall layer where they later open; biflagellate zoospores with one pulsating vacuole, no stigma (Figures 8a to 8i). Occurs on (in?) earth.

Myrmecia biatorellae **lichenized** — Cells are single, in general smaller than in culture; cell walls are not partly thickened. Propagation is only by autospores. In some lichens (e.g.,

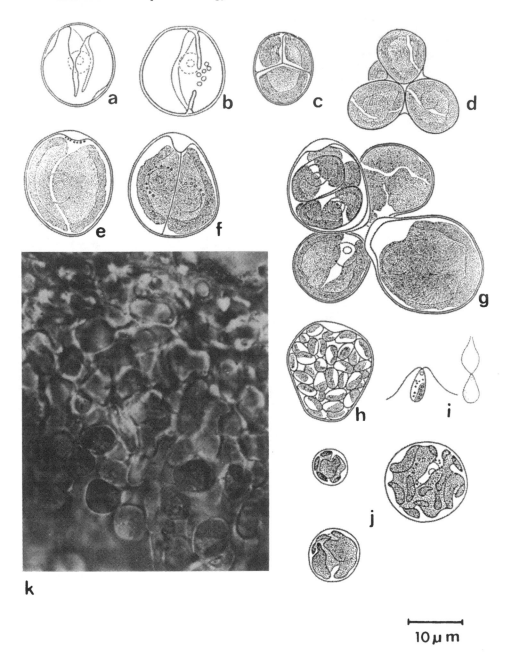

10 μm

FIGURE 8.   *Myrmecia biatorellae*. a to i = free-living in culture; a and b = vegetative cells; c = young, d = older autospores; e, f, and left upper cell in g = developing zoosporangia; h = ripe zoosporangium; i = zoospores; j = from the thallus of *Lecidea berengeriana*, k = lichenized in *Lobaria linita*. (Figures a to i from Tschermak-Woess, E. and Plessl, A., *Osterr. Bot. Z.*, 95, 194, 1948. Figure j from Geitler, L., *Osterr. Bot. Z.*, 109, 41, 1962. Figure k from Tschermak-Woess, E., *Nova Hedwigia*, 35, 63, 1981. With permission.)

Lecidea berengeriana[129]), chloroplast is complexly elaborated and folded which makes identification difficult; in others, chloroplast is as described (Figures 8j and 8k).

Stichococcus bacillaris free-living (in nature and in culture) — Filamentous alga with cylindrical, round-ended cells [(1.8) 2 to 4 × 5 to 10 (18) μm]; chains of cells which mostly fall apart. Some strains with mucous colonies, mucilage diffusing. Cells with one chloroplast,

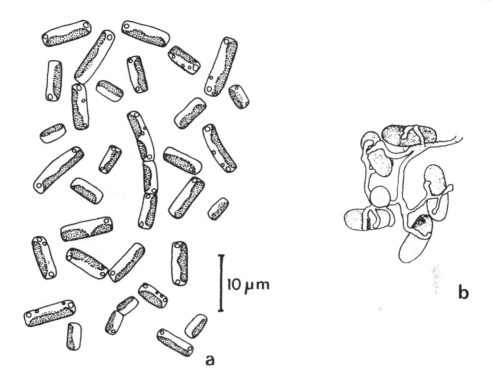

FIGURE 9. *Stichococcus bacillaris.* a = free-living in culture; b = lichenized with *Chaenotheca stemonea.* (Figure b from Raths, H., *Ber. Schweiz. Bot. Ges.*, 48, 329, 1938. With permission.)

usually lining the longitudinal wall; without pyrenoid. Propagation is solely by bipartition (Figure 9a). Occurs aerophytically on bark, rocks, and earth.

Stichococcus bacillaris lichenized — Apparently is little changed compared to free-living or cultured alga (Figure 9b).*

Trentepohlia **sp. free-living (in culture, from** *Cystocoleus ebeneus)* — Filamentous alga with cylindrical, barrel-, and irregularly shaped cells; filaments are irregularly branched, adhering to the substrate and erect (2 to 3 mm). Numerous disk-shaped chloroplasts without pyrenoids, with more or less starch; conspicuous storage product (according to Czygan and Kalb,[130] produced in response to N-deficiency), orange-colored hematochrome (carotenoids dissolved in oil); characteristic pits in crosswalls (Figure 10a). Propagation is by fragmentation of threads and by biflagellate zoospores formed in sessile spherical sporangia (Figure 10b). In other cases also ellipsoidal and stalked sporangia and quadriflagellate zoospores. Sexual fusion of biflagellate swarmers reported. Identity of the depicted species with species from nature is problematical. The latter occurs aerophitically on rocks, bark, wood, etc., in more or less damp habitats.

Trentepohlia **sp. lichenized (in** *Cystocoleus ebeneus)* — Filaments are less branched than in free-living state, but are not broken up in short pieces or single cells as in other lichens (Figures 10 c,d); no organs of reproduction are reported.

B. Phycobionts Newly Described or Encountered as Lichen Partners Since 1967

Trebouxia crenulata **Archibald free-living (in culture)** — Isolated spherical cells (5 to 18 (24) μm) with axial, massive chloroplast with comparatively shallow incisions; around peripheral nucleus, deeper indentation. One central rounded to angular pyrenoid, no con-

---

* Examples of the most common trebouxioid phycobionts that may be expected next are given in part B, because older descriptions in general are insufficient (compare also comment 55).

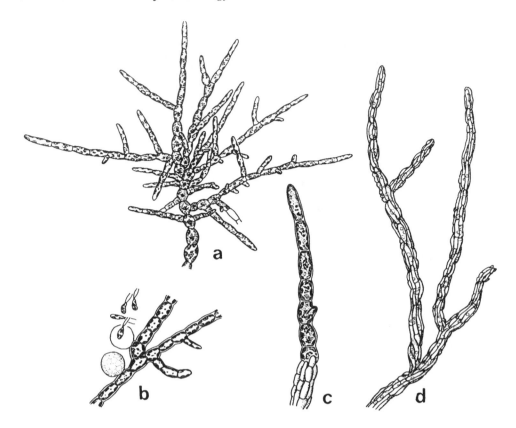

FIGURE 10.   *Trentepohlia* sp. a and b = free-living in culture; c = growing out of the fungal ensheathment in culture; d = fully lichenized with *Cystocoleus ebeneus*. (Figures a to d redrawn from Skuja, H. and Ore, M., *Acta Horti Bot. Univ. Latviensis*, 8, 21, 1933.)

tinuous starch sheath present according to Gärtner[76] (Figures 11a and 11b). The statement[89] that cells are multinucleate is corrected to uninucleate for the whole genus (*Pseudotrebouxia* included[94]) (compare also Reference 76). Reproduction is by zoo- or aplanospores (in the sense of arrested zoospores), the former biflagellate, with stigma and posterior nucleus; no autospores.* Known only from culture, isolated from *Xanthoria calcicola* Oxner (= *X. aureola* auct.), *Parmelia acetabulum* (Neck.) Duby, and *P. caperata*.[76]

Trebouxia pyriformis **Archibald free-living (in culture)** — Cells are solitary, spherical, or pyriform when young, pyriform when fully grown, infrequently also oviform, up to 18 × 20 (18 × 25) μm. Chloroplast with deeper and apparently more irregular incisions than in *T. crenulata*, with a single angular, naked pyrenoid (Figure 11). Nucleus is as described for *T. crenulata*. Some cells with unipolarly thickened walls. Reproduction is by zoo- and aplanospores; zoospores with a median to anterior nucleus and stigma; no autospores. Presently there are two known strains differing in some minor details.[76,89,97] Isolated from some *Cladonia* spp., *Stereocaulon dactyllophyllum* var. *occidentale* (Magn.) Grumm., and *St. pileatum*.[21,76,89]

Pseudotrebouxia decolorans **(Ahm.) Archibald free-living (in culture)** — Cells are solitary, generally spherical (diameter is up to 20 (30) μm). Axial chloroplast is coarsely lobed, with distinct, spherical, naked pyrenoid, relatively large grains of starch in stroma or near pyrenoid; one peripheral nucleus (Figure 12a). Reproduction is by 4, 8 (16) autospores, biflagellate zoo- and aplanospores, both of the latter probably arising in higher number

---

*   Some authors do not differentiate between autospores and aplanospores.[76,94]

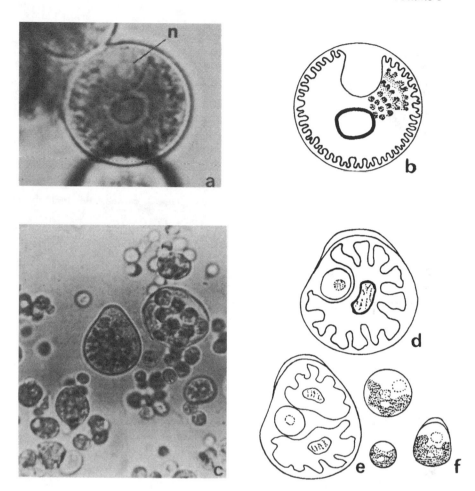

FIGURE 11. *Trebouxia*, free-living in culture. a and b = *T. crenulata;* c to f = *T. pyriformis;* c = vegetative pyriform cell and aplanosporangium; d = vegetative cell; e = after first division of chloroplast in young sporangium; f = spherical and pyriform young cells. (Figures a and b from Ettl, H. and Gärtner, G., *Plant Syst. Evol.*, 148, 135, 1984. Figure c from Hildreth, K. C. and Ahmadjian, V., *Lichenologist*, 13, 65, 1981. Figures d to f from Gärtner, G., *Arch. Hydrobiol. Suppl. 71, Algol. Stud. 41*, 495, 1985. With permission.)

(32, 64); zoospores with posterior nucleus and anterior stigma; loss of color when exposed to direct light.[76,88,89] Isolated from *Xanthoria parietina* and *Buellia punctata*.

Pseudotrebouxia simplex **(Tsch.-Woess) Tsch.-Woess free-living (in culture)** — Solitary spherical and relatively small cells (diameter is up to 14 μm). Axial chloroplast is mostly not indented; when actively growing, lobation is more pronounced. Chloroplast with one naked pyrenoid containing regularly dispersed light microscopically visible pyrenoglobuli which give it a sieve-like appearance (Figure 12b); starch grains when present in the stroma near pyrenoid or also more dispersed. Chloroplasts with a yellow-green tinge; in stationary phase cultures secondary carotenoids dissolved in oil or crystalloid in peripheral cytoplasm. Reproduction is by 4, 8 (16) autospores; chloroplast during division not or only slightly flattened (Figures 12c to 12f). Reproduction is by biflagellate zoo- and aplanospores only at low temperatures (around 10°C), not studied intensively;[12,131] alga sensitive to relative high temperatures (30°C), but not to light.

Pseudotrebouxia simplex **lichenized with** *Chaenotheca chrysocephala* **and** *C. subros-*

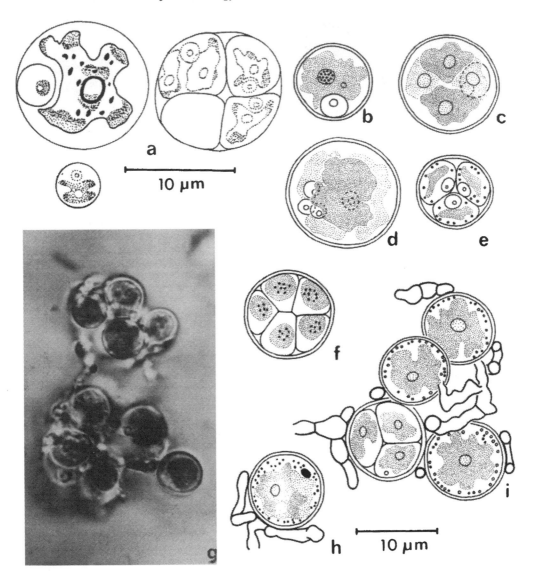

FIGURE 12. *Pseudotrebouxia.* a to f = free-living in culture. a = *P. decolorans,* young and fully grown vegetative cell and autosporangium; b to f = *P. simplex;* b = vegetative cell; c to f = developing and fully developed autosporangia; g to i = *P. simplex* lichenized in *Chaenotheca chrysocephala,* note secondary carotinoids (in h) dissolved in oil in peripheral cytoplasm. (Figure a from Gärtner, G., *Arch. Hydrobiol. Suppl. 71, Algol. Stud. 41,* 495, 1985. Figures b to i from Tschermak-Woess, E., *Plant Syst. Evol.,* 129, 185, 1978. With permission.)

cida — Chloroplast is simple (Figures 12g to 12i); pyrenoglobuli are not observed, starch and carotenoids are occasionally present; reproduction is only by autospores.

***Elliptochloris bilobata* Tsch.-Woess free-living (in culture)** — Fully grown cells are ellipsoidal (up to 10.5 × 13 μm), ovoid, occasionally spherical (up to 13 μm), or flattened along one side. No individual or confluent gelatinous sheaths around cells, but cultures on agar are slimy. One parietal cup-shaped chloroplast with two deep indentations, no pyrenoid, small flat starch grains found in stroma; no secondary carotenoids. Nucleus mostly near base of chloroplast. Cytoplasm often with many vacuoles. Reproduction is by two or four relatively large autospores (formed by successive division) or by 16 or 32 small bacilliform spores (2 × 5 to 3.5 × 6 μm); no flagellate states (Figures 13a to 13j).[132] Occurrence free-living in nature not reported (but see the following).

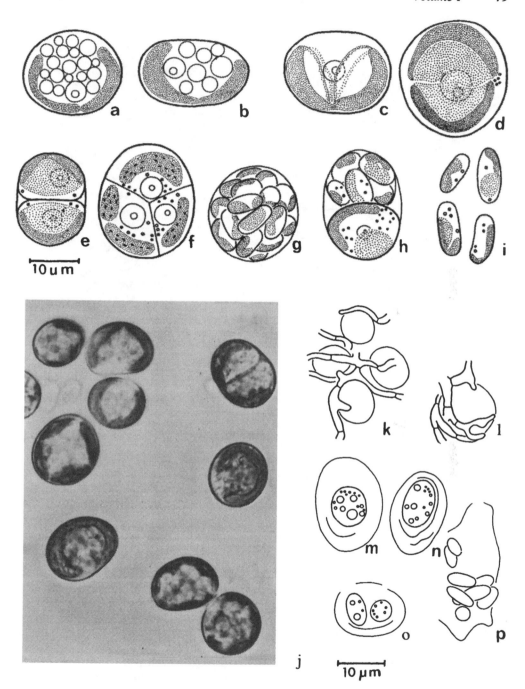

FIGURE 13. *Elliptochloris bilobata.* a to j = free-living in culture; a to d and j = large vegetative cells; e and f = autosporangia; g to i = more or less rodlike small daughter cells in sporangia and free; k to p = lichenized in *Protothelenella corrosa* (k and l), respectively free-living (m to p) from natural habitat (chloroplasts omitted). (Figures a to p from Tschermak-Woess, E., *Plant Syst. Evol.*, 136, 63, 1980; Tschermak-Woess, E., *Herzogia*, 7, 105, 1985. With permission.)

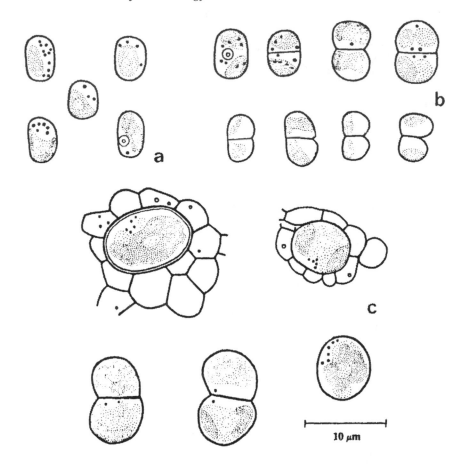

FIGURE 14.   *Nannochloris normandinae.* a and b = free-living in culture; c = from thallus of *Normandina pulchella.* (Figures a to c from Tschermak-Woess, E., *Nova Hedwigia,* 35, 63, 1981. With permission.)

Elliptochloris bilobata* lichenized in *Baeomyces rufus, Catolechia wahlenbergii,* and *Protothelenella — In thalli of *Catolechia* and *Baeomyces,* cells are usually large, without sheaths. Reproduction is by two or four relatively large autospores; in some thallus parts where fungus loses contact with alga (*Baeomyces*), small rod-like daughter cells occur in higher number per sporangium, and also without sheaths. In *Protothelenella corrosa,* in intensely lichenized parts, large spherical cells are without sheaths; in loosely and not lichenized parts, smaller ellipsoidal cells and bacilliform spores are embedded in mucilage (Figures 13k to 13p).[132,133]

***Nannochloris normandinae* Tsch.-Woess free-living (in culture)** — Single rod-like, occasionally more rounded cells (2.5 to 3 × 3.5 to 5 μm) with thin walls; one plate-like chloroplast mostly along longitudinal part of wall, with naked pyrenoid and tiny starch grains in stroma; secondary carotenoids in old cultures. Bipartition perpendicular to longitudinal axis is the sole mode of reproduction (Figures 14a and 14b, compare comment 66).[134] Not observed free-living in nature.

Nannochloris normandinae* lichenized in *Normandina pulchella — Ellipsoidal cells in central region of thallus markedly enlarged (6 × 8 to 10 × 15 μm), in peripheral parts somewhat less enlarged; chloroplast is more compact than in free-living state, eventually with flat lobes (Figure 14c).[134]

***Leptosira obovata* Vischer free-living (in culture)** — Thallus formed by irregularly branching filaments of cylindrical, irregularly inflated, spherical, ellipsoidal, or pear-shaped cells (cylindrical cells 6 to 10 × 11.5 to 42 (60) μm); in basal parts, filaments are readily falling apart into isolated rounded cells. Single, thick, elongate chloroplast, irregularly plate-like or folded and with lobes; one (occasionally two) lenticular naked pyrenoid consisting of several subunits which can be observed in the light microscope, and by electron microscopy are shown to be isolated by lamellae of thylakoids.[135,137] Starch throughout the stroma, no sheath around pyrenoid. Reproduction is by elongate autospores seen only for a short time after isolation from lichen. After development of filaments, propagation by zoo- and aplanospores originating in older parts of the inoculum in sporangia which become markedly yellowish in color and develop a local thickening of the cell wall where they later open. Zoospores biflagellate, markedly or slightly flattened, with stigma and anterior nucleus (Figures 15a and 15b).[136,137] Has been isolated from bogs of water near Basel, Switzerland, probably occurred free-living.

Leptosira obovata **lichenized in** *Vezdaea aestivalis* — Drastically changed, only with isolated spherical or ellipsoidal relatively small cells (diameter 4 to 12.5 μm, resp. 6 to 12.5 × 7.5 to 14.5 μm); chloroplast of various shapes (Figures 15c to 15g).[137]

VI. OCCURRENCE IN THE FREE-LIVING STATE

As can be seen in Table 1, many phycobionts can be found not lichenized (free-living) in nature. In a number of cases, they occur on the same substrates and in the same ecological niches as the lichen harboring them. Examples include: *Gloeocapsa sanguinea* and *Synalissa symphorea*, both growing on sun-exposed rocks; *Hyella caespitosa* and *Arthopyrenia halodytes*, occurring on marine rocks and shells of molluscs; and *Dilabifilum arthopyreniae* and *Verrucaria adriatica*, on maritime rocks in the splash zone (see Table 1). The ecological amplitude of free-living and lichenized algae may be equal, a little, or widely different. Lichens and their free-living algae may share the same macrohabitats; however, with respect to microhabitat slight differences occur. *Gloeocapsa* grows not only on rocks, but also on and in tufts of rock inhabiting mosses; *Synalissa* apparently cannot invade the latter. In *Verrucaria adriatica* the lichen tends to grow higher up in the splash zone than its free-living *Dilabifilum* partner. *Lichina confinis* lives in the supralittoral, while its algal partner *Calothrix* occurs at the upper border of the intertidal zone. The two free-living *Dictyochloropsis* species seem to prefer habitats with some degree of humidity, and most lichens harboring them show oceanic tendencies or preferences for a humid microhabitat.[80,92]

Most of these examples demonstrate small variations in the ecological demands of free-living algae and lichens containing them. In many other instances, however, the ecological breadth and developmental possibilities (biomass) of algae are distinctly broadened by lichenization. This applies, above all, to trebouxioid algae which, until now, have been reported only sporadically and in low cell numbers in nature.[20,80,82,138-143] Nakano[139,140] isolated trebouxioid algae from soil samples, but these may have been present in soredia. He also isolated apparently free-living trebouxioid algae from bark of *Nothofagus* in Patagonia. In a recent study,[141] *Trebouxia glomerata* was isolated from spikes of *Typha* in Japan. Bubrick et al.[20] found free-living cells or small groups of cells of a *Pseudotrebouxia* (or *Trebouxia*) species in the vicinity of *Xanthoria parietina* thalli which, on the basis of morphological and immunological criteria studied in culture, were identical with two phycobionts isolated from the lichen (also germinating spores of *Xanthoria* were observed in the vicinity of the free-living algal cells suggesting that resynthesis in nature can occur). Formerly, Werner[144] had reported the occurrence of free-living cystococcoid algal cells (probably *Pseudotrebouxia*) near thalli and germinating spores of *Caloplaca* and *Lecania* sp. Ahmadjian (Reference 2, p. 192) has argued that free-living microcolonies of trebouxioid algae (including *Pseudotrebouxia*) probably originated from zoospores which escaped from

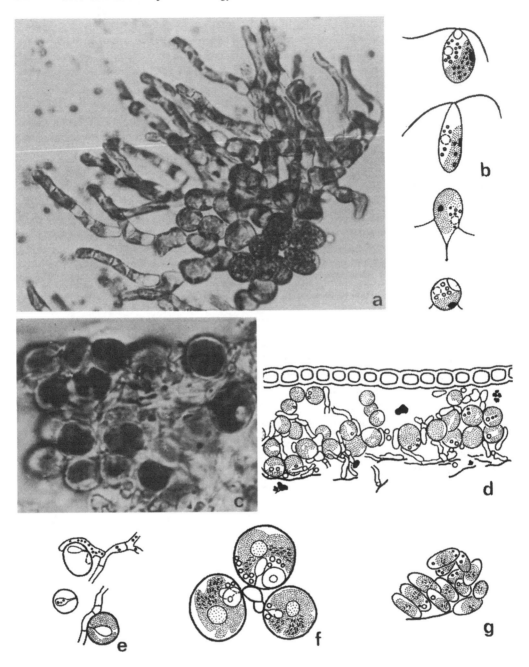

FIGURE 15.   *Leptosira obovata.* a and b = free-living in culture (note zoosporangia in old parts of filaments in a and characteristic behavior of flagella in b); c to f = lichenized in *Vezdaea aestivalis* [d = cross-section through leaf of moss (*Camptothecium*) and lichen, c = remnants of the cuticula of the moss on which Vezdaea grows subcuticularly]; g = elongate spores formed in young inoculum. (Figures a to g from Tschermak-Woess, E. and Poelt, J., in *Lichenology: Progress and Problems,* Brown, D. H., Hawksworth, D. L., and Bailey, R. H., Eds., Academic Press, New York, 1976, 89. With permission.)

nearby lichen thalli. However, in certain cases (Botanical Garden in a district of Vienna with high air pollution) no lichen thalli or soredial traces were present nearby, so that observed colonies can be considered as members of the long-term, free-living algal community.[80] Other statements for the presence or absence of trebouxioid algae free-living in nature have been summarized by Ahmadjian.[145]

VII. REPRODUCTION AND PROPAGATION OF THE ALGAL PARTNER WITHIN THE LICHEN THALLUS

When lichenized, certain modes of propagation commonly encountered in the free-living alga may be absent or highly restricted. *Gloeocapsa*, a common cyanophycean phycobiont, reproduces by bipartition when free-living and lichenized; however, akinetes and nannocytes have not been reported from lichenized *Gloeocapsa*. In free-living Chroococcaceae, nannocytes develop when rates of division are more rapid than cell enlargement. In the lichen thallus, conditions for rapid algal division are probably unfavorable due to high demands of photosynthate transfer to the fungus and consequent lack of available carbon for algal cell multiplication.[57,103,124,146,147] From *Nostoc* no reports on hormogonia exist. In *Placynthium nigrum* (phycobiont probably *Dichothrix orsiniana*), the cyanophycean biont develops hormogonia after periods of excess rainfall which can escape from loosely organized thallus parts.[103] In other lichenized Rivulariaceae and from *Stigonema* and *Scytonema*, formation of hormogonia in the thallus has not been reported. In some cases (e.g., *Ephebe*), it may have simply been overlooked. Akinetes in free-living *Stigonema* and *Scytonema* seldomly occur and until now have not been found in the lichen association.

Heterococcus caespitosus (Tribophyceae = Xanthophyceae), lichenized in *Verrucaria funckii* and *V. laevata*, reproduces mainly by bipartition, exceptionally by the formation of autospores and does not develop zoospores; in rapidly growing cultures of the isolated alga zoospores are abundantly produced.[62] In *Petroderma maculiforme* (Fucophycee = Phaeophyceae) unilocular zoosporangia are suppressed in the lichenized state.[148]

Among the Chlorophyceae, *Coccomyxa* and *Gloeocystis*, when lichenized or free-living, reproduce solely by autospores. *Trebouxia* De Pulm. emend. Arch. in culture multiplies by zoo- and aplanospores (arrested zoospores); in the lichen thallus only aplanospores are regularly observed. Ahmadjian[11] thinks that in the thallus, zoospores may be sporadically produced and released, and that these may even escape from thalli to form colonies of the free-living alga. In the thallus of *Parmelia caperata*, Slocum et al.[149] found daughter cells within sporangia which had developed flagella (compare also Scheidegger[150] for similar behavior of *Asterochloris* in the thallus of *Anzina*). They interpreted this as evidence for the potential of zoospore release and escape. The process of escape, which was not observed, is probably far more complex and deserves further investigation. In contrast to *Trebouxia*, *Pseudotrebouxia* when free-living in nature or culture can produce autospores as well as zoo- and aplanospores. In numerous lichens containing *Pseudotrebouxia* the present author has observed reproduction solely by autospores; in these cases, some aspects of lichenization apparently prevent formation of even aplanospores. A similar situation can be found in *Dictyochloropsis reticulata*, *D. symbiontica*, and *Myrmecia biatorellae*; in culture they form auto-, zoo-, and aplanospores whereas only autospores are formed in lichen thalli. In free-living *D. reticulata* on trees and rocks, until now only autospores were observed (with one exception).

Elliptochloris in culture develops two types of daughter cells without flagella; 2 or 4 relative large ellipsoidal, rarely spherical autospores and 16 or 32 small rod-like spores. In the lichenized state, formation of the latter is suppressed;[132] it occurs only in thallus parts where the algal cells are not yet encircled by fungal hyphae or have secondarily become free of the fungal influence.[133]

Chaetophoralean Chlorophyceae which produce zoospores in culture (*Dilabifilum arthopyreniae*, *Leptosira obovata*, *L. thrombii*) do not generally produce them when lichenized. Under certain natural conditions (long periods of rainfall), *Leptosira thrombii* can sporadically produce zoosporangia in the lichen thallus.[240] These occur in parts where fungal encirclements of the algal cells are less tight. In the lichen, the palmelloid state of the alga reproducing by autospores prevails, while under optimal cultural conditions filaments develop. Under suboptimal conditions (e.g., storage at stationary phase) palmelloid cells are formed and may be surrounded by confluent gelatinous sheaths. The latter may also be observed on natural substrates, where algal cells have escaped from the thallus.[151]

Reproduction and propagation of free-living *Stichococcus* species are uniform and consist of bipartition and falling apart of daughter cells. This cannot be further reduced by lichenization.

With respect to the existence of reproductive structures (zoosporangia, gametangia) in lichenized *Trentepohlia*, reports are contradictory. According to Uyenco,[117] they do not normally occur; the present author also did not find any traces of them in *Graphis scripta*, *Gyalecta jenensis*, and *Opegrapha rufescens*. However, McCoy (according to Meier and Chapman[152]) reported the frequent occurrence of sporangia in lichenized *Trentepohlia*, and Meier and Chapman[152] found fungal hyphae near some gametangia-bearing filaments of *Trentepohlia* in *Coenogonium interplexum*. In *Coenogonium moniliforme* lichenized with *Trentepohlia monilia* de Wild. (= *Physolinum monilia* (De Wild.) Printz), Santesson[7] found typical aplanosporangia in corticolous specimens, and sporangia (aplano- or zoo-?) in a foliicolous specimen. Further investigations paying particular attention to the degree of lichenization of reproductive organ-bearing structures are needed.

Reproductive structures from other Trentepohliaceae have been reported. According to Santesson,[7] *Phycopeltis expansa* lichenized with *Microtheliopsis ulcana* "fairly regularly forms zoosporangia as well as gametangia" and *Cephaleuros virescens* as a phycobiont of *Strigula complanta* sometimes develops stalked zoosporangia. It would be interesting to determine whether the filaments bearing these sporangia are lichenized or have grown out of the thallus. In free-living Trentepohliaceae, the stalked sporangia are known to break off and to be windborne before the spores are released. Thus, in the *Strigula* and *Microtheliopsis* association, the possibility of effective, separate propagation of the algal and fungal partner exists.

In summary, as far as can be judged from the restricted data available, it appears that, in certain Tribo-, Fuco-, and Chlorophyceae, production of zoospores is suppressed in the fully lichenized state. In other Chlorophyceae (*Trebouxia*) presumptive zoospores probably are mostly transformed into aplanospores, while in the Trentepohliaceae a number of different reproductive structures may be present.

VIII. OUTLOOK FOR THE FUTURE

Future investigations certainly will reveal further genera and species of phycobionts new to science as well as other algae not presently known to occur as phycobionts. New species are especially expected in the genera *Trebouxia* and *Pseudotrebouxia*. Until now 11 well-characterized species of *Trebouxia* and 12 of *Pseudotrebouxia* have been described as algal partners of lichens (besides many others described incompletely).[76] The number of free-living forms identified as to species in 1960 was two; nowadays it is still only two or three;[76,88] even in these, uncertainties exist. Ahmadjian[88] pointed to the possibility that the many trebouxioid forms might have developed as a consequence of lichenization. It also must be kept in mind that it is much easier to find certain lichenized algae than to detect solitary free-living cells of them.

The study of phycosymbiodemes[8] (different morphotypes, chimeroid associations) has shown that different algae play a decisive role in morphogenesis and physiological potentialities of one fungal species. It, therefore, seems justified to request that more attention be paid to phycobionts than is presently done.

The author thanks Prof. Dr. A. Henssen, Prof. Dr. H. Hertel, Doc. Dr. L. Tibell, and especially Prof. Dr. J. Poelt for support in nomenclatoral and taxonomical problems of lichens. Dr. G. Gärtner kindly placed the proofs of his publication on the genus *Trebouxia* at the author's disposal. Special thanks are due to Dr. P. Bubrick for correcting the English style and for other essential improvements.

REFERENCES

1. **Scott, G. D.,** Lichen terminology, *Nature (London)*, 179, 486, 1957.
2. **Ahmadjian, V.,** Algal/fungal symbiosis, in *Progress in Phycological Research*, Round, F. and Chapman, D. J., Eds., Elsevier, Amsterdam, 1982, 179.
3. **Ahmadjian, V.,** A guide to the algae occurring as lichen symbionts: isolation, culture, cultural physiology and identification, *Phycologia*, 6, 127, 1967.
4. **Ahmadjian, V.,** *The Lichen Symbiosis*, Blaisdell Publ., Waltham, Mass., 1967.
5. **Ahmadjian, V.,** Methods of isolating and culturing lichen symbionts and thalli, in *The Lichens*, Ahmadjian, V. and Hale, M. E., Eds., Academic Press, New York, 1973, 653.
6. **Tschermak-Woess, E.,** Über den *Chlorella*-Phycobionten von *Trapelia coarctata, Plant Syst. Evol.*, 130, 253, 1978.
7. **Santesson, R.,** Foliicolous lichens, I., *Symb. Bot. Ups.*, 12, 1, 1952.
8. **James, P. W. and Henssen, A.,** The morphological and taxonomic significance of cephalodia, in *Lichenology: Progress and Problems*, Brown, D. W., Hawksworth, D. L., and Bailey, R. H., Eds., Academic Press, New York, 1976, 27.
9. **Brodo, I. M. and Richardson, H. S.,** Chimeroid associations in the genus *Peltigera, Lichenologist*, 10, 157, 1978.
10. **Renner, B. and Galloway, D. J.,** Phycosymbiodemes in *Pseudocyphellaria* in New Zealand, *Mycotaxon*, 16, 197, 1982.
11. **Ahmadjian, V.,** Separation and artificial synthesis of lichens, in *Cellular Interactions in Symbiotic and Parasitic Associations*, Cook, C. B., Pappas, P. W., and Rudolph, E. D., Eds., Ohio State University Press, Columbus, 1980, 3.
12. **Tschermak-Woess, E.,** Über die Phycobionten der Sektion *Cystophora* von *Chaenotheca*, insbesondere *Dictyochloropsis splendida* und *Trebouxia simplex*, spec. nova, *Plant Syst. Evol.*, 129, 185, 1978.
13. **Tibell, L.,** The lichen genus *Chaenotheca* in the northern hemisphere, *Symb. Bot. Ups.*, 23, 1, 1980.
14. **Tibell, L.,** Caliciales of Costa Rica, *Lichenologist*, 14, 219, 1982.
15. **Tibell, L.,** A reappraisal of the taxonomy of Caliciales, *Nova Hedwigia Beih.*, 79, 597, 1984.
16. **Tibell, L.,** in *Flora of New Zealand: Lichens*, Galloway, D. J., Ed., P.D. Hasselberg Printer, Wellington, New Zealand, 1985.
17. **Tschermak-Woess, E.,** *Chaenothecopsis consociata* — kein parasitischer oder parasymbiontischer Pilz, sondern lichenisiert mit *Dictyochloropsis symbiontica*, spec. nova, *Plant Syst. Evol.*, 136, 287, 1980.
18. **Lamb, I. M.,** On the morphology, phylogeny, and taxonomy of the lichen genus *Stereocaulon, Can. J. Bot.*, 29, 522, 1951.
19. **Archibald, P. A.,** Physiological characteristics of *Trebouxia* (Chlorophyceae, Chlorococcales) and *Pseudotrebouxia* (Chlorophyceae, Chlorosarcinales), *Phycologia*, 16, 295, 1977.
20. **Bubrick, P., Galun, M., and Frensdorff, A.,** Observations on free-living *Trebouxia* and *Pseudotrebouxia*, and evidence that both symbionts from *Xanthoria parietina* can be found free-living in nature, *New Phytol.*, 97, 455, 1984.
21. **Meisch, J.-P.,** Beiträge zur Isolation, Kultur und Systematik von Flechten-Algen, Dissertation, Universität Innsbruck, Austria, 1981.
22. **Ahmadjian, V.,** A guide for the identification of algae occurring as lichen symbionts, *Bot. Not.*, 111, 632, 1958.
23. **Letrouit-Galinou, M. A.,** Les algues des lichens, *Bull. Soc. Bot. Fr. Mem. 1968, Colloque sur les Lichens 1967*, 35, 1968.
24. **Lewin, R. A.,** Naming the blue-greens, *Nature (London)*, 259, 360, 1976.
25. **Bourrelly, P.,** Les Cyanophycees, algues ou bacteries?, *Rev. Algol.*, 14, 5, 1979.
26. **Geitler, L.,** Einige kritische Bemerkungen zu neuen zusammenfassenden Darstellungen der Morphologie und Systematik der Cyanophyceen, *Plant Syst. Evol.*, 132, 153, 1979.
27. **Golubić, S.,** Cyanobacteria (blue-green algae) under the bacteriological code? An ecological objection, *Taxon*, 28, 387, 1979.
28. **Kondratyeva, N. V.,** On difference of opinions of phycologists and bacteriologists concerning the nomenclature of Cyanophyta, *Arch. Protistenkd.*, 126, 247, 1982.
29. **Friedmann, E. I. and Borowitzka, L. J.,** The symposium on taxonomic concepts in blue-green algae: toward a compromise with the bacterial code?, *Taxon*, 31, 673, 1982.
30. **Friedmann, E. I.,** Cyanophycota, in *Synopsis and Classification of Living Organisms*, Vol. 1, Parker, S. P., Ed., McGraw Hill, New York, 1982, 45.
31. **Drouet, F. and Daily, W. A.,** Revision of the coccoid Myxophyceae, *Butler Univ. Bot. Stud.*, 12, 1, 1956.
32. **Wetmore, C. M.,** The lichen family Heppiaceae in North America, *Ann. Mo. Bot. Gard.*, 57, 158, 1970.
33. **Bubrick, P. and Galun, M.,** Cyanobiont diversity in the Lichinaceae and Heppiaceae, *Lichenologist*, 16, 279, 1984.

34. **Jaag, O.,** Untersuchungen über die Vegetation und Biologie der Algen des nackten Gesteins in den Alpen, im Jura und im schweizerischen Mittelland, *Beitr. Kryptogamenflora Schweiz,* 9(3), 3, 1945.
35. **Zahlbruckner, A.,** Lichenes, in *Die natürlichen Pflanzenfamilien,* Vol. 8, 2nd ed., Engler, A., Ed., Wilhelm Engelmann, Leipzig, 1926.
36. **Henssen, A.,** Drei neue Arten der Flechtengattung *Phylliscum, Sven. Bot. Tidskr.,* 57, 145, 1963.
37. **Henssen, A. and Jahns, H. M.,** *Lichenes,* Georg Thieme Verlag, Stuttgart, 1974.
38. **Poelt, J.,** *Bestimmungsschlüssel Europäischer Flechten,* J. Cramer, Lehre, 1969.
39. **Wirth, V.,** *Flechtenflora,* Eugen Ulmer, Stuttgart, 1980.
40. **Henssen, A.,** Eine Revision der Flechtenfamilien Lichinaceae und Ephebaceae, *Symb. Bot. Ups.,* 18, 1, 1963.
41. **Büdel, H. and Henssen, A.,** *Chroococcidiopsis* (Cyanophyceae), a phycobiont in the lichen family Lichinaceae, *Phycologia,* 22, 367, 1983.
42. **Brodo, I. M. and Hertel, H.,** The lichen genus *Amygdalaria* (Porpidiaceae) in North America, *Herzogia,* 7, 493, 1987.
43. **Poelt, J.,** personal communication.
44. **Schiman, H.,** Beiträge zur Lebensbeschichte homoeomerer und heteromerer Cyanophyceen-Flechten, *Osterr. Bot. Z.,* 104, 409, 1958.
45. **Lamb, I. M.,** The lichen genus *Argopsis* Th. Fr., *J. Hattori Bot. Lab.,* 38, 447, 1974.
46. **De Bary, A.,** Morphologie und Physiologie der Pilze, Flechten und Myxomyceten, in *Handbuch der Physiologischen Botanik,* Vol. 2, Hofmeister, W., Ed., Wilhelm Engelmann, Leipzig, 1866.
47. **De Toni, J. B.,** *Sylloge Algarum,* Vol. V, Patavii, 1807.
48. **Livingstone, D. and Whitton, B. A.,** Influence of phosphorus on morphology of *Calothrix parietina* (Cyanophyta) in culture, *Br. Phycol. J.,* 18, 29, 1983.
49. **Jordan, W. P. and Rickson, F. R.,** Cyanophyte cephalodia in the lichen genus *Nephroma, Am. J. Bot.,* 58, 562, 1971.
50. **Hawksworth, D. L. and Hill, D. J.,** *The Lichen-Forming Fungi,* Methuen, New York, 1984.
51. **Danilov, A. N.,** Le *Nostoc* en état de symbiose, *Russ. Arch. Protist.,* 6, 83, 1927.
52. **Geitler, L.,** Cyanophyceae, in *Rabenhorst's Kryptogamenflora,* Vol. 14, Kolkwitz, R., Ed., Akademische Verlagsges, Leipzig, 1932.
53. **Degelius, G.,** The lichen genus *Collema* in Europe, *Symb. Bot. Ups.,* 13, 2, 1954.
54. **Arvidsson, L.,** A monograph of the lichen genus *Coccocarpia, Opera Bot.,* 67, 5, 1982.
55. **Parmasto, E.,** The genus *Dictyonema* ("Thelophorolichenes"), *Nova Hedwigia,* 29, 99, 1978.
56. **Bornet, M. E.,** Recherches sur les gonidies des lichens, *Ann. Sci. Nat. Ser. 5 Bot.,* 17, 45, 1873.
57. **Tschermak-Woess, E.,** Das Haustorialsystem von *Dictyonema* kennzeichnend für die Gattung, *Plant Syst. Evol.,* 143, 109, 1983.
58. **Geitler, L.,** Zur Kenntnis der Flechte *Thermutis velutina, Osterr. Bot. Z.,* 112, 263, 1965.
59. **Fujita, M.,** Fine structure of lichens, *Misc. Bryol. Lichenol.,* 4, 157, 1968.
60. **Christenssen, T.,** *Algae, A Taxonomic Survey,* Fasc. 1, Aio Try, as, Odense, 1980.
61. **Tschermak, E.,** Untersuchungen über die Beziehungen von Pilz und Alge im Flechtenthallus, *Osterr. Bot. Z.,* 90, 233, 1941.
62. **Zeitler, I.,** Untersuchungen über die Morphologie, Entwicklungsgeschichte und Systematik von Flechtengonidien, *Osterr. Bot. Z.,* 101, 453, 1954.
63. **Tschermak-Woess, E.,** unpublished data, 1974/1975.
64. **Kohlmeyer, J.,** Higher fungi as parasites and symbionts of algae, *Veroeff. Inst. Meeresforsch. Bremerhaven Suppl.,* 5, 339, 1974.
65. **Kohlmeyer, J. and Kohlmeyer, E.,** *Marine Mycology: The Higher Fungi,* Academic Press, New York, 1979.
66. **Silva, P. C.,** Chlorophyta, in *Synopsis and Classification of Living Organisms,* Vol. 1, Parker, S. P., Ed., McGraw-Hill, New York, 1982, 133.
67. **Komárek, J. and Fott, B.,** Chlorophyceae (Grünalgen), Ordnung Chlorococcales, in *Das Phytoplankton des Süsswassers,* Vol. 7 (Part 1), Huber-Pestalozzi, G., Ed., E. Schweizerbartsche Verlagsbuchhandlung, Stuttgart, 1983.
68. **Jaag, O.,** *Coccomyxa* Schmidle, Monographie einer Algengattung, *Beitr. Kryptogamenflora Schweiz,* 8(1), 1, 1933.
69. **Hindák, F.,** The genus *Gloeocystis* (Chlorococcales, Chlorophyceae), *Preslia,* 50, 3, 1978.
70. **Hedlund, T.,** A contribution to the knowledge of the development of the aerophile Chlorophyceae, *Bot. Not.,* 43, 173, 1949.
71. **Hue, A.,** Description de deux espèce de lichens et de céphalodies nouvelles, *Bot. Centralbl.,* 99, 34, 1905.
72. **Kupffer, K. R.,** *Stereonema chthonoblastes,* eine lebende Urflechte, *Naturf. Ver. Riga,* 58, 111, 1924.
73. **Jaag, O. and Thomas, E.,** Neue Untersuchungen über die Flechte *Epigloea bactrospora* Zukal, *Ber. Schweiz. Bot. Ges.,* 43, 77, 1934.

74. **Döbbeler, P.**, Symbiosen zwischen Gallertalgen und Gallertpilzen der Gattung *Epigloea* (Ascomycetes), *Nova Hedwigia*, 79, 203, 1984.

75. **Scheidegger, C.**, Systematische Studien zur Krustenflechte *Anzina carneonivae*, *Dipl. Arbeit am System. Geobot. Inst. Univ. Bern*, 1983.

76. **Gärtner, G.**, Die Gattung *Trebouxia* Puymaly (Chlorellales, Chlorophyceae), *Arch. Hydrobiol. Suppl. 71, Algol. Stud. 41*, 4, 495, 1985.

77. **Yoshimura, I.**, The genus *Lobaria* of eastern Asia, *J. Hattori Bot. Lab.*, 34, 231, 1971.

78. **Brodo, I. M.**, The lichen genus *Coccotrema* in North America, *Bryologist*, 76, 260, 1973.

79. **Geitler, L.**, Notizen über wenig bekannte Grünalgen und eine neue Chytridiale, *Osterr. Bot. Z.*, 112, 603, 1965.

80. **Tschermak-Woess, E.**, *Myrmecia reticulata* as a phycobiont and free-living — free-living *Trebouxia* — the problem of *Stenocybe septata*, *Lichenologist*, 10, 69, 1978.

81. **Geitler, L.**, Über Flechtenalgen, *Schweiz. Z. Hydrol.*, 22, 131, 1960.

82. **Raths, H.**, Experimentelle Untersuchungen mit Flechtengonidien aus der Familie der Caliciaceen, *Ber. Schweiz. Bot. Ges.*, 48, 329, 1938.

83. **Tibell, L.**, Comments on Caliciales exsiccatae. III, *Lichenologist*, 17, 189, 1985.

84. **Trenkwalder, H.**, Untersuchungen zur Bodenalgenflora verschiedener Föhrenwaldtypen im Raum von Brixen (Südtirol, Italien), Thesis, University of Innsbruck, Austria, 1975.

85. **Vinatzer, G.**, Untersuchungen über die Bodenalgen in der alpinen Stufe des Pitschberges (2300 m), Südtirol, Thesis, University of Innsbruck, Austria, 1975.

86. **Watanabe, S.**, New and interesting green algae from soils of some Asian and Oceanian regions, *Arch. Protistenkd.*, 127, 223, 1983.

87. **Warén, H.**, Reinkulturen von Flechtengonidien, *Ofvers. Finska Vetensk. Soc. Forh.*, 61(14), Afd. A, 1, 1918/1919.

88. **Ahmadjian, V.**, Some new and interesting species of *Trebouxia*, a genus of lichenized algae, *Am. J. Bot.*, 47, 677, 1960.

89. **Archibald, P. A.**, *Trebouxia* de Puymaly (Chlorophyceae, Chlorococcales) and *Pseudotrebouxia* gen. nov. (Chlorophyceae, Chlorosarcinales), *Phycologia*, 14, 125, 1975.

90. **Tschermak-Woess, E.**, Über die Abgrenzung der Chlorosarcinales von den Chlorococcales, *Plant Syst. Evol.*, 139, 295, 1982.

91. **Mattox, K. R. and Stewart, K. D.**, Classification of the green algae: a concept based on comparative cytology, in *Systematics of Green Algae*, Irvine, D. E. G. and John, D. M., Eds., Academic Press, New York, 1984, 29.

92. **Tschermak-Woess, E.**, Über die weite Verbreitung lichenisierter Sippen von *Dictyochloropsis* und die systematische Stellung von *Myrmecia reticulata* (Chlorophyta), *Plant Syst. Evol.*, 147, 299, 1984.

93. **Gärtner, G.**, Taxonomische Probleme bei den Flechtengattungen *Trebouxia* und *Pseudotrebouxia* (Chlorophyceae, Chlorellales), *Phyton (Horn, Austria)*, 25, 101, 1985.

94. **Ettl, H. and Gärtner, G.**, Über die Bedeutung der Cytologie für die Algentaxonomie, dargestellt an *Trebouxia* (Chlorellales, Chlorophyceae), *Plant Syst. Evol.*, 148, 135, 1984.

95. **König, J. and Peveling, E.**, Cell walls of the phycobionts *Trebouxia* and *Pseudotrebouxia*: constituents and their localization, *Lichenologist*, 16, 129, 1984.

96. **Peveling, E. and König, J.**, Differences in formation of vegetative cells and their walls in *Trebouxia* and *Pseudotrebouxia* as further evidence for the classification of these genera, *Lichenologist*, 17, 281, 1985.

97. **Hildreth, K. C. and Ahmadjian, V.**, A study of *Trebouxia* and *Pseudotrebouxia* isolates from different lichens, *Lichenologist*, 13, 65, 1981.

98. **Werner, R.-G.**, La gonidie marocaine du *Xanthoria parietina* (L.) Beltr., *Bull. Soc. Sci. Nancy N.S.*, 13, 8, 1954.

99. **Jaag, O.**, Recherches Expérimentales sur les Gonidies des Lichens Appartenant aux Genres *Parmelia* et *Cladonia*, Thèse Fac. Sci., University of Geneva, Geneva, 1929.

100. **Tschermak-Woess, E.**, unpublished data, 1984.

101. **Wang-Yang, J.-R. and Ahmadjian, V.**, A morphological study of the algal symbionts of *Cladonia rangiferina* (L.) Web. and *Parmelia caperata* (L.) Ach., *Taiwania*, 17, 170, 1972.

102. **Zehnder, A.**, Über Einfluss von Wuchsstoffen auf Flechtenbildner, *Ber. Schweiz. Bot. Ges.*, 59, 201, 1949.

103. **Geitler, L.**, Beitrage zur Kenntnis der Flechtensymbiose. IV, V, *Arch. Protistenkd.*, 82, 51, 1934.

104. **Chodat, R.**, Monographie d'algues en culture pure, *Beitr. Kryptogammenflora Schweiz*, 4, 1, 1913.

105. **Lamb, I. M.**, A conspectus of the lichen genus *Stereocaulon* (Schreb.) Hoffm., *J. Hattori Bot. Lab.*, 43, 191, 1977.

106. **Brown, L. M. and Elfman, B.**, Is autosporulation a feature of *Nannochloris?*, *Can. J. Bot.*, 61, 2647, 1983.

107. **Bourrelly, P.**, *Les Algues d' eau Douce, Algues Vertes*, N. Boubée & Cie, Paris, 1966.

108. **Vischer, W.,** Reproduktion und systematische Stellung einiger Rinden- und Bodenalgen, *Schweiz. Z. Hydrol.,* 212, 330, 1960.

109. **Stahl, E.,** *Beiträge zur Entwicklungsgeschichte der Flechten, II. Über die Bedeuntung der Hymenialgonidien,* Arthur Felix, Leipzig, 1877.

110. **Letellier, A.,** Etude de quelques gonidies de lichens, *Bull. Soc. Bot. Geneve Ser. 2,* 9, 373, 1917.

111. **Tibell, L.,** The Caliciales of boreal North America. Taxonomy, ecological and distributional comparisons with Europe, and ultrastructural investigations in some species, *Symb. Bot. Ups.,* 21, 1, 1975.

112. **Tibell, L.,** Lavordningen Caliciales i Sverige. Slaktena *Chaenotheca* och *Coniocybe, Sven. Bot. Tidskr.,* 72, 171, 1978.

113. **Ahmadjian, V. and Heikkilä, H.,** The culture and synthesis of *Endocarpon pusillum* and *Staurothele clopima, Lichenologist,* 4, 259, 1970.

114. **Printz, H.,** *Die Chaetophoralen der Binnengewasser,* Junk, Den Haag, 1964.

115. **Reed, M.,** Two new ascomycetous fungi parasitic on marine algae, *Univ. Calif. Publ. Bot.,* 1, 141, 1902.

116. **Norris, J. N.,** Observations on the genus *Blidingia* (Chlorophyta) in California, *J. Phycol.,* 7, 145, 1971.

117. **Uyenco, F. R.,** Studies on some lichenized *Trentepohlia* associated in lichen thalli with *Coenogonium, Trans. Am. Microsc. Soc.,* 84, 1, 1965.

118. **Skuja, H. and Ore, M.,** Die Flechte *Coenogonium nigrum* (Huds.) Zahlbr. und ihre Gonidien, *Acta Horti Bot. Univ. Latviensis,* 8, 21, 1933.

119. **Vainio, E. A.,** Lichenes insularum Philippinarum. III, *Ann. Acad. Sci. Fenn. Ser. A,* 15(6), 1, 1921.

120. **Reinke, J.,** Abhandlungen über Flechten. V, *Jahrb. Wiss. Bot.,* 29, 171, 1896.

121. **Hérisset, A.,** Démonstration expérimentale du rôle du *Trentepohlia umbrina* (Kg.) Born. dans la synthese des Graphidees corticales, *C. R. Acad. Sci.,* 222, 100, 1946.

122. **Geitler, L.,** Schizophyceen. II. Umgearbeitete Auflage, in *Encyclopedia of Plant Anatomy,* Zimmermann, W. and Ozenda, P., Eds., Borntraeger, Berlin, 1960.

123. **Tschermak-Woess, E.,** Haustorienbefall und inaquale Teilung des *Nostoc*-phycobionten von *Lempholemma botryosum* (Lichinaceae), *Plant Syst. Evol.,* 137, 317, 1981.

124. **Tschermak-Woess, E., Bartlett, J., and Peveling, E.,** *Lichenothrix riddlei* is an ascolichen and also occurs in New Zealand — light and electron microscopical investigations, *Plant Syst. Evol.,* 143, 293, 1983.

125. **Jahns, H. M.,** Die Entwicklung von Flechten-Cephalodien aus *Stigonema*-Algen, *Ber. Dtsch. Bot. Ges.,* 85, 615, 1972.

126. **Skuja, H.,** Taxonomie des Phytoplanktons einiger Seen in Uppland, Schweden, *Symb. Bot. Ups.,* 9(3), 1, 1948.

127. **Fott, B.,** *Choricystis,* eine neue Gattung der Chlorococcales (Chlorophyceae), *Arch. Hydrobiol. Suppl. 49, Algol. Stud. 17,* 382, 1976.

128. **Brunner, U.,** Ultrastrukturelle und chemische Zellwanduntersuchungen an Flechtenphycobionten aus 7 Gaattungen der Chlorophyceae (Chlorophytina) unter besonderer Berücksichtigung sporopollenin-ähnlicher Biopolymere, Dissertation, University of Zurich, Zurich, 1985.

129. **Geitler, L.,** Über die Flechtenalge *Myrmecia biatorellae, Osterr. Bot. Z.,* 109, 41, 1962.

130. **Czygan, F.-C. and Kalb, K.,** Untersuchungen zur Biogenese der Carotinoide in *Trentepohlia aurea, Z. Pflanzenphysiol.,* 55, 59, 1966.

131. **Tschermak-Woess, E.,** unpublished data, 1984.

132. **Tschermak-Woess, E.,** *Elliptochloris bilobata,* gen. et spec. nov., der Phycobiont von *Catolechia wahlenbergii, Plant Syst. Evol.,* 136, 63, 1980.

133. **Tschermak-Woess, E.,** *Elliptochloris bilobata* kein ganz seltener Phycobiont, *Herzogia,* 7, 105, 1985.

134. **Tschermak-Woess, E.,** Zur Kenntnis der Phycobionten von *Lobaria linita* und *Normandina pulchella, Nova Hedwigia,* 35, 63, 1981.

135. **Wujek, D. E.,** Light and electron microscope observations on the pyrenoid of the green alga *Leptosira, Mich. Acad.,* 3, 59, 1971.

136. **Vischer, W.,** Über einige kritische Gattungen und die Systematik der Chaetophorales, *Beih. Bot. Zentralbl.,* 51, 1, 1933.

137. **Tschermak-Woess, E. and Poelt, J.,** *Vezdaea,* a peculiar lichen genus, and its phycobiont, in *Lichenology: Progress and Problems,* Brown, D. H., Hawksworth, D. L., and Bailey, R. H., Eds., Academic Press, New York, 1976, 89.

138. **Nakano, T.,** Subaerial algae of Patagonia, South America. I, *Bull. Biol. Soc. Hiroshima Univ.,* 38, 2, 1971.

139. **Nakano, T.,** Some aerial and soil algae from the Ishizucki Mountains, *Hikobia,* 6, 139, 1971.

140. **Nakano, T.,** Flora and distribution of edaphic algae on Mt. Daisen, Tottori Prefecture, *Hikobia Suppl.,* 1, 111, 1981.

141. **Nakano, T.,** Some epiphytic algae growing on the spikes of *Typha, J. Jpn. Bot.,* 60, 5, 1985.

142. **Galun, M., Bubrick, P., and Frensdorff, A.,** Initial stages in fungus-alga interaction, *Lichenologist,* 16, 103, 1984.

143. **Tschermak-Woess, E.,** unpublished data, 1985.

144. **Werner, R.-G.,** Histoire de la synthèse lichénique, *Mem. Soc. Sci. Nat. Maroc,* 27, 7, 1931.

145. **Ahmadjian, V.,** The lichen association, *Bryologist,* 63, 250, 1960.

146. **Geitler, L.,** Beiträge zur Kenntnis der Fletchensymbiose. I-III, *Arch. Protistenkd.,* 80, 378, 1933.

147. **Geitler, L.,** Beiträge zur Kenntnis der Flechtensymbiose. VI. Die Verbindung von Pilz und Alge bei den Pyrenopsidaceen *Synalissa, Thyrea, Peccania* und *Psorotichia, Arch. Protistenkd.,* 88, 161, 1937.

148. **Wynne, M. J.,** Life history and systematic studies of some pacific North American Phaeophyceae (brown algae), *Univ. Calif. Publ. Bot.,* 50, 1, 1969.

149. **Slocum, R. D., Ahmadjian, V., and Hildreth, K. C.,** Zoosporogenesis in *Trebouxia gelatinosa:* ultrastructural potential for zoospore release and implications for the lichen association, *Lichenologist,* 12, 173, 1980.

150. **Scheidegger, C.,** Systematische Studien zur Krustenflechte *Anzina carneonivea* (Trapeliaceae, Lecanorales), *Nova Hedwigia,* 41, 191, 1985.

151. **Tschermak-Woess, E.,** Über wenig bekannte und neue Flechtengonidien. III. Die Entwicklungsgeschichte von *Leptosira thrombii* nov. spec., der Gonidie von *Thrombium epigaeum, Osterr. Bot. Z.,* 100, 203, 1953.

152. **Meier, J. L. and Chapman, R. L.,** Ultrastructure of the lichen *Coenogonium interplexum* L., *Am. J. Bot.,* 70, 400, 1983.

153. **Ettl, H. and Komárek, J.,** Was versteht man unter dem Begriff "coccale Grünalgen"? (Systematische Bemerkungen zu den Grünalgen II), *Arch. Hydrobiol. Suppl.,* 60, 345, 1982.

154. **Tschermak-Woess, E.,** unpublished data, 1967.

155. **Henssen, A. and Büdel, B.,** *Phylliciella,* a new genus of the Lichinaceae, *Nova Hedwigia,* 79, 381, 1984.

156. **Henssen, A., Büdel, B., and Wessels, S.,** New or interesting members of the Lichinaceae from southern Africa. I. Species from northern and eastern Transvaal, *Mycotaxon,* 22, 169, 1985.

157. **Marton, K. and Galun, M.,** *In vitro* dissociation and reassociation of the symbionts of the lichen *Heppia echinulata, Protoplasma,* 87, 135, 1976.

158. **Duvigneaud, P.,** Les *Stereocaulon* des hautes montagnes du Kivu, *Lejeunia Mem.,* 14, 1, 1955.

159. **Henssen, A.,** New or interesting cyanophilic lichens, *Lichenologist,* 5, 444, 1973.

160. **Hertel, H.,** *Amygdalaria* Norm., in *Bestimmungsschlüssel Europaischer Flechten, Ergänzungsheft II,* Poelt, J. and Vězda, A., Eds., J. Cramer, Vaduz, 1981, 111.

161. **Hertel, H.,** personal communication, 1985.

162. **Paran, N., Ben-Shaul, Y., and Galun, M.,** Fine structure of the blue-green phycobiont and its relation to the mycobiont in two *Gonohymenia* lichens, *Arch. Mikrobiol.,* 76, 103, 1971.

163. **Forssell, K. B. J.,** Beiträge zur Kenntnis der Anatomie und Systematik der Gloeolichenen, *Nova Acta Regiae Soc. Sci. Ups. C,* 13, 1, 1885.

164. **Snyder, J. M. and Wullstein, L. H.,** Nitrogen fixation on granite outcrop poineer ecosystems, *Bryologist,* 76, 196, 1973.

165. **Köfaragó-Gyelnik, V.,** Lichinaceae, Heppiaceae, Pannaraiaceae, Stictaceae, Peltigeraceae, in *Rabenhorst's Kryptogammenflora,* Vol. 9, Kolkwitz, R., Ed., Akademische Verlagsges, Leipzig, 1940, 2.

166. **Tschermak-Woess, E.,** Weitere Untersuchungen zur Frage des Zusammenlebens von Pilz und Alge in den Flechten, *Wien. Bot. Z.,* 92, 15, 1943.

167. **Forssell, K. B. J.,** Lichenologische Untersuchungen, *Flora,* 67, 177, 1884.

168. **Brodo, I. M.,** personal communication, 1984.

169. **Tschermak-Woess, E.,** unpublished data, 1984.

170. **Schiman, H.,** unpublished data, 1985.

171. **Henssen, A. and Büdel, B.,** *Peccania cerebriformis* und *Psorotichia columnaris.* Zwei neue Lichinaceen von Lanzarote, *Int. J. Mycol. Lichenol.,* 1, 261, 1984.

172. **Bubrick, P.,** Studies on the Phycobionts of Desert Cyanolichens, Thesis, Florida State University, Tallahassee, 1978.

173. **Büdel, B.,** Blue-green phycobionts in the family Lichinaceae, *Arch. Hydrobiol. Suppl. 71 Algol. Stud.,* 38/39, 335, 1985.

174. **Swinscow, T. D. V.,** The marine species of *Arthopyrenia* in the British Isles, *Lichenologist,* 3, 55, 1965.

175. **Klement, O.,** Eine Flechte auf lebenden Schnecken, *Ber. Naturhist. Ges. Hannover,* 106, 57, 1962.

176. **Tschermak-Woess, E.,** Algal taxonomy and the taxonomy of lichens: the phycobiont of *Verrucaria adriatica,* in *Lichenology: Progress and Problems,* Brown, D. H., Hawksworth, D. L., and Bailey, R. H., Eds., Academic Press, New York, 1976, 79.

177. **Jahns, H. M.,** Remarks on the taxonomy of the European and North American species of *Pilophorus, Lichenologist,* 4, 199, 1970.

178. **Schwendener, S.,** *Die Algentypen der Flechtengonidien,* C. Schultze, Basel, 1869.

179. **Ozenda, P. and Clauzade, G.,** *Les Lichens,* Masson et Cie, Paris, 1970.

180. **Lamb, I. M.**, Keys to the species of the lichen genus *Stereocaulon* (Schreb.) Hoffm., *J. Hattori Bot. Lab.*, 44, 209, 1978.

181. **Galloway, D. J., Lamb, I. M., and Bratt, G. C.**, Two new species of *Stereocaulon* from New Zealand and Tasmania, *Lichenologist*, 8, 61, 1976.

182. **Henssen, A.**, Hyphomorpha als Phycobiont in Flechten, *Plant Syst. Evol.*, 137, 139, 1981.

183. **Henssen, A.**, *Hertella*, a new lichen genus in the Peltigerales from the southern hemisphere, *Mycotaxon*, 22, 381, 1985.

184. **Henssen, A.**, Three non-marine species of the genus *Lichina*, *Lichenologist*, 4, 88, 1969.

185. **Ahmadjian, V.**, Lichens, in *Physiology and Biochemistry of Algae*, Lewin, R. A., Ed., Academic Press, New York, 1962, 817.

186. **Coppins, B. J. and James, P. W.**, New or interesting British lichens. V, *Lichenologist*, 16, 241, 1984.

187. **Wetmore, C. M.**, The lichen genus *Nephroma* in North and Middle America, *Publ. Mus. Mich. State Univ. Biol. Ser.*, 1, 369, 1960.

188. **Jørgensen, P. M.**, The lichen family Pannariaceae in Europe, *Opera Bot.*, 45, 1, 1978.

189. **James, P. W. and Henssen, A.**, A new species of *Psoroma* with sorediate cephalodia, *Lichenologist*, 7, 143, 1975.

190. **Forssell, K. B. J.**, Über den Bau und die Entwicklung des Thallus bei *Lecanora (Psoroma) hypnorum* (Hoffm.) Ach., *Flora*, 67, 187, 1884.

191. **Henssen, A. and James, P. W.**, The lichen genus *Steinera*, *Bull. Br. Mus. Nat. Hist. Bot.*, 10, 227, 1982.

192. **Lamb, I. M.**, A monograph of the lichen genus *Placopsis* Nyl., *Lilloa*, 13, 151, 1947.

193. **Jordan, W. P.**, Erumpent cephalodia, an apparent case of phycobial influence on lichen morphology, *J. Phycol.*, 8, 112, 1972.

194. **Linkola, K.**, Kulturen mit *Nostoc*-Gonidien der *Peltigera*-Arten, *Ann. Soc. Zool. Bot. Fenn. Vanamo*, 1, 1, 1920.

195. **Lamb, I. M.**, *Compsocladium*, a new genus of lichenized ascomycetes, *Lloydia*, 19, 157, 1956.

196. **Arvidsson, L. and Galloway, D. J.**, *Degelia gayana* new genus new combination new lichen in the Pannariaceae, *Lichenologist*, 13, 27, 1981.

197. **Slocum, R. D. and Floyd, G. L.**, Light and electron microscopic investigations in the Dictyonemataceae (Basidiolichenes), *Can. J. Bot.*, 55, 2565, 1977.

198. **Slocum, R. D.**, Light and electron microscopic investigations in the Dictyonemataceae (Basidiolichenes). *Dictyonema irpicinum*, *Can. J. Bot.*, 58, 1005, 1980.

199. **Roskin, P. A.**, Ultrastructure of the host-parasite interaction in the basidiolichen *Cora pavonia* (Web.) E. Fries, *Arch. Mikrobiol.*, 70, 176, 1970.

200. **Oberwinkler, F.**, Fungus-alga interactions in basidiolichenes, *Nova Hedwigia*, 79, 739, 1984.

201. **Oberwinkler, F.**, Symbiotic relationships between fungus and alga in basidiolichens, in *Endocytobiology*, Schwemmler, W. and Schenk, H. E., Eds., Walter de Gruyter, New York, 1981, 305.

202. **Henssen, A.**, *Placynthium arachnoideum*, a new lichen from Patagonia, and notes on other species of the genus in the southern hemisphere, *Lichenologist*, 16, 265, 1984.

203. **Riddle, L.**, *Pyrenothrix nigra*, gen. et sp. nov., *Bot. Gaz. (Chicago)*, 64, 513, 1917.

204. **Henssen, A.**, Was ist *Pyrenothrix nigra?*, *Ber. Dtsch. Bot. Ges.*, 77, 317, 1964.

205. **Eriksson, O.**, The families of bitunicate Ascomycetes, *Opera Bot.*, 60, 1, 1981.

206. **Henssen, A.**, The genus *Zahlbrucknerella*, *Lichenologist*, 9, 17, 1977.

207. **Parra, O. O. and Redon, J.**, Aislamento de *Heterococcus caespitosus* Vischer ficobionte de *Verrucaria maura* Wahlenb., *Bol. Soc. Biol. Concepcion*, 51, 219, 1977.

208. **Skuja, H.**, Ein Fall von fakultativer Symbiose zwischen operculatem Discomycet und einer Chlamydomonade, *Arch. Protistenkd.*, 96, 365, 1943.

209. **Pott, L.**, Struktur und Vermehrung des Phycobionten *Coccomyxa*, *Staatsarbeit Munster*, 1972; as cited in **Peveling, E. and Galun, M.**, *New Phytol.*, 77, 713, 1976.

210. **Peveling, E. and Galun, M.**, Electron-microscopical studies on the phycobiont *Coccomyxa* Schmiddle, *New Phytol.*, 77, 713, 1976.

211. **Geitler, L.**, *Clavaria mucida*, eine extratropische Basidiolichene, *Biol. Zentralbl.*, 74, 145, 1955.

212. **Geitler, L.**, Ergänzende Beobachtungen über die extratropische Basidiolichene *Clavaria mucida*, *Osterr. Bot. Z.*, 103, 164, 1956.

213. **Plessl, A.**, Über die Beziehungen von Haustorientypus und Organisationshöhe bei Flechten, *Osterr. Bot. Z.*, 110, 194, 1963.

214. **Acton, E.**, *Coccomyxa subellipsoidea*, a new member of the Palmellaceae, *Ann. Bot. (London)*, 23, 573, 1909.

215. **Acton, E.**, *Botrydina vulgaris* Brébisson, a primitive lichen, *Ann. Bot. (London)*, 23, 579, 1909.

216. **Geitler, L.**, *Botrydina*-keine Symbiose einer Alge mit einem Moosprotonema, *Osterr. Bot. Z.*, 103, 469, 1956.

217. **Jaag, O.**, *Botrydina vulgaris* Bréb. eine Lebensgemeinschaft von Moosprotonemen und Grünalgen, *Ber. Schweiz. Bot. Ges.*, 42, 169, 1933.

218. **Poelt, J.**, Eine Basidiolichene in den Hochalpen, *Planta*, 52, 600, 1959.

219. **Vězda, A.**, Flechtensystematische Studien. II. *Absconditella*, eine neue Flechtengattung, *Preslia*, 37, 237, 1965.

220. **Tschermak-Woess, E.**, *Asterochloris phycobiontica*, gen. et spec. nov., der Phycobiont der Flechte *Varicellaria carneonivea*, *Plant Syst. Evol.*, 135, 279, 1980.

221. **Tshcermak-Woess, E.**, Über wenig bekannte und neue Flechtengonidien. II. Eine neue Protococcale, *Myrmecia reticulata*, als Algenkomponente von *Catillaria chalybeia*, *Osterr. Bot. Z.*, 98, 412, 1951.

222. **Tschermak-Woess, E.**, unpublished data, 1983.

223. **Tschermak-Woess, E.**, Über wenig bekannte und neue Flechtengonidien. IV. *Myremcia reticulata*-der Algenpartner in *Phlyctis argena* und seine systematische Stellung, *Osterr. Bot. Z.*, 116, 167, 1969.

224. **Tschermak-Woess, E.**, unpublished data, 1970.

225. **Řeháková, H.**, Lišejnikové řasy rodu *Trebouxia*, *Diplosphaera* a *Myrmecia*, C.Sc. thesis, Katedra botaniky Přirodovědecké fakulty, University of Karlovy, Praze, 1968.

226. **Tschermak-Woess, E. and Plessl, A.**, Über zweierlei Typen der sukzedanen Teilung und ein auffallendes Teilungsverhalten des Chromatophors bei einer neuen Protococcale, *Myrmecia pyriformis*, *Osterr. Bot. Z.*, 95, 194, 1948.

227. **Galun, M., Ben-Shaul, Y., and Paran, N.**, The fungus-alga association in the Lecideaceae: an ultrastructural study, *New Phytol.*, 70, 483, 1971.

228. **Geitler, L.**, Über Haustorien bei Flechten und uber *Myrmecia biatorellae* in *Psora globifera*, *Osterr. Bot. Z.*, 110, 270, 1963.

229. **Weber, W. A.**, The lichen flora of Colorado. II. Pannariaceae, *Univ. Colo. Stud. Ser. Biol.*, 16, 1, 1965.

230. **Ahmadjian, V., Russell, L. A., and Hildreth, K. C.**, Artificial reestablishment of lichens. I. Morphological interactions between the phycobionts of different lichens and the mycobionts *Cladonia cristatella* and *Lecanora chrysoleuca*, *Mycologia*, 72, 73, 1980.

231. **Quispel, A.**, The mutual relations between algae and fungi in lichens, *Recl. Trav. Bot. Neerl.*, 40, 413, 1943—1945.

232. **Giudici de Nicola, M. and Di Benedetto, G.**, Ricerche preliminari sui pigmenti nel ficobionte lichenico *Trebouxia decolorans* Ahm. III. Clorofille e carotinoidi, *Boll. Ist. Bot. Univ. Catania Ser. 3*, 3, 22, 1962.

233. **Jaag, O.**, Ueber die Verwendbarkeit der Gonidienalgen in der Flechtensystematik, *Ber. Schweiz. Bot. Ges.*, 42, 724, 1933.

234. **Ahmadjian, V.**, The Taxonomy and Physiology of Lichen Algae and Problems of Lichen Synthesis, Ph.D. dissertation, Harvard University, Cambridge, 1959; as cited in **Archibald, P. A.**, *Phycologia*, 14, 125, 1975.

235. **Tschermak-Woess, E.**, New and known taxa of *Chlorella (Chlorophyceae)*: occurrence as lichen phycobionts and observations on living dictyosomes, *Plant Syst. Evol.*, 159, 1988.

236. **Tschermak-Woess, E.**, Über wenig bekannte und neue Flechtengonidien. I. *Chlorella ellipsoidea* Gerneck, als neue Flechtenalge, *Osterr. Bot. Z.*, 95, 341, 1948.

237. **Tschermak-Woess, E.**, Über wenig bekannte und neue Flechtengonidien. V. Der Phycobiont von *Verrucaria aquatilis* und die Fortpflanzung von *Pseudopleurococcus arthopyreniae*, *Osterr. Bot. Z.*, 118, 443, 1970.

238. **Binz, A. and Vischer, W.**, Zur Flora des Rheinlaufs bei Basel, *Verh. Naturforsch. Ges. Basel*, 67, 195, 1956.

239. **Tschermak-Woess, E.**, unpublished data, 1974.

240. **Schiman, H.**, Über die Entwicklungsmöglichkeiten von *Leptosira thrombii* Tsch.-Woess als Algenkomponente in der Flechte *Thrombium epigaeum*, *Osterr. Bot. Z.*, 108, 1, 1961.

241. **Werner, R.-G.**, Etude biologique de la gonidie hymeniale de l'*Endocarpon pallidum* Ach., *Bull. Soc. Sci. Nancy*, 1960, 212, 1960.

242. **Tschermak, E.**, Beitrag zur Entwicklungsgeschichte und Morphologie der Protococcale *Trochiscia granulata*, *Osterr. Bot. Z.*, 90, 67, 1941.

243. **Geitler, L.**, Beiträge zur Kenntnis der Flechtensymbiose. VII. Über Hymenialgonidien, *Arch. Protistenkd.*, 90, 489, 1938.

244. **Ahmadjian, V. and Jacobs, J. B.**, The ultrastructure of lichens. III. *Endocarpon pusillum*, *Lichenologist*, 4, 268, 1970.

245. **Feldmann, J.**, Le *Blodgettia confervoides* Harv. est-il un lichen?, *Rev. Bryol. Lichenol.*, 11, 155, 1938.

246. **Chapman, R. L.**, Ultrastructural investigation on the foliicolous pyrenocarpous lichen *Strigula elegans* (Fée) Mull. Arg., *Phycologia*, 15, 191, 1976.

247. **Marche-Marchad, J.**, Some ecological data about *Cephaleuros virescens* and some lichens of which it is the symbiotic alga, *Cryptogam. Algol.*, 2, 289, 1981.

248. **Sérusiaux, E.**, Goniocysts, goniocystangia and *Opegrapha lambinonii* and related species, *Lichenologist*, 17, 1, 1985.

249. **Sérusiaux, E.**, Two new foliicolous lichens from tropical Africa, *Lichenologist*, 11, 181, 1979.
250. **Hawksworth, D. L., Coppins, B. J., and James, P. W.**, *Blarneya*, a lichenized hyphomycete from southern Ireland, *Bot. J. Linn. Soc.*, 79, 357, 1980.
251. **Henssen, A., Vobis, G., and Renner, B.**, New species of *Roccellinastrum* with an emendation of the genus, *Nord. Jordbrugsforsk. Bot.*, 2, 587, 1982.
252. **Ellis, E. A.**, Observations on the ultrastructure of the lichen *Chiodecton sanguineum*, *Bryologist*, 78, 471, 1975.
253. **Follmann, G. and Huneck, S.**, Über das Vorkommen von Confluentinsäure in *Enterographa crassa* (de Cand.) Fé und die Stellung von *Herpothallon sanguineum* (Swans.) Tobler, *Willdenowia*, 5, 3, 1968.
254. **Withrow, K. and Ahmadjian, V.**, The ultrastructure of lichens. VII. *Chiodecton sanguineum*, *Mycologia*, 75, 337, 1983.
255. **Coppins, B. J. and James, P. W.**, New or interesting British lichens. III, *Lichenologist*, 11, 27, 1979.
256. **Ahmadjian, V.**, Further studies on lichenized fungi, *Bryologist*, 67, 87, 1964.
257. **Henssen, A., Renner, B., and Vobis, G.**, *Sagenidium patagonicum*, a new South American lichen, *Lichenologist*, 11, 263, 1979.
258. **Jørgensen, M. and Vězda, A.**, *Topelia*, a new Mediterranean lichen genus, *Nova Hedwigia*, 79, 501, 1984.
259. **Lambright, D. D. and Tucker, S. C.**, Observations on the ultrastructure of *Trypethelium eluteriae* Spreng., *Bryologist*, 83, 170, 1980.
260. **James, P. W. and Coppins, B. J.**, Key to the British sterile crustose lichens with *Trentepohlia* as phycobiont, *Lichenologist*, 11, 253, 1979.
261. **Verseghy, K.**, A *Graphis scripta* Ach. (Lichenes) gonidiumára vonatkozó vizsgálatok, *Bot. Kozl.*, 49, 95, 1961.
262. **Koch, W.**, Die Gonidie von *Racodium repestre* Pers., *Vortr. Gesamtgeb. Bot. Dtsch. Bot. Ges. N. F.*, 1, 61, 1962.
263. **Watanabe, A. and Kiyohara, T.**, Symbiotic blue-green algae of lichens, liverworts and cycads, *Studies in Microalgae and Photosynthetic Bacteria*, Japan Soc. Plant Physiol., *Tokyo*, 1963, 189.
264. **Kele, R. A.**, Isolation of the Mycobiont and Phycobiont of an Underwater Lichen, *Hydrothria venosa*, Honors thesis, Clark University, Worcester, 1964; as cited in **Ahmadjian, V.**, *The Lichens*, Ahmadjian, V. and Hale, M. E., Eds., Academic Press, New York, 1973, 653.
265. **Santesson, R.**, *The Lichens of Sweden and Norway*, Swedish Museum of Natural History, Stockholm, 1984.
266. **Zahlbruckner, A.**, *Catalagus Lichenum Universalis*, Borntraeger, Leipzig, 1922.
267. **Lamb, I. M.**, *Index Nominum Lichenum*, Ronald Press, New York, 1963.
268. **Puymaly, A., de**, Le *Coenogonium ebeneum* (Thwaites) A. L. Smith dans les Pyrénées, aux environs de Cauterets; l'écolgie de ce lichen et de son algae gonidiale, le *Trentepohlia aurea* Mart., *Botaniste*, 27, 1, 1935.
269. **Henssen, A., Büdel, B., and Wessels, D.**, New or interesting members of the Lichinaceae from Southern Africa. I. Species from northern and eastern Transvaal, *Mycotaxon*, 22, 169, 1985.
270. **Henssen, A.**, *Edwardella mirabilis*, a holocarpous lichen from Marion island, *Lichenologist*, 18, 51, 1986.
271. **Henssen, A. and Kantvilas, G.**, *Wawea fruticulosa*, a new genus and species from the southern hemisphere, *Lichenologist*, 17, 85, 1985.
272. **Henssen, A.**, The North American species of *Placynthium*, *Can. J. Bot.*, 41, 1687, 1963.
273. **Andreeva, V. M., Czaplygina, O., and Strelkova, L. A.**, Chlorococcales et Chlorosarcinales terrestres in parte Ucrainiae Polessie dicta distributae, *Nov. Syst. Plant. non Vasc.*, 22, 3, 1985.
274. **Reisigl, H.**, Zur Systematik alpiner Bodenalgen, *Osterr. Bot. Z.*, 111, 402, 1964.
275. **Nováček, F.**, De statibus *Gloeocapsae sanguineae* Ag. notula, *Sb. Klubu Prirdovedeckeho Brne*, 13, 1, 1930.
276. **Nováček, F.**, Epilithické sinice serpentinu mohelenských, *Arch. Verb. Natur. Heimatschutz Mahrisch-schlesischen Lande*, 3(a), 1, 1934.
277. **Tupa, D. D.**, An investigation of certain chaetophoralean algae, *Beih. Nova Hedwigia*, 46, 1, 1974.
278. **Tschermak-Woess, E.**, unpublished data.
279. **Poelt, J.**, unpublished data.
280. **Swinscow, T. D. V.**, personal communication to Wynne, M. J.
281. **Bergman, B. and Hällbom, L.**, *Nostoc* of *Peltigera canina* when lichenized and isolated, *Can. J. Bot.*, 60, 2092, 1982.
282. **Spector, D. L. and Jensen, T. E.**, Fine structure of *Leptogium cyanescens* and its cultured phycobiont *Nostoc commune*, *Bryologist*, 80, 445, 1977.
283. **Friedmann, E. I.**, personal communication.

Section III: The Lichen Thallus

Chapter III

THE LICHEN THALLUS

Hans Martin Jahns

I. THE SIGNIFICANCE OF THE LICHEN THALLUS

Lichens are a specialized group of fungi with which algae are permanently associated in a symbiotic relationship. The fungus is the dominant element in the taxonomic definition of a lichen species, despite the importance of the algal component in supplying photosynthates to the fungus. Lichens include a diverse collection of fungi so that, while they have their own well-established position in classical systematics, they are essentially an artificial grouping linked by their specialized mode of metabolism and nutrition rather than by a taxonomic similarity.

The taxonomy of lichens largely depends on the structure and development of generative fruit bodies,[1] as they are formed by the mycobiont alone and in their development they hardly differ from those of the nonlichenized fungi (see also Chapter X). However, it is the formation of the thallus which is the true characteristic of a lichen, as nonlichenized fungi are incapable of differentiation into different kinds of thalline tissues.

Thallus structures were previously studied only in a descriptive manner without an analytical approach, with various tissues being described and named, and different growth forms distinguished. This cataloging of the structures and forms is still of great importance and so they will be described here in detail. However, there are some problems in the customary ways of describing and naming lichen thalli and tissues.

II. PRINCIPLES OF FUNCTIONAL MORPHOLOGY

A problem with all lichen descriptions to date is that most anatomical and morphological terms have been taken from higher plant science. Such a transfer of terms has the advantage of the implied similarity with well-known higher plant structures, but also has the disadvantage of almost unconsciously transferring ideas on similarity of function and development, often leading to misconceptions of typical lichen reactions. For example, a "parenchymatic cortex" of a lichen is easily visualized, but in suggesting a differentiated and apparently determined surface tissue gives a false picture of the special functions of the lichen cortex and its enormous morphogenetic plasticity.

A further problem arises from the fact that the description and, particularly, the comparison of structures is only meaningful when linked to their function. This aspect was neglected in the past so that one can often only guess at the function of numerous structures. Where possible in the following, structures are discussed in terms of the demands placed on them by their progression from the nonlichenized to the lichenized state. The following points are of special importance.

A. Contact between Fungus and Alga

A lichen can only be formed when the association of the symbionts is a permanent feature in their life cycle, an accidental contact being insufficient.[2] The anatomy and morphology of the thallus must facilitate the coexistence of both partners and there must be exchange of metabolites (see Chapters IV, VI, and VIII).

B. Simultaneous Dispersal of the Symbionts

Since the presence of both partners is a prerequisite to the existence of a lichen, the contact between the symbionts has to be assured not only for the individual lichen but also in the course of the hologeny. For dispersal either both partners must be distributed simultaneously or certain adaptations must ensure contact and relichenization after separate dissemination (see Chapter VIII.B).

C. Adaptations to Autotrophy

Photosynthesis of the algae provides the energy source for the lichen and the structures of the thallus are primarily determined by adaptations to autotrophic nutrition. The following aspects are of importance:

1. Adaptation of the Growth Form

For an optimal utilization of the sunlight, the algae must be distributed in a flat layer or in some similar horizontal arrangement of branches of a fruticose thallus. Some of the structures developed by lichens are analogous to the leaves of higher plants, but other constructions also occur. In many cases, the lichen develops a fruticose thallus for better exploitation of its environment. This growth form, and others as well, can be explained to a certain extend as adaptations to energy gain.

The twigs and branches that are developed for this purpose require special anatomical structures for their stability, and the highly differentiated thalli have to be anchored to the substrate, so tissue specialization is to be expected (e.g., supporting tissues, assimilating tissues, cortex tissues, growth zones).

To be effective, photosynthesis not only requires a more or less leaf-like shape of the thallus but also a suitable arrangement of the algae inside the thallus. On one hand, the algae must be protected from excessive irradiation while, on the other hand, the position and thickness of the algal layer must facilitate an effective utilization of the available light.

2. Adaptation of Water Relations

A sufficient water supply is a central problem for lichens, as for most of the other lower terrestrial plants. Until recently, the water economy of lichens was thought to be controlled purely by passive physical processes. The lichens were considered to have no influence on uptake, transport, and loss of water, so that periods of drought could only be survived in a state of suspended life. This simplification is no longer justified since we now know that certain vegetative structures influence uptake, storage capacity, and loss of water, even if control is not complete[3] (see Chapter VII.B).

3. Adaptations to Gaseous Exchange

As in higher plants, photosynthesis and respiration require gas exchange in the tissues. In both lichens and phanerogams there is a competitive situation between gas exchange and water relations. The function of thallus tissues is adjusted to cope with the contradicting requirements to minimize evaporation and to allow free CO_2 exchange in the thallus tissues.

4. Adaptation to Nutrient Acquisition

The algae in the lichen thallus must be provided with inorganic nutrients from the substrate, air, and water. Special structures, such as cephalodia are involved in the nitrogen metabolism of the lichen (see Chapter VI.B).

D. Adaptation to Longevity

In most nonlichenized fungi compact structures are differentiated only when the reproductive organs develop, but these fruit bodies are relatively short-lived. The thalli of the lichens, on the other hand, must continue their assimilatory functions over a longer period of time. A short life span of the thallus, on which the fungus depends for its nutrition, would require its instant regeneration which would lead to an enormous waste of material. Lichens must, therefore, be long-lived, and this necessity is manifested in many structural characteristics of the thalli.

III. PHYLOGENY OF THE LICHEN THALLUS

The thallus of a lichen is the structure which differentiates between lichenized and non-lichenized fungi. Therefore, it is of interest to speculate on the phylogeny of this important structure. On the one hand, the lichens are of polyphyletical origin, since lichenized representatives are found in very different fungal taxa. On the other hand, the majority of species of the Lecanorales, one of the largest groups, developed a similar kind of lichen thallus so that, for this group at least, lichenization has apparently been a very successful adaptation.

Several theories try to explain the origin of the lichen thallus. An acceptable explanation is difficult to find, since compact plectenchymatous tissues in nonlichenized fungi are usually only found in connection with fruit body development, while the vegetative phase of the fungal life cycle covers the substrate with a loose hyphal network. Even in those fungi which develop a stroma, this often serves as the base for the differentiation of reproductive stages.

Since the compactly built thallus resembles the stroma of several nonlichenized fungi, the lichen could perhaps be related to some stromatic fungi. The stroma could have incorporated some algae and developed into a lichen thallus. Such utilization of free-living algae, as we shall see, occurs also in mature lichen thalli. This evolutionary theory has the advantage that it is based on a postulated process in fossil fungi that occurs in a similar form in contemporary species.

Another theory starts from the fruit bodies of the lichens.[4,5] Their fundamental structure corresponds to that of the nonlichenized fungi. In many cases, the lichen fruit bodies contain algae in their marginal zones. It is phylogenetically possible that compact fruit bodies of nonlichenized fungi have incorporated some algae, and, to enlarge the photosynthetically active area, these algae-containing tissues of the generative stage were enlarged until a vegetative thallus had evolved. This theory is strengthened by the fact that the *thallus verticalis* of certain lichens (e.g., *Cladonia*) is ontogenetically a part of the fruit body since it develops from the primordium of the ascocarp.[6]

Both theories start with the same prerequisites. They postulate that the initiation for development of a compact thallus comes only from the mycobiont and that, therefore, compact structures must have been present in the nonlichenized ancestors of the recent lichens. Algae would have been incorporated into these existing structures.

Recently, it has been shown that the fungus does not dominate the consortium alone, but that the algae are morphogenetically relevant for the development of a thallus.[7,8] If both partners are of equal importance then the ability to develop a thallus could also have been coevolutionary parallel to the development of the symbiosis. Phylogenetically, the mycobionts only need to have had the ability to develop compact tissues, but it is irrelevant whether these have been stromata or fruit bodies. The actual vegetative lichen thallus can then be traced back to the loosely interwoven hyphal mycelium of nonlichenized fungi. This hyphal network, which served originally for saprophytic or parasitic nutrition, could have come into contact with algae and, in the course of phylogeny, algae and fungi could, with many intermediate developmental stages, finally have formed the lichen thallus. This implies that the morphogenetic influence of the algae triggered the inherent ability of the fungi to develop compact tissue so that a highly organized lichen thallus replaced the more primitive undiffererntiated mixture of hyphae and algae.

Such a concept has the advantage that it allows for more variability and, therefore, is in accord with a polyphyletic origin of the lichen thallus. The two other theories mentioned above can be considered as possible developmental variations and added to the phylogenetical scheme.

IV. FORM AND STRUCTURE OF THE LICHEN THALLUS

In some primitive lichens, the thallus resembles that of free-living fungi or algae. If the mycobiont dominates, the thallus consists of a loosely interwoven network of fungal hyphae, in which the algae are more or less regularly embedded (Figure 1*). The algae can also dominate, in that case the lichen consists of a gelatinous algal thallus, traversed by fungal hyphae. In both cases there is a relatively low degree of anatomical differentiation into separate tissues, and this type of unstratified lichen is termed *homoeomerous*. Most lichens, however, are divided into definite tissue layers and these stratified thalli are termed *heteromerous*. These thalli have a characteristic habit and differ completely in form, color, and structure from free-living fungi and algae.

The appearance of the thallus is mostly determined by fungal tissues, as the greater part of the lichen is formed by the mycobiont. Only in a few lichens does the algal partner exceed the fungus in volume. Even if one or the other of the symbionts dominates the shape of the lichen and builds the greater part of the thallus, this does not lead to any conclusions on the morphogenetic influence of mycobiont or photobiont. In many lichens, where the fungus dominates quantitatively, the algae can still give the decisive impulses for the morphogenetic control of thalline development.

A. The Growth Forms of the Lichens

In the descriptive morphology, the lichens are divided on the basis of their habit into three large groups that are called *fruticose* (Figures 27, 28, 29, and 30), *foliose* (Figures 15, 16, and 118), and *crustose* (Figures 2, 3, 5, 7, and 8). There are several other special types, such as fruticose lichens with thin cylindrical lobes, which are known as *beard lichens* (Figure 23) or *hair lichens* (Figures 24, 25, and 26). Some lichens can absorb large amounts of water and attain a gelatinous consistency and, therefore, are called *gelatinous lichens* (Figure 21). One should also remember that all lichens are brittle when dry and all expand on moistening, becoming soft and pliable.

The simple division into three growth forms, does not adequately take into account the many variations in habit. In general, crustose lichens are flattened to the substrate, adhered by their entire lower surface and cannot be loosened without damage. Foliose and fruticose species, which are collectively called *macrolichens,* are typically rather loosely attached and can be easily removed. Such a simple definition is bound to be inadequate, since it specifies only a few typical conditions out of the almost continuously grading developmental series from primitive to highly differentiated thalli, but it serves for general orientation.

Over and over again, the same growth forms have developed from different systematic groups of the lichens, and these forms show an astonishing convergence not only in habit but also in the differentiation and arrangement of their tissues. However, it seems inadequate to draw conclusions from this convergence about the phylogenetics or systematics as has often been done in the past. It has been proposed that the crustose lichens are the most primitive representatives within any given group,[9] but this statement is not generally acceptable. Like many foliose and fruticose lichens, many crustose species have developed special organs and do not, in this respect, differ from the macrolichens.

Within the taxa, converging development has occurred and retrogradic developments can also be postulated, so that the habit of a lichen gives no clues to its phylogenetic or systematic position (see Chapter X). The great flexibility and lack of determination of the lichen tissue

* All figures can be found at the end of this chapter.

permits a morphological change in a shorter time span than in the phanerogams. Since the morphology of the lichen thallus usually represents an obvious functional adaptation to certain environmental conditions, the concept of a generally acceptable evolutionary trend from crustose, via foliose, to fruticose forms should, therefore, be discarded.

1. Crustose Lichens

The homoeomerous crustose lichens cover the substrate with a more or less even layer. They develop a granulate, scurfy, or varnish-like layer on stone, soil, wood, moss, or plant remains. Their habit may be determined in part by the shape of the substrate. This often happens in young thalli while the older lichens develop a continuous layer (Figures 4 and 5).

There are simply constructed crustose lichens that grow inside their substrate, for example, in wood or porous, soft stone. Species that grow inside rocks are called *endolithic* (Figure 3), and those that grow in wood are called *endophloeodic*. The algae, as well as the hyphae, of these lichens penetrate the substrate to a depth of several millimeters. The way of life of the endolithic lichens depends on their ability to dissolve rock with their lichen acids prior to the penetration of hyphae and algae. Some lichens can even dissolve quartz, but real colonization and thallus formation occurs only in soft rock, such as sandstone and limestone[10,11] (see Chapter IX.A).

Species living in calcareous rock develop peculiar oil-hyphae, of which the irregularly swollen cells are filled with oil droplets.[11,12] The colonization of stone or wood by endolithic or endophloeodic lichens, respectively, is often recognizable as a discoloration of the substrate, but frequently the erumpent fruit bodies are clearly visible.

Most of the crustose lichens are more or less heteromerous and can attain a high degree of differentiation. Numerous special organs, such as soralia, isidia, and pseudocyphellae, may develop. In a typical crustose lichen, the thallus consists of small areoles (Figure 2) in which the beginnings of stratification can be seen, as algae are more numerous in the upper part. The areoles can reach several millimeters in thickness. On the surface they possess a more or less distinct cortex. Often, a false cortex of dead and gelatinized fungal cells of the outermost layer of the thallus develops.[13] This *necral layer* (Figure 55) is continuously shedded and renewed from within by the growth of the thallus. The areoles of the thallus can form a continuous layer or be separated by cracks. In the second case, they may be scattered singly or in small groups on the *prothallus* (Figure 6), an algal-free mycelium of fungal hyphae. This growth from is very interesting in connection with the questions of growth and lichenization during reproduction.

A number of crustose lichens show forms transitional to the foliose lichens. Areoles are replaced by either elongated and narrow or scale-like lobes. These forms sometimes possess a well-developed cortex, which is either restricted to the upper side of the thallus or binds it on all sides. When the entire thallus consists of such leaf-like lobes that are closely adnate with their entire lower side to the substratum, they are called *placoid* (Figure 7). Often, the development is not so advanced and the thallus center consists of the typical areoles while only the marginal zone is formed by elongated lobes. In this case, one speaks of a crustose lichen with an *effigurate margin* (Figure 8).

The described forms present a clear transition to the foliose type, but are still adnate to the substrate. Other crustose lichens in which the margin of the individual squamules are not attached to the substrate, but raised above it, are called *squamulose* (Figure 9). Often the squamules may be arranged crowded together in a rosette.

When squamulose thallus lobes fuse laterally, another growth form results, where a larger shield-like structure is attached by a holdfast in the center of the lower side. Such *peltate* thalli, as found in *Lecanora*, may even show a complete habitual resemblance to a foliose umbilicate lichen (Figures 10 and 11).

Crustose lichens not only form transitions to the foliose lichens but also, although seldom, to the fruticose growth form. Small lobes may grow upwards and in some species they may be inflated. Such thalli are termed *pulvinate* (Figure 13). Numerous transition forms of this type are found in the lichen genera *Peltula* and *Toninia*.

2. Foliose Lichens

The thallus of the leaf-like lichens consists of flattened, dorsiventrally constructed lobes that are more or less closely attached to the substrate. In several cases the underside of the thalli is inseparable from the substrate and in these lichens there is hardly any difference from placoid crustose lichens (Figure 14). The division of the growth forms is here quite arbitrary. The thalli of other foliose lichens are attached to the substrate only with a part of their lower surface while the margins are raised upwards.

Often, the lower surface of foliose lichens is covered by rhizinae, cilia, or tomentum which may attach the thallus to the substrate or, what is of equal importance, may also raise it slightly above the surface. Some foliose lichens, characterized by lobes, more or less strongly involuted, are attached to the substrate with only a small part of their thallus, while the rest of the lobes grow upright and may die off at the base. Such forms represent the transition to the fruticose lichens. The boundaries between the types are again arbitrary.

A special type of the foliose growth form is the *umbilicate* lichen (Figure 17). Its shield-like thallus is only attached centrally by a disk-like holdfast, corresponding to a navel-like depression on the upper side (Figures 18 and 19). From this "umbilicus" the genus *Umbilicaria* takes its name. The umbilicate type is not restricted to this genus but also occurs in *Dermatocarpon* and several other genera. It is interesting that these genera are not closely related so that, with regard to the umbilicate growth form, there prevails an astonishing degree of convergence.

The function of the different growth forms of the foliose lichens can only be understood when one considers their different contacts with the substrate and also takes into account the mutual arrangements of the different lobes. It is rare for foliose lichens to develop single, flat leaves or flat, squamulose rosettes. Generally, the marginal lobes or neighboring thalli partly overlap. Often, systems of thalli originate with a tile-like arrangement of the lobes (Figure 12) or a cushion of intergrowing lobes. These arrangements create a specific microclimate which is important for the lichen.

The lower surface of a lichen can incorporate particles of the substrate into its lower tissue layers or it can penetrate the substrate with its hyphae or with a gelatinous substance. In this way an unstable substrate (for example, sand) becomes fixed and colonized. Often, a lichen thallus grows on dead remnants of an earlier lichen generation. The resulting structures are important for water relations, regeneration, and competition of the species.

The anatomy of the foliose lichens is often very complicated. The function of the numerous accessory organs is often closely related to the growth form. The development of an upper and lower cortex, rhizinae, and tomentum in connection with different growth forms leads to differing adaptations of the water economy (see Chapter VII.B). The foliose lichens also have further accessory organs that are important for metabolism (cyphellae, pseudocyphellae, and cephalodia) or for the vegetative propagation (soralia, isidia, etc.).

3. Fruticose Lichens

The thallus of the fruticose lichens can be strap-shaped (Figures 27 and 28) or cylindrical (Figures 30 and 31). The strap-shaped thalli resemble those of some of the foliose lichens. While cylindrical thalli, in most cases, show a radial construction, the strap-shaped lichens are built radially as well as dorsiventrally. These lichens are generally attached at their base to the substrate where a distinct holdfast can develop (Figure 75). Some fruticose lichens are not attached to the substrate at all but hang loosely with their filamentous thallus draped over the branches of trees (Figure 23). Others lie on the ground and can be dispersed by the wind (Figure 31).

Fruticose lichens grow either singly or in bush-like cushions on the ground, or they hang from trees or rocks. Hanging and upright thalli are exposed to different mechanical stresses. The necessary strengthening elements are arranged either in the form of a tubular sheath at the periphery of the thallus (Figure 57) or in a central strand (Figures 61 and 62). The tubular supporting tissue gives a high degree of resistance to bending and is, therefore, suitable for erect thalli, while a central medullar strand gives pendant lichens good resistance to stretching. For further increase in stability diagonal supports are added inside the thallus (Figure 60).[14] Lichens from different systematic groups have developed similar types of supporting tissue.

Fruticose lichens (for example, of the genus *Usnea*) that possess very thin, circular thalli and hang down from the substrate, are often called beard lichens. In regions with high humidity they represent a considerable part of the epiphytic vegetation of forests. Species of the genus *Usnea* can reach a length of several meters and belong to the largest known lichens. The hair lichens are similar to the beard lichens although their thalli reach only a length of a few millimeters. In the latter, the symbiotic alga determines the growth form and consist of filamentous green algae or cyanobacteria, which are more or less completely enclosed by an envelope of hyphae (Figures 24 to 26).

As already described, the fruticose lichens often develop thick lawns or cushions (Figure 29) in which the lower part of the thalli dies off, while the terminal regions continue their growth. Cushion-like growth form and dead thallus remnants influence the microclimate and, thereby, the growing conditions of the lichen. This growth form is also of importance for regeneration and reproduction. The filamentous nature of the thallus of the beard lichens can also be explained as a certain kind of ecological adaptation.

Anatomically, the fruticose lichens demonstrate the same differentiations as do the foliose lichens.

4. Lichens with a Twofold Thallus

The thallus of some lichens consists of a horizontal part (*thallus horizontalis*) and of an erect part (*thallus verticalis*). The horizontal thallus can be crustose (Figure 33) or foliose (Figure 32). In some lichens, it contains the algae and fulfills the photosynthetic requirements while the fruticose *thallus verticalis* bears the fruit bodies (e.g., in *Baeomyces*). Often, the erect thallus dominates and the horizontal thallus is reduced. In these cases, the photosynthetic apparatus is located in the *thallus verticalis*. The existence of this developmental process is phylogenetically significant.

The *thallus verticalis* can develop in two different ways.[6] In the genus *Cladonia* and related genera, the *thallus verticalis* is ontogenetically a part of the fruit body and is derived from its primordium. This also applies when the *thallus verticalis*, as is often the case, does not develop apothecia and remains sterile. The *thallus verticalis* that develops from the primordium of the fruit body is called *podetium* (Figures 32 and 33).

In other lichens, e.g., the genus *Stereocaulon* (Figure 34), the *thallus verticalis* is a thalloid outgrowth that is not connected with fruit-body development. This structure is called a *pseudopodetium*. On the pseudopodetium, primordia can originate which may develop into fruit bodies.

5. Gelatinous Lichens

In the gelatinous lichens we find crustose, foliose (Figure 21), and fruticose thalli, but most species are quite small. Only a few foliose species can attain a size of several square centimeters. The lichens of this group are black, gray black, or greenish, and they lack the bright colors of other lichens since the lichen substances that cause this coloration are not produced. A light reddish color, which is found infrequently, is caused by the gelatinous sheaths of certain cyanobionts.

Gelatinous lichens are generally less differentiated than other lichens. Simple crustose lichens consists of fungal mycelium in which algae are loosely embedded, whereas, in the gelatinous lichens the hyphae generally run inside the jelly of the algae, but often do not touch the algal cells at all (Figure 139). The characteristic intumescence on imbibition is mostly caused by the swelling of the gelatinous sheaths of the cyanobionts that can absorb large amounts of water.

The thallus of simple gelatinous lichens is homoeomerous. In other species, however, the fungus develops an outer envelope of hyphae or a cortical layer round the thallus (Figure 35), nevertheless, a sharply delimited algal layer is always absent. Other differentiations, for example, the accessory organs, are rarely found in gelatinous lichens when compared to other growth forms.

B. Ecological Significance of the Growth Forms

The significance of the different growth forms of lichens becomes apparent when one sees them as functional adaptations to certain environmental conditions.

If the crustose thallus is accepted as the initial phylogenetic form, then the question arises, what is the advantage of developing complicated foliose and fruticose structures? Basically, the crustose lichens can only develop a relatively small biomass in relation to a large surface area. Fruticose lichens at the other extreme, can develop a large amount of tissue in a small area. This could be most important in the competitive situation existing between lichens and phanerogams. Crustose lichens are generally weak in competition since they can easily be overgrown by other lichens, mosses, or phanerogams. The crustose species, therefore, only dominate on rocks in extreme environments. Foliose and fruticose lichens have a better chance to maintain their competitive position.

The most important and basic requirement for the autotrophic nutrition of the lichen is a thallus that possesses a sufficiently thick algal layer, which is exposed to the light. It is obvious that crustose as well as foliose lichens fulfill this prerequisite. In the fruticose lichens, the situation is more complicated. On one side, a finely branched, erect, or pendulous thallus can arrange the neighboring twigs and branches in a tightly packed three-dimensional system that can absorb the total amount of irradiation. On the other side, in cushions of *Cladonia*, the lower and inner branches are shielded from light.[15] The upper branches contain most of the algae and chlorophyll while the lower stems consist almost entirely of fungal hyphae.[16] The photosynthetically active part of the lichen must, therefore, supply a tissue with mainly supporting functions. This presents a disadvantage which is compensated for by advantages in water relations.

It has been assumed that the change in the water content of lichens is a purely passive physical process and that the thallus has no influence on the uptake, storage, and loss of water. This is correct insofar as all lichens quickly loose their water under unfavorable conditions and survive dry periods in a state of suspended life.[17] However, the desiccation rate is influenced by the anatomy of the thallus. On sunny days many lichens can only reach a positive photosynthetic rate for a few hours of the morning, so that a delay in desiccation for even one hour is already important.[18] Of all the growth forms, the crustose thalli (and several foliose forms) are, without doubt, the best protected against water loss, as their entire lower surface is attached to the substrate and no water loss can occur at this side. As we shall see from the description of the anatomy of the thallus, the upper surface may have an extra protective layer against evaporation. Even better protected are endolithic and endophloeodic lichens. Therefore, it is not surprising to find more crustose lichens than any other kind in exposed extreme localities.

Some lichens are poorly protected against water loss. The finely divided branches of the beard lichens dehydrate quickly in dry air. This is one reason for their occurrence in areas with high humidity, such as montane forests. Other fruticose lichens are protected against evaporation by their growth form, such as cushions of *Cladonia*, within which pockets of air with higher humidity decrease evaporation.

Water storage is mainly regulated by the inner tissues of the thallus, but the growth form also plays a part. All lichens can regenerate and new thalli can be built on the old tissues. The dead remnants underneath the living lichen represent an effective water storage compartment (Figure 50).[19] Also, the photosynthetically inactive lower branches of lichen cushions are important for water storage.

The morphology of water-storing tissues sometimes shows ecologically linked variations.[20]

For water uptake, the thallus form can also be important. For many lichens, air humidity is an important water source and is effectively used, for instance, by the finely branched beard lichens. This growth form occurs in humid regions not only because dehydration is less but also because an optimal water supply is at hand. The ability to use air humidity is important to these lichens, as liquid water is not always available (for example, inside the tree-crown).

Certain growth forms are adpated to the uptake of liquid water. After a short rain, lichens that are completely attached to their substrate with their lower side are at a disadvantage when compared to foliose thalli whose underside is raised a bit above the substrate (Figure 20). Between thallus underside and substrate, a layer of water is retained which can be absorbed slowly by the lichen.[21] This procedure may be amplified by accessory organs such as rhizinae or tomentum[22,23] (see also Chapter VII.B).

The different growth forms are also exposed to external mechanical influences. Some crustose and cushion-like forms can be seen as adaptations to sites with high wind speed.

V. CELLS, HYPHAE, AND TISSUES

A. Types of Cells

The spores of a lichen fungus germinate by germ tubes, which then elongate to form branching and septate hyphae (Figure 135). All lichen fungi have a filamentous organization; meristematic tissues, which originate by three-dimensional divisions of a single cell, never occur. The original component of a mycelium is the branched, transversely divided hypha with cylindrical segments and thin walls (Figure 135). The laterals normally branch off at an acute angle to the parent hyphae, only rarely at a right angle. The hyphae may run parallel side by side or, in their subsequent growth, spread out laterally. In this case, they diverge radially, giving rise to a tissue which is expanded in flabellate or fountainhead formation.

With these relatively simple basic structures, through a process of differentiation, the lichen can produce very different tissues. High stability can be attained with different methods. Hyphae of the same layer are almost never arranged in parallel but instead are interwoven (Figures 36 and 41). Individual cells of different hyphae may fuse together secondarily. Through these *anastomoses* (Figures 39 and 40) a three-dimensional hyphal network is formed with great mechanical stability.

The processes of differentiation lead to an alteration in the form of the cell and to secondary growth of the wall.[24] Cylindrical cells may swell up and become rounded off, so that short cells become spherical (Figures 37 and 54), and long cells more ellipsoidal. Clavate cells are produced if only one end of the cell becomes swollen (Figure 38). Uneven swelling of this kind is often typical of the terminal cells of paraphyses. Outside the hymenium, numerous unevenly thickened cells occur. Their lumina are often polyhedral and are provided with thin extensions (Figure 136). Such cells are triangular or multiangular in transverse section. Apart from the cell connections by anastomoses, the secondary changes of the cell wall are of importance for hyphal contact. Through secondary growth and impregnations, the wall can become thicker than the diameter of the cell lumen. Cell walls can agglutinate and fuse together to form a homogeneous mucilaginous substance in which the cell lumina are embedded (Figures 43 and 44).

The space between the hyphae is often filled with a gelatinous substance that is secreted by the cells. Such gel-like substances cannot always be separated from the gelatinized cell walls. Very thick cell walls are compactly built close to the cell lumen, while the structure loosens further outwards. In this case, it is difficult to say where the hyphal wall ends and the interhyphal substance starts.[25] The hyphae can be encrusted by lichen substance (Figure 41).

B. Types of Tissue

A tissue with basically reticulate structure and branching, anastomosing hyphae is characteristic of lichens and nonlichenized fungi. These woven tissues, to which the term *plectenchyma* (Figure 36) is applied, are formed by the aggregation and intertexture of individual hyphae. Genuine meristematic tissues occur perhaps very rarely in muriform spores and in the ascostromata of certain ascolocular fungi.[1]

Next to networks made of loosely interwoven hyphae that are immediately recognizable as plectenchyma, tissues are formed, as a result of conglutinization of hyphae, which are more difficult to analyze. Various types of plectenchyma may be distinguished that have a deceptive resemblance to certain tissues of higher plants and are named after them.[26,27] For example, the terms pseudoparenchymatical and prosoplectenchymatical tissue are applied to structures which are similar to parenchyma and prosenchyma of phanerogams. In such lichen tissues, the direction of the individual hyphae can no longer be discerned.

A *paraplectenchyma (pseudoparenchyma)* originates, when hyphae with tightly packed spherical cells become flattened against each other and conglutinate (Figure 42). If the cells are thin-walled, the solid paraplectenchyma resembles a parenchyma with isodiametric cells. While in such a tissue the direction of the individual hyphae is not recognizable, in other paraplectenchyma tissues the hyphae are usually separated from each other to a varying extent (Figure 137). These tissues do not show such a distinct isodiametric structure and the individual hyphae can still be discerned.

Prosoplectenchyma of lichens consist of long-celled hyphae in parallel arrangement, the walls of which are swollen and completely fused (Figure 43). This structure resembles the prosenchyma of the phanerogams. In the lichenological literature, the term prosoplectenchyma is frequently used for all interwoven complexes which form solid tissues by the swelling and conglutination of hyphal walls (Figures 44 and 137).[28] Here the term "*scleroplectenchyma*" would be more suitable. The cells may be shorter or longer, and the arrangement of the hyphae parallel or reticulate.

Other conspicuous types of tissue also occur. For example, the single-layer bounding tissue of the gelatinous lichens consists either of a pattern of contiguous isodiametrical cells (Figure 47) or one in which the walls are interlocked with jigsaw-like projections (Figure 45). Tissues can be characterized by the orientation of the hyphae in the thallus. A tissue of anticlinal hyphae with cylindrical cells and conglutinated walls is called *pallisade plectenchyma* (Figure 46). In other cases, the hyphae may run periclinally, i.e., parallel to the surface.

A more detailed classification of the described tissues and their numerous varieties has been attempted, but only little additional understanding can be attained by such schematic distinctions because the structures are very variable. The degree of determination is very low and the structures are often influenced by changes in their function and by the environment. For example, a prosoplectenchymatic tissue can change its structure according to the light or water conditions,[29] lose its compact structure, and become a net-like plectenchyma. Boundary tissues can resume their growth and, in the course of ontogeny, participate in the development of special organs.[30,31] One must always keep in mind that the naming of lichen tissues were strongly influenced by their superficial resemblance to the tissues of higher plants, whereas their function corresponds only rarely, and the degree of determination is not always comparable.

VI. STRUCTURE OF THE VEGETATIVE THALLUS

A. Thallus Layers

1. Algal Layer and Medulla

The heteromerous species have a stratified thallus with an *algal layer* and a subtending *medulla* (Figures 48 and 49) in contrast to the undifferentiated thallus construction of the homoeomerous lichens. The difference between the two layers is small, as the algae lie in that part of the tissue where optimal light intensities prevail, and, if, for example, the thallus is turned over, the algal layer and the medulla can change position after some time.[6] The loose hyphal tissue of the medulla facilitates gas exchange. The hyphae, especially when impregnated with lichen substances (Figure 41), can be resistent to wetting, which has an advantage for gas exchange. The lichen substances often cover the cell walls with a dense crystal layer.

2. Cortex

A thallus that consists only of a medulla and algal layer is not very durable and is very sensitive to environmental influences. During a dry period all the water is evaporated quickly and during high irradiation the algae may be damaged. To overcome these difficulties, a cortex often develops either on the upper side of the thallus only, or it bounds the thallus on all sides (Figure 48).

The gradual stages in the formation of a cortex can be particularly well observed in the homoeomerous thalli of certain gelatinous lichens. Many of them are ecorticated, but sometimes hyphae which reach the surface, bend at a right angle and continue their growth, still inside the jelly, along the surface (Figure 139). Other species show all transitions from a primitive incomplete cortex of irregular cells, which are not closely contiguous (Figure 141), to a regular cortex of isodiametric cells (Figure 140).

In the cortex of foliose and fruticose lichens, all types of pseudoparenchymatical or prosoplectenchymatical tissue occur. A double cortex may also be present, consisting of two different types of tissue (Figure 138). The cortex on the upper side of the thallus may have a different structure from that on the underside.

The thickness of the thallus may vary between habitats. In *Xanthoria parietina*, for instance, thalli are much thinner in shady locations than in those exposed to full sunshine. This has the effect of protecting those algae which cannot tolerate high light intensities. The cortical layer of lichens absorbes 26 to 43% of the light falling on it.[32] Lichen substances, especially pigments, which are deposited in the cortex and give some lichens their colorful appearance also absorb a large part of the incident light. Parietin in the Teloschistaceae, for instance, gives a deep yellow or orange-red coloration. Green and yellow shades are common, but red, blue, and violet colors are of rarer occurrence. Lichens exposed to full sunshine generally contain more pigment than individuals of the same species from shaded locations.[33]

Some lichen substances of the cortex do not have a protective function. Atranorin, a very widely distributed colorless depside, may have the opposite effect.[34] It absorbs ultraviolet light and emits a part of the energy thus taken up by fluorescing in the form of light of a longer wave-length; this reemitted light is of a wavelength which can be utilized for photosynthesis. In this way, atranorin in the cortex of a lichen could supplement photosynthesis of the phycobiont.

The protective function of the cortex may also have disadvantages. The cortex, with its strongly conglutinated hyphae, offers a serious obstacle to gaseous exchange.[35,36] When moist, the cells are pressed so tightly together that no intercellular spaces are present. Only in dried thalli is the cortex perforated by minute pores that are filled with air. Dry lichens have a gray appearance because of the total reflection of sunlight on these cracks. These pores are hardly sufficient for gas exchange and for this reason the algal layer of some lichens can protude into the cortex and even reach the surface of the lichen. Of more importance for gas exchange are special respiratory openings of the thallus.

In some lichens, the actual cortex is covered by another membraneous protective layer that resembles the cuticle of the phanerogams and is called an *epicortex* (Figure 51). This structure may represent extra protection from evaporation, but for gaseous exchange it has to be punctured with pores.[37-39]

Lichens which are not attached to the substrate by their lower surface and which do not have a lower cortex have no problems with gas exchange, but, naturally, such a thallus loses larger amounts of water from the exposed medulla (Figure 52). On the other hand, water uptake is easier for the sponge-like medulla than for the compact cortex. Especially foliose lichens that grow close to the substrate and can, with their lower surface, easily absorb liquid water that gathers underneath them. It is obvious that a structural advantage for one trait often is a disadvantage for another. Which combination of constructural characteristics is most favorable to the lichen depends on environmental conditions.

B. Tissues of the Fruticose Lichens

The division of the thallus into the subsequent layers of the upper cortex, the algal layer, the medulla, and the lower cortex naturally only exists in foliose thalli, but the radial thalli of the fruticose lichens are, in principle, not different in construction from the flattened thalli. Here, the layers formed by the algal zone and the cortex surround the centrally located medulla in the form of a cylinder (Figures 57, 61, and 62).

Foliose lichens growing close to the substrate need no special supporting tissues. Only the largest foliose lichens possess a network of *veins* (Figure 58) or *ridges* (Figure 59) that give stability to the lobes. Fruticose lichens cannot exist without a well-developed supporting tissue. Lichens that are erect must have tensile strength. The supporting elements are, therefore, found in the tubular thalline cortex (Figures 46 and 57). Its hyphae are interconnected in a reticulate fashion, run parallel to the surface, or form a palisade parenchyma. The medulla of these lichens is cottony, gelatinous, or hollow. A system of diagnonal supports may provide further stabilization (Figure 60).

Pendulous lichens need special resistance to stretching that is achieved by a central medullary strand of densely aggregated hyphae, which is formed by conglutination of the longitudinal running, thick-walled medullary hyphae. The best known example of this type is the genus *Usnea*, with its thread-like, elastic central strand (Figure 61). In other species, several central strands may become confluent (Figure 62).

C. Superficial Structures

The surface of a corticated thallus is often smooth, especially when the lichen possesses an epicortex (Figure 51). In other cases, the uppermost cells of the cortical hyphae grow out in irregular groups, giving rise to a lumpy-areolate surface (Figures 53 and 54). Frequently, individual cortical cells elongate into branched or unbranched *hairs* (Figures 48, 52, and 84) which give the lichen a felty appearance. The hairs can be decumbent or form upright bunches, with either elongated or spherical cells, giving the hairs a beaded form (Figure 37). A dense felt of short hairs, not serving for attachment, is called a *tomentum*, while hairs that attach the thallus to the substrate are termed *rhizoid hyphae*. A peculiar form of tomentum, termed here *spongiostratum* (Figure 63), is found on the lower surface of the thallus of the genus *Anzia* where it takes on the structure of a coarse reticulum. All the structures described here may occur on the upper surface as well as on the lower surface.

Hairs and tomentum can fulfill different functions. They may protect against evaporation or too strong irradiation,[40] a function also fulfilled by similar structures in higher plants. The water uptake is probably a more important function since the tomentum absorbs water like a wick. There are transitions between rhizoid hyphae and rhizinae (see below), both with the function of anchoring the lichen.

The upper or lower sides of lichens may be covered with reticulate ridges or veins that serve as supporting tissues and can develop from the cortex or the medulla (Figures 58 and 59). The surface of thallus and fruit bodies may be covered by minute particles called pruina formed by the scaling off of dead cortical cells (Figure 55) or derived from the deposition of lichen substances. A crystal layer may develop round the pores of the cortex (Figure 56).

D. Appendages of the Thallus[41]

As discussed above, the thallus can be covered with hairs that, as in the spongiostratum, may be arranged in a close network. In other cases, adjacent hairs form small bushes consisting of either a sheaf of hyphae which are not connected with each other (*rhizoptae*) (Figure 64) or hyphae interconnected by anastomoses. Firmly connected strands of hyphae which may serve as anchorage for the thallus are called *rhizinae*. The structure of the rhizinae is quite different in the lichen genera. They consist either of loosely packed bundles of hyphae that may spread out in the form of a brush (Figure 58) or may be more compactly constructed with a smooth cortex (Figure 65). They develop from the cortex or the medulla of the lichen or from both.

When mature rhizinae touch the substrate, a broadly expanded foot, which forms the attachment to the substrate, may originate in the course of subsequent superficial growth at their tips (Figures 65 and 66). The development of rhizinae can sometimes be initiated by contact of the thallus with the substrate, but, in most cases, numerous rhizinae are developed by the lichen without any external induction (Figure 67). In some lichens, the first rhizinae grow from the primoridum of the lichen, even before the tissues of the thallus have been differentiated (Figure 125). From the numerous rhizinae developed, only a few touch the substrate and serve to attach the thallus.

Rhizinae are very variable in their form. They may be simple (Figure 65) or branched (Figure 68). The branches, if present, are either irregular or, rarely, dichotomously forked. Particularly conspicious are those organs which are bulbously swollen at their lower end.

Many lichens possess protuberances on the lower surface that are similar to rhizinae but do not anchor the lichen. These accessory organs often occur in umbilicate lichens and are called *rhizinomorphs* (Figure 71). Other rhizinae-like structures which proliferate from the thallus margin or from the upper surface only seldomly serve to attach the lichen. They are called *cilia* (Figures 72 and 73) or, when they are in fact tiny lateral branches, *fibrillae* (Figure 74).

Many fruticose lichens and the umbilicate lichens are attached to the substrate by a *holdfast* disk (Figures 18 and 75) that is called an umbilicus in the umbilicate lichens. It consists of a dense strand of conglutinated hyphae which originate from the medulla.

Hyphae from the lower surface of closely attached thalli sometimes grow more quickly than the actual thallus of crustose and foliose lichens and surround them with a margin, which is different in color and structure from the rest of the thallus. This *prothallus* may have a felted-filamentous (Figures 76, 128, and 130) or cartilaginous structure (Figure 79). It is also termed a *hypothallus,* especially when it remains persistently between the areoles of the thallus in its continued growth. Occasionally a continuous prothallus does not develop, but individual hyphal strands radiate from the thallus (Figure 77).

Nearly all accessory organs described above do not only serve for the attachment of the thallus but they may also function in water and mineral uptake.[22,23] This does not imply that rhizinae function like the roots of higher plants in which a stream of water and inorganic nutrients are transported, but provide a surface enlargement which absorbs liquid water like a candle wick making it available to the lichen. Numerous accessory organs are covered by small hairs or thin-walled cells which may have a function in water uptake (Figures 67, 69, 70, and 73). The prothallus may be important for growth and dispersal.

E. Organs of Gaseous Exchange

By describing the cortex, the difficulties this strongly conglutinated structure presents for gas exchange in and to the underlying tissues have already been pointed out. Even though the cortical layer is perforated by minute pores,[37] many lichens with an upper and lower cortex seem to be dependent upon developing extra respiratory openings. The best known are those found in the genus *Sticta* which are called *cyphellae* (Figure 81). They originate as small pits on the lower side of the thallus. The lower cortex in these places forms a circular prominence which is depressed in the center in the form of a crater. At the bottom of the depression the cortex is replaced by detached medullar hyphae with spherical terminal cells (Figure 82). The organ thus produced is reminiscent of the lenticels of higher plants. Old cyphellae achieve a diameter of several millimeters, but are no longer functional as the spherical cells become densely aggregated and conglutinated by a mucilaginous substance (Figure 83).

The respiratory pores on the underside of the thallus of *Nephroma resupinatum* are similarly filled with spherical terminal cells of the hyphae (Figure 85). Beneath the pores, a cavity can be observed. Here, in contrast to *Sticta*, the respiratory apertures are not in the form of a crater-like depression, but are proliferations of the cortex, known as *tubercles* (Figure 84).

The cyphellae with their regular appearance occur in only one genus, while many other genera form somewhat irregular respiratory pores which are called *pseudocyphellae* (Figure 78). In the genus *Pseudocyphellaria*, so named on their account, they form irregular lumps on the underside of the thallus, are corticate on their periphery, and filled in the center with short-celled medullary hyphae (Figure 80). Generally, the form of the pseudocyphellae shows many variations in different genera.

In addition to cyphellae and pseudocyphellae, one finds in many lichens respiratory pores to which no particular name has been assigned. For example, wart-like protuberances on the thallus upperside may have a central punctiform depression. The cortex in such places is reduced and individual medullary hyphae are exposed (Figure 86).

Naturally, all organs, regardless of their main function, may assist in gas exchange if they connect the medulla with the outside. Every soralium and places left by the erosion of isidia are significant in connection with gas exchange. In some lichen apothecia, a fissure can develop between hymenium and the surrounding thallus margin through which gas exchange can also take place.[42] In lichens with a thick cortex, the algal layer sometimes protrudes into the cortical layer nearly reaching the surface and facilitating gas exchange (Figure 87).

While, on the one hand, the respiratory pores are needed for gas exchange, they present, on the other hand, especially in xeric environments with strong irradiation, a disadvantage for the water economy as water can easily evaporate through these openings. For example, in species of *Caloplaca* found in such habitats, the elongated pseudocyphellae are found at the side of the lobes (not on the surface) and in gaps between the long and narrow thallus lobes where they are not directly exposed to irradiation.[42,43]

F. Cephalodia

Some of the lichen fungi form a symbiotic association with cyanobionts, but most of them associate with green algae (see Chapter II.B). The coexistence with a cyanobacterium has the advantage for the fungus that at least some species of them are capable of fixing atmospheric nitrogen. Certain lichens which contain a green alga as phycobiont achieve the same nutritional advantage by taking up a cyanobiont in certain parts of their thallus as a third partner in the symbiosis. The cyanobionts only rarely develop a second algal stratum underneath the layer of green algae in the thallus. Usually, the cyanobacteria are placed in distinctly delimited parts of the thallus which are called *cephalodia* (Figures 16 and 145).

Cephalodia in lichens have evolved independently in repeated instances and they occur in diverse systematic groups. Cephalodia occur in about 520 species. They are either internal, in the medulla, or external, on the upper or lower surface of the thallus. The cyanobiont of the cephalodia is most often *Nostoc*, but also *Calothrix, Gloeocapsa, Scytonema*, and *Stigonema* have been reported. All genera belonging in the Stereocaulaceae have cephalodia; all species of the family Stictaceae with a eukaryotic primary phycobiont (*Sticta, Pseudocyphellaria*, and *Lobaria*), and the family Peltigeraceae (*Peltigera, Nephroma*, and *Solorina*) have cephalodia. Other genera which have species with cephalodia are *Placopsis* and *Psoroma* and also some species of Caliaceae and Lecidaceae.

In all studied cases, the lichen thalli display a high degree of selectivity in the acquisition of the third partner (the cyanobiont of the cephalodia). Even when various cyanobacteria or green algae are growing epiphytically on the thallus, only the preferred symbiont is incorporated and the cephalodium formed (see Chapter VIII.C).

These "selected" symbionts are trapped by hairs on the upper surface or by rhizines[44a] on the lower surface and apparently elicit an active proliferation of fungal hyphae which begin to envelop the cyanobiont. The cyanobiont grows rapidly at this stage and forms a thick colony, which becomes engulfed by the mycobiont to form the cephalodium (see Figures 4D to 4F in Chapter II.B).[31] The cortex and algal layer beneath the cephalodia of the upper thallus surface dissociate and a direct contact between the cephalodia and the medulla is formed (Figure 145). In the case of cephalodia on the lower surface, the lower cortex disintegrates to form the direct contact with the medulla, or, in the case of internal cephalodia,[44b] the cephalodia finally become positioned inside the medulla.

For nitrogen fixation in the cephalodia and the transport of the nitrogen metabolism products to the main thallus, see Chapter VI.B.

Next to the well-developed cephalodia, some lichen fungi can be loosely associated with cyanobionts, a contact that seems to be important for their juvenile development. Fungal hyphae and several kinds of cyanobionts can colonize a substrate, covering it with an unstructured network, before the fungus acquires its special green alga and develops a normal thallus, free from cyanobionts.[44] This mechanism is interesting in connection with the dispersal strategies of lichens.

VII. VEGETATIVE REPRODUCTION

The reproduction and propagation of a symbiotic organism is very complex, because, by separate dissemination of the two partners, the symbiosis has to be reestablished each time (see Chapter VIII.C). The lichens have developed methods to optimize the success of separate distribution, but the best response to the problem are special diaspores by which both bionts are distributed simultaneously. Mutual dispersal can occur by fragmentation of the thallus or by special organs for vegetative propagation.

The vegetative reproductive organs are classified, according to their structure, as *isidia, soralia*, and *hormocystangia*. The formation of isidia and soralia is widespread in crustose, foliose, and fruticose lichens, while gelatinous lichens form only isidia. Hormocystangia are, up until now, known only from one genus.[45] Soralia and hormocystangia are organs in which many soredia and lichenized hormocysts are formed as diaspores. Isidia are outgrowths of the thallus, which break off in one piece. Each isidium is, therefore, a single diaspore.

A. Regeneration and Distribution by Fragmentation

Most lichen thalli become very brittle when dry and, especially the large foliose and fruticose lichens, can easily crumble in this condition. From the fragments new thalli develop by regeneration, either at the same locality or at a new site after dispersal. The same regeneration process occurs also in an undamaged lichen when it has reached old age and dies. In some lichens, this final stage of the ontogeny is reached after they have developed sexual reproductive organs. Since the differentiation of lichen tissues is not finally determined, regeneration can start from nearly every part of the thallus. Most often, parts of the algal layer protrude from the degenerated lichen or the fragment of the thallus and differentiate into a new lichen.

In other cases, undispersed isidia (e.g., in *Pseudevernia* and *Parmelia*) (Figure 88),[19] pycnidia (in *Cetraria*) (Figure 89),[46] or even paraphyses of the ascocarp (in *Cladonia*)[47] resume growth and change into small thalline lobules. These regeneration processes seem to require a minimum size of the thallus fragments. Very small fragments and propagated isidia cannot directly develop into a new thallus but must follow a more complicated way of differentiation.

B. Isidia

1. Definition and Function of the Isidia

Isidia are small outgrowths of the thallus which contain algae and correspond in their anatomy with the structure of the thallus. According to general opinion, they serve for vegetative propagation. This definition seems to be clear but in reality it is not always easy to distinguish isidia from other outgrowths and appendages of the thallus. Many lichens possess, for example, warts and papillae on their surface that are similar to young isidia. Old isidia may vary only gradually from foliose outgrowth that are called lobuli. There are also transitions to fibrils and small lateral branches, and frequently old isidia, which have not broken off, change into other thalline structures.[19] Only in some lichens are isidia more or less actively dispersed by degeneration of their cortical base (Figure 90). Particularly, the isidia of many crustose lichens, e.g., the Pertusariaceae, break off easily. In others, the isidia are liberated only by the dying off of the thallus. In many gelatinous and foliose lichens, the isidia remain permanently attached to the thallus. In such lichens, it is possible that they serve primarily to increase the surface area.

In *Leptotrema* and *Graphina*, the isidia are important for the aeration of the thallus.[48] The hyphal tissue at their apex may be loosened and porous so that they represent a transition to a respiratory pore. When isidia break off, a crateriform respiratory pore remains (Figure 91) which may be closed by subsequent development of a secondary cortex.[49]

2. Form and Development of the Isidia

Isidia can be very variable in appearance. Frequently, one finds spherical (Figure 92), cylindrical (Figure 93), squamulose (Figure 98), or branched-coralloid (Figure 94) forms. These formations are usually of solid construction, but a few lichens have isidia which are hollow and appear as if inflated. Transitions between isidia and soralia can occur. At the tips of isidia, soralia may develop (Figure 96) and, likewise, isidia can be formed inside a soralium (Figure 97). In the latter case a compact outgrowth develops out of the loosely packed network of the soralium and forms a cortex secondarily.

The development of isidia is just as diverse as their form.[49] Their formation can be initiated by almost all the participating tissues. Medullary hyphae or hyphae from the algal layer may intrude between the cortical hyphae, carrying algal cells upwards with them.[50] The differentiation of a cortex on the protuberance originating in this way occurs only secondarily. This developmental type originates often in pseudocyphellae because the outgrowing hyphae do not have to penetrate a cortex (Figure 100).[19]

In *Peltigera praetextata*, the isidial structures only develop in places where the cortex has been injured, beneath the ruptured and partly separated cortical layer (Figure 99). The formation of isidia on thalli of *Peltigera praetextata* may be induced experimentally by cutting of the thallus. It is doubtful whether these dorsiventrally constructed foliose outgrowths should be named isidia. Perhaps they should be described as lobuli, with regenerative function.

In another mode of formation, the isidium develops from a protuberance of the thalline cortex. The tissue of the algal layer grows upwards simultaneously into the originating outgrowth (Figure 144). In some cases an isidium can develop exclusively from cortical hyphae by trapping free-living algae, which fall accidentally onto the surface of the thallus. In this process the algae are surrounded and enclosed by outgrowing hyphae of the cortex (Figure 101).[51] A connection with the inner layers of the thallus is established secondarily. The same process is repeated in the formation of cephalodia and again shows the ability for functional change of lichen tissues.[30]

In the gelatinous lichens, the development of isidia (Figure 95 and 143) is initiated by an active division of the algal cells at the thallus edge (Figure 142). On ecorticate thalli, a small protuberance forms which is then invaded by the hyphae. In corticate thalli the cortical cells are similarly stimulated to increased divisions by the multiplication of the algae. Division of the cortical cells keeps pace with the growth of the algae, so that the thalline protuberance is provided with a regular cortical layer from the beginning.

3. Distribution of the Lichens by Isidia

As already mentioned above, many isidia do not break off and, therefore, obviously do not serve the purpose of vegetative propagation. Nevertheless, they are important for regeneration since they can, on an old thallus, grow directly into new lobes (Figure 88). This direct development to a complete thallus seems to be impossible for isidia after dissemination.

Isidia are not easily dispersed due to their relatively large weight, and reproduction of lichens by isidia is not as efficient as reproduction by soredia.[52] The only advantage of isidia when compared to soredia could be that they are more resistant to harsh envrionmental conditions due to their cortex. For rapid regeneration of lichens, larger fragments are most effective, but, of course, their great weight makes dispersal difficult.

C. Soralia

In contrast to isidia, which are solid and corticated, soralia are open excrescences of the thallus. Small globulose bodies, called *soredia*, which consist of a few algal cells, more or less densely enmeshed by hyphae, are formed in these fissures. The soredia are disseminated as diaspores and grow into a new thallus. In the soralia, the massed soredia form a mealy or coarsely granulose mass which is water repellent. Since the soralia are openings in the thalline cortex, they may play a part in the aeration of the thallus.

1. Types of Soralia

Soralia are divided into various types on the basis of their form.[53-55] *Maculiform soralia* (Figure 104) represent the simplest type of soralia. They consist of rounded or elongated flat depressions scattered over the thallus surface. In the case of isolateral lobes they often develop at the margins of the thallus. In other lichens, the soredia form spherical clumps called *spherical soralia* (Figure 105). Spherical soralia which occur at the ends of the lobes are also called *capitate soralia* (Figure 106). *Cuff-shaped soralia* (Figure 107) resemble capitate soralia in appearance. They are raised above the thallus by lobe-like thalline protuberances. As they grow older, they become perforated in the center and resemble the cuff of a sleeve.

Many dorsiventral thalli bear soralia not only at the ends, but are bordered by soralia along their entire margin (Figure 108). In such cases they are called *marginal soralia* or *border soralia*.

Some soralia are not formed on the surface but develop only in cracks. These *fissure soralia* are elongated and partly branched (Figure 109).

On a dorsiventral thallus, soralia frequently originate not on the upper side but at the lower terminal edge of the lobes. Here the upper layers of the thallus become split and reflex upward. This type is called a *labriform soralium* (Figure 110) since the upwardly turned cortex, with the soralium developed below it, resembles a lip. Often, the ends of the thallus which bear a soralium are also arched into a dome-like form, so that they give the appearance of being inflated. Those soralia which are situated on the inner side of this dome are called *vault soralia* (Figure 111).

The formation of soredia is not always confined to soralia, they may be produced diffusely on the surface of lobes (Figure 112), especially in noncorticated thalli. Moreover, soralia which are originally formed as maculiform soralia may become secondarily confluent and give an evenly sorediose appearance to a large part of the thallus. Different types of soralia may occur together on one lichen thallus.

2. Development of Soralia and Soredia

Soralia are formed by the combination of several developmental steps.[49] One step is the degeneration of the thalline cortex. In another step, individual groups of algal cells become enmeshed by hyphae and the connection between the newly formed soredia and the tissue of the algal layer becomes loosened. The developmental stages may occur in different sequences. For example, the thallus cortex deteriorates first, followed by the development of soredia from the uppermost part of the algal layer. The production of soredia then penetrates into deeper parts of the tissue until the entire algal layer is used up.

In other cases, the development starts at the border between algal layer and medulla. The first soredia originate in the interior of the thallus and are, under certain circumstances, collected in a medullar cavity. The continued production of soredia and the subsequent increased pressure finally burst open the cortex and the soredia can be dispersed (Figure 113).

In many cases new soredia continue to be produced at the base of a soralium. Thus, the older soredia are gradually pushed out of the thallus (Figure 114).

Labriform soralia positioned at the edge of a thallus lobe show movement due to swelling of the cortical tissue at water uptake or water loss, so that the soredia are loosened and scattered.[56]

The structure of the soredia varies in different systematic groups. In the Parmeliaceae, the hyphal envelope which surrounds the relatively large algal cells is rather loose (Figure 116). The hyphae of the Stictaceae, on the other hand, tightly enclose the algal cells from the beginning (Figure 117). Sometimes soredia are formed nearly exclusively by algae enclosing a few hyphae.[57]

3. Influences on Soralium Development

The development of specific types of soralia was claimed to be genetically determined and species-specific,[54] but environmental influences may also be important for the formation of soredia.[58] In general, soralia develop with special luxuriance in moist and shady surroundings. Shady rock surfaces are almost exclusively colonized by sorediose lichens. *Parmelia saxatilis* at sunny sites develops isidia, the same lichen in shady localities develops loosely packed structures that can be called neither typical isidia nor soredia.[19]

The formation of soredia sometimes depends on the age of the thallus. In *Peltigera spuria* only young specimens bear maculiform soralia (Figure 115) which gradually diminish as soon as apothecia originate at the margin of the lobes.[59]

Certain species which normally do not possess soredia may show a spontaneous production of these bodies in geographically widely separated regions. Such aberrant forms, however, seldom seem to extend their range appreciably. An exception to this is *Cetraria delisei*, a lichen of the Boreal-Arctic tundra; it frequently produces extensive populations of a sorediate form. Apparently, we have here a case of development of a genotypically modified parallel species from the normal apothecia-bearing form. The development of such related types is known as the phenomenon of paired species.[60]

D. Hormocystangia

Hormocystangia are, up until now, known only from the genus *Lempholemma*.[45] They develop as vesicular swellings on the fruticose thalli (Figure 102). In the hormocystangia, the cyanobionts develop hormocysts, which are surrounded by a heavy gelatinous sheath. Hyphae of the mycobiont penetrate into the mucilaginous envelope of the hormocyst and these lichenized hormocysts are liberated by the disintegration of the hormocystangium.

E. Other Vegetative Dispersal Mechanisms

Next to isidia, soredia, and hormocystangia, other vegetative dispersal mechanisms exist. In *Baeomyces* and other lichens, small portions of the upper layers of the thallus become detached. Underneath these scales, which are called *schizidia* (Figure 103),[61] a cupular layer develops in which increased cell division occurs.[62] At this boundary the scales are shed. Schizidia usually are distributed over short distances only. Thread-shaped cords made of closely appressed hyphae growing below the ground surface may function as reproductive structures; they are called *rhizomorpha*.[63]

F. Efficiency of the Different Types of Propagation

1. Maintaining Sites against Competition

The exceptional regenerative ability of many lichen thalli enables them to maintain colonized sites. In the cushions of *Cetraria islandica*, the individual plants die after developing apothecia, but, at the same time, new thalli grow out from the pycnidia at the thallus edge.[46] In this way, the site is continuously colonized by the same species and competitors have no chance. Similar mechanisms are found in many other fruticose species.[6,64]

In the foliose lichen *Parmelia saxatilis*, older isidia on the thallus surface develop into new thalline scales. These new lobes grow tile-like on the old thalli, covering them and renewing the growth at the site. The tissue of the lower, dead lobes serves for water storage.[19]

In other foliose and crustose lichens, another type of regeneration plays an important role in maintaining a site. Thalli of *Parmelia centrifuga* grow radially from a starting point (Figure 118). After some time, the oldest parts of the thallus die off in the center, while the peripheral zone continues its centrifugal growth. As the old parts of the thallus die, small vital fragments remain and regenerate so that a second thallus ring eventually develops inside the first. This process is repeated so that a system of concentric rings is developed and enlarged without the colonized area ever being abandoned by the lichen. Similar, although less regular, are the annular growth processes of the crustose lichen *Lecanora muralis* (Figure 120).

2. Efficiency of the Different Dispersal Mechanisms

It was and to some extent still is generally assumed that the distribution by vegetative diaspores is much more favorable than the propagation by ascospores, which must establish new contact with suitable algae in order to be capable of forming a lichen. The formation of a new lichen by this means was thought to occur only in one out of several million cases[2] (see Chapter VIII.C).

The distribution through soredia and isidia has the advantage that the mycobiont stays in association with the correct algae. This may be advantageous for the rapid colonization of new sites and may explain why the sorediate species of a species-pair is generally widely distributed, whereas the fertile, esorediate species is restricted to a limited geographical area.[60] For a meaningful comparison between sexual and asexual distribution methods, the mechanisms of propagation of more species have to be investigated.

The larger mass of an isidium does not present a developmental advantage when compared to the smaller soredium. A direct development of the isidium into a thallus is impossible. In a new thallus, a large quantity of cortical and medullar tissue must be developed that is made up exclusively of fungal hyphae and is not photosynthetically active. The photosynthates produced by the few algae of the isidium are insufficient for this, so an isidium attaches to the substrate and degenerates to an undifferentiated mass of tissue after dispersal (Figure 121).[19] Inside this primordium the algae multiply and the further development follows the pattern that is characteristic of the growth of young lichens from soralia or spores.

Probably, the isidia can survive better under unfavorable conditions (e.g., a drought period) than the soredia. Soredia can be distributed over greater distances than isidia and, moreover, can easily attach themselves to the substrate by hyphae growing from their loosely constructed surface.

A lichen thallus hardly ever develops from a single isidium or soredium. For example, soredia from *Baeomyces rufus* may cover mosses in large numbers (Figure 119) and, since the germinating soredia quickly fuse, the growth of the new thallus starts not from one place but from numerous points.[65] This principle is adopted by most species with reproduction by soredia. Probably, the heavy isidia are transported only a very short distance from the parental thallus and, after germination, help to enlarge the thallus. They probably serve more for growth and regeneration than for distribution.

In some lichens, the different means of dispersal are used alternately in the various stages of the life-cycle and, thereby, may fulfill the differing requirements of enlarging the young lichen, dispersal of the adult thallus, and regeneration of the old lichen.[65]

VIII. JUVENILE DEVELOPMENT, GROWTH, AND DEATH OF THE THALLUS

A. Thallus Development from Vegetative Diaspores and Ascospores

All lichens that do not directly develop from the tissue of a dying thallus through regeneration, begin their juvenile development with an undifferentiated network of algae and fungal hyphae.[3,66] The starting point of the development can be soredia, isidia, or separately distributed fungal spores and algae.

After distribution, soredia become attached to the substrate by outgrowing hyphae and gelatinous secretions. More hyphae reach out in all directions (Figure 122) and take up contact with other soredia.[3,66,67] Together, they develop a hyphal net which covers the substrate and incorporates suitable free-living algae (Figure 123). Through cell divisions and strong gelatinous secretions, the loosely connected mesh becomes a tightly constructed basal tissue. From this basal tissue, young thallus lobes emerge (Figure 124) which show a stratified organization. In other cases, an undifferentiated basal tissue can change completely into a layered thallus.[21] Of special importance is the formation of a closed cortical surface by gelatinous secretion that usually is only completed with the final development of normal thalline lobes. The gelatinous secretion can be damaged by air pollution. The resulting thalli have no closed cortex (Figure 127) and their water uptake seems to be deficient.[3]

The first accessory organs or supporting tissues can develop very early during the ontogeny, sometimes even the undifferentiated basal tissue may form rhizinae (Figure 125).[21] The differentiation of a thallus from soredia has been investigated only in a few species. Only small ontogenetic variations have been discovered, but further investigations are necessary.

Isidia have a higher degree of organization than soredia, which is lost after distribution. They are reduced to an undifferentiated basal tissue which then develops into a thallus by similar stages as in soredial development.

Germinating sexual spores of many macrolichens have difficulties in establishing a symbiotic relationship, since the necessary *Trebouxia*-algae only rarely occur in a free-living state.[68-70] The mycobiont of *Xanthoria parietina* can alternatively live in contact with other algae in an unorganized form and colonize the substrate without developing a real thallus.[70a] Thallus development starts directly at those points where the expanded network comes into contact with the suitable algae. The specific algae can also be supplied by soredia of other lichens as well as by fragments of dead thalli. The further differentiation of the thallus follows the same principles as were described above for soredia. The resynthesis of the lichen thallus has been successfully studied in vitro.[71,72]

B. Growth of the Thallus

Mature, fully differentiated thalli of foliose and fruticose lichens usually possess growth zones at the apical ends of the lobes.[73,74] Here, the gelatinous layer of the cortex often is pierced by pores indicating elongation processes (Figure 126).[21] In general, the foliose lichens show very little growth in the vertical direction; in most species, the upper cortex merely scales off slowly and is regenerated from below by the underlying layer of the thallus (see Chapter VII.A).

In old thalli, the hyphae of the medulla and cortex are capable of reverting at any time to a meristematic condition and produce new proliferations. We know almost nothing about the inductive effects, factors, and mechanisms controlling these types of development.

In some lichens, the growth of the thallus margins is not due to an increase of the complete tissue, but, instead, the fungal hyphae alone grow ahead and form a peripheral zone which contains no algae. This prothallus has a filamentous or cartilaginous structure, and the growing peripheral zone must then acquire algae secondarily. In some lichens the fungal hyphae of the thalline margins become concentrated into fascicles of thrusting hyphae orientated in parallel formation, which force the algae further into the peripheral zone.[75] The filamentous prothallus of *Placynthium nigrum* (Figure 128) obtains its algae by a different method. When the thallus becomes moist, hormogonia of the cyanobiont (Figure 129) or lichenized particles of the thallus detach themselves from the margin of the thalline squamules and alight on the prothallus, where they become enmeshed by the hyphae (Figure 130).[76] Possibly free-living algae are also incorporated. In any case, the new squamules are differentiated in scattered formation in the marginal zone of the thallus, and only later become secondarily concrescent into a continuous crust.

C. Aging and Death of the Thallus

The thalli of many species have a prolonged continuous growth period. The growth rates differ in different lichens (see Chapter VII.A). However, the longevity of lichens is often overestimated. Above all, important differences exist between the growth forms. Crustose lichens grow slowly and can reach an age of several hundred years in extreme environments.[77]

In the foliose lichens the situation is different. The cartilaginous thalli of foliose lichens, like *Parmelia centrifuga,* are composed of small individual lobes. It is very doubtful whether the thallus as a whole can be considered as one individual plant, since every lobe grows and branches independently. The contact of the neighboring lobes is slowly lost due to branching and the dying of the older parts of the thallus. The oldest living part of the lobes is only 30 to 50 years of age.

Some large foliose lichens that consist of a single leaf, as, for example, specimens of the genus *Peltigera*, are relatively short-lived and reach an age of ten years at the most.[78] About the same age is reached by some fruticose lichens such as *Cladonia*.[79] This does not apply to sterile fruticose thalli of the subgenus *Cladina*. These lichens branch apically and die off at their base, so that finally lateral branches become separate plants which are potentially immortal.

Some lichens die after developing sexual stages.[79,80] This happens to the reindeer mosses after they produce apothecia, which only occurs under favorable conditions. In *Baeomyces rufus*, propagation by ascospores, schizidia, and soredia alternate during the ontogeny of the thallus and faciliate in this way an optimal distribution and colonization of new substrates. In this lichen, the thallus only reaches an age of four years.[65]

IX. INDIVIDUALITY OF THE LICHENS

Numerous ascospores and algae take part in the formation of a new lichen. The thallus development from vegetative diaspores involves several soredia. These vegetative diaspores probably have been produced by different parental thalli. The tissue of the lichen in this instance, therefore, is not genetically homogenous.[81] The phenotype of the lichen is determined by the united morphogenetic influence of the genetically different hyphae and algae. New soredia produced by this heterogeneous thallus might well contain only some of the genetically different hyphae occurring in the thallus. In the next generation, soredia from different paternal thalli again are responsible for the formation of a new thallus and, thereby, genetically controlled features are intermixed and interchanged. Even mutations that change the habit of the lichen can be dispersed in this way without sexual processes.

The vegetative transformation of characteristics without sexual reproduction can also occur between thalli of adult lichens. In the dense clumps formed by many fruticose species, e.g., the cushion-like thalli of *Cladonia*, the lichens frequently become concrescent with each other (Figure 131). This may occur between plants of the same species as well as between different taxa. When a podetium of *Cladonia squamosa*, covered with small squamules (Figure 133), fuses with a strongly branched *Cladonia rangiferina* (Figure 132), the growth form of the resulting lichen resembles the branched *Cladonia rangiferina* but is covered with squamules (Figure 134). A frequent occurrence of this intermingling would make the delimitation of species impossible, but, in mixed cushions formed by several species of *Cladonia*, interspecific fusions are found only rarely while fusions between specimens of the same species occur ten times more frequently. Therefore, while it is certain that an interspecific transmission of characteristics may occur, there seems to exist a partial incompatibility between the tissues of different species.

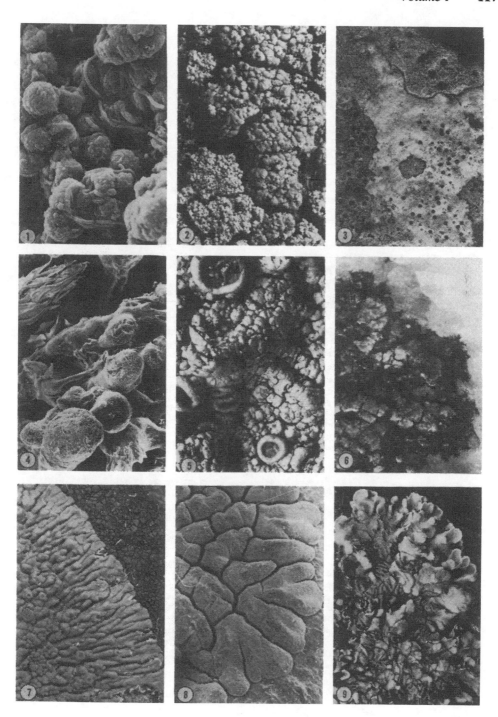

Figure 1. *Lecanora conizaeoides*; thallus consisting of loosely interwoven hyphae and algae. (Magnification × 510.) **Figure 2.** *Candelariella coralliza*; granulate crustose thallus. (Magnification × 5.) **Figure 3.** *Verrucaria*; endolithic thalli with immersed fruit bodies. (Magnification × 1.5.) **Figure 4.** *Ochrolechia frigida*; young thallus growing over mosses. (Magnification × 14.) **Figure 5.** *Ochrolechia frigida*; old areolated thallus with apothecia. (Magnification × 2.) **Figure 6.** *Rhizocarpon geographicum*; areolate thallus with black prothallus. (Magnification × 2.5.) **Figure 7.** *Caloplaca thallincola*; placoid thallus. (Magnification × 4.) **Figure 8.** *Dimelaena oreina*; crustose thallus with an effigurate margin. (Magnification × 8.) **Figure 9.** *Squamarina cartilaginea*; squamulose thallus. (Magnification × 2.)

Figure 10. *Rhizoplaca chrysoleuca*; upper side of areolated peltate thallus. (Magnification × 1.5.) **Figure 11.** *Rhizoplaca chrysoleuca*; lower side of peltate thallus with umbilicus. (Magnification × 1.5.) **Figure 12.** *Hypocenomyce scalaris*; thalline squamules with tile-like arrangement and sorediate margins. (Magnification × 7.) **Figure 13.** *Toninia caeruleonigricans*; pulvinate thallus. (Magnification × 4.) **Figure 14.** *Physciopsis adglutinata*; foliose thallus closely attached to the substrate, with maculate soralia. (Magnification × 16.) **Figure 15.** *Lobaria pulmonaria*; reticulate foliose thallus. (Magnification × .55.) **Figure 16.** *Peltigera aphthosa*; foliose thallus with small black cephalodia on its surface. (Magnification × .30.)

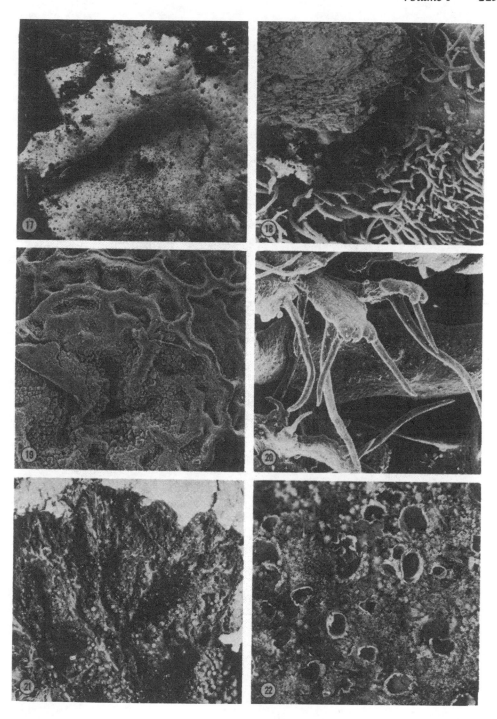

Figure 17. *Umbilicaria vellea*; upper side of umbilicate thallus with central umbilicus. **Figure 18.** *Umbilicaria hirsuta*; lower side with rhizinae and part of the large holdfast. (Magnification × 8.) **Figure 19.** *Umbilicaria proboscidea*; umbilicus with veins and pruina. (Magnification × 8.) **Figure 20.** *Physcia tenella*; thalli raised above the substrate by rhizinae. (Magnification × 22.) **Figure 21.** *Collema nigrescens*; gelatinous thallus with numerous apothecia. (Magnification × 1.5.) **Figure 22.** *Solorina spongiosa*; apothecia borne on circular squamules containing green algae, the squamules are surrounded by tiny granules containing cyanobacteria and green algae.

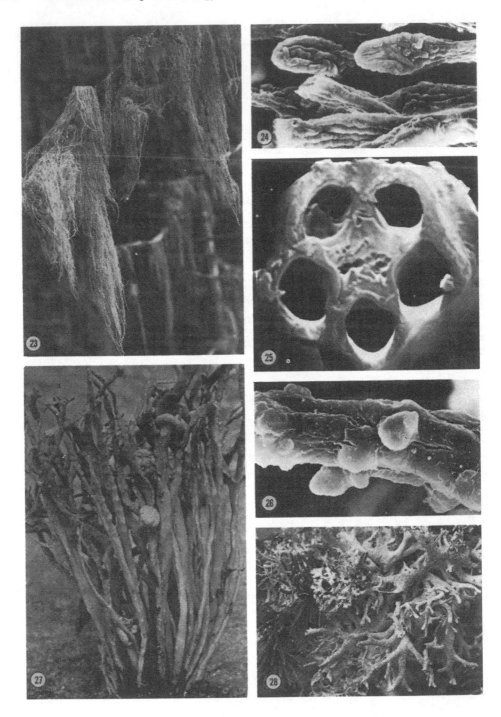

Figure 23. *Usnea*; thalli hanging loosely from branches of a tree. (Magnification × 3.) **Figure 24.** *Cystocoleus niger*; filamentous lichen with *Trentepohlia* as symbiotic alga. The alga is surrounded by about 10 hyphae. (Magnification × 325.) **Figure 25.** *Racodium rupestre*; section through filamentous lichen, the central *Trentepohlia* alga is surrounded by 5 hyphae. (Magnification × 1760.) **Figure 26.** *Racodium rupestre*; filamentous lichen. (Magnification × 880.) **Figure 27.** *Ramalina curnowii*; fruticose strap-shaped thallus growing on rocks. (Magnification × 3.) **Figure 28.** *Pseudevernia furfuracea*; fruticose strap-shaped thallus growing on trees.

Figure 29. *Cladonia rangiferina*; thalli with a cushion-like growth form. **Figure 30.** *Sphaerophorus fragilis*; cylindrical thalli with fruit bodies. **Figure 31.** *Cornicularia aculeata*; fruticose thallus without attachment to the substrate. (Magnification × 1.5.) **Figure 32.** *Cladonia fimbriata*; lichen with a foliose thallus horizontalis and a cup-like thallus verticalis. (Magnification × 1.5.) **Figure 33.** *Baeomyces rufus*; lichen with a crustose thallus horizontalis and stalked apothecia. (Magnification × 9.) **Figure 34.** *Stereocaulon vesuvianum*; fruticose lichen with pseudopodetia. (Magnification × 1.5.)

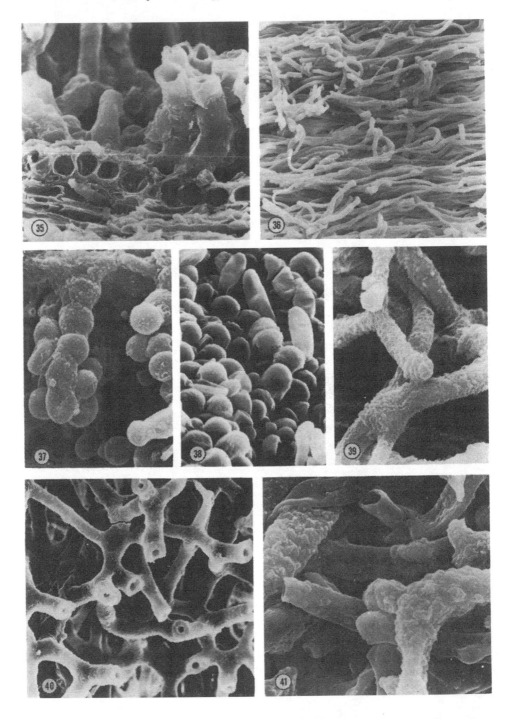

Figure 35. *Leptogium hildenbrandii*; cortical layer with hairs. (Magnification × 730.) **Figure 36.** *Sphaero-phorus globosus*; interwoven hyphae from the medulla. (Magnification × 206.) **Figure 37.** *Leptogium hibernicum*; tomentum of beaded hyphae. (Magnification × 1100.) **Figure 38.** *Coenogonium interplexum*; clavate ends of paraphyses. (Magnification × 1180.) **Figure 39.** *Ramalina siliquosa*; hyphae with anasto-moses. (Magnification × 1180.) **Figure 40.** *Anzia ornata*; netlike hyphae with anastomoses in the spon-giostraum. (Magnification × 375.) **Figure 41.** *Solorina crocea*; hyphae covered with lichen substances. (Magnification × 1250.)

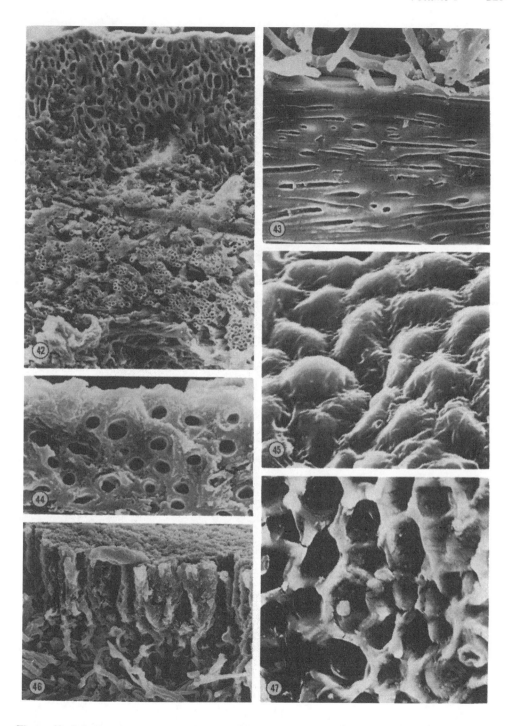

Figure 42. *Solorina crocea*; section through the thallus with pseudoparenchymatic cortex, algal layer, and medulla. (Magnification × 175.) **Figure 43.** *Usnea ceratina*; prosoplectenchymatous tissue of the central strand. (Magnification × 450.) **Figure 44.** *Heterodermia hypoleuca*; scleroplectenchymatous tissue. (Magnification × 1470.) **Figure 45.** *Leptogium diffractum*; cells of the irregular cortex. (Magnification × 1680.) **Figure 46.** *Roccella fuciformis*; pallisade plectenchyma. (Magnification × 265.) **Figure 47.** *Leptogium sinuatum*; cortex of isodiametric cells. (Magnification × 1100.)

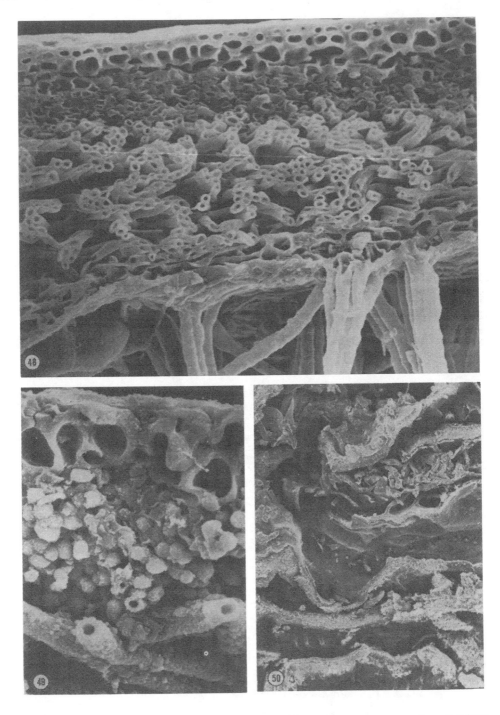

Figure 48. *Sticta canariensis*; section through the thallus with parenchymatous upper cortex, algal layer, medulla, lower cortex, and hairs of the tomentum. (Magnification × 365.) **Figure 49.** *Peltigera aphthosa*; algal layer with a few cells of the cortex and of the medulla. (Magnification × 715.) **Figure 50.** *Parmelia saxatilis*; dead thalli under the living lichen serving for water storage. (Magnification × 8.)

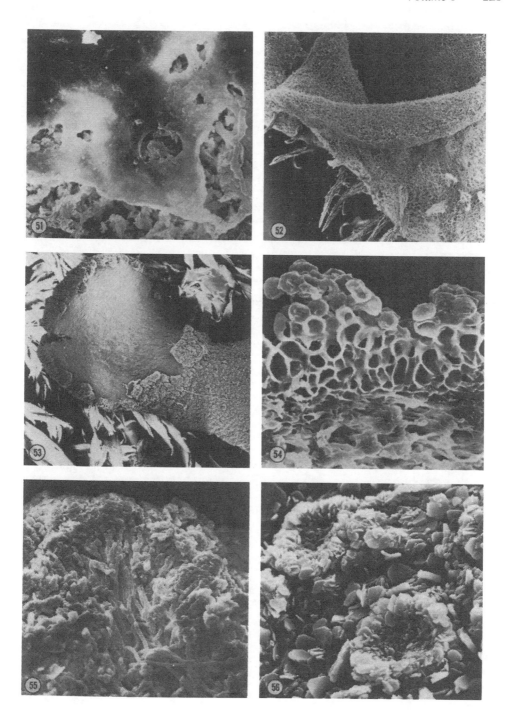

Figure 51. *Parmeliopsis aleurites*; pored epicortex. (Magnification × 450.) **Figure 52.** *Peltigera canina*; felty tomentous surface. (Magnification × 8.) **Figure 53.** *Peltigera scabrosa*; apothecium and thallus with areolated surface. (Magnification × 10.) **Figure 54.** *Peltigera scabrosa*; section through the thallus with lumpy areolated part of the cortex. (Magnification × 265.) **Figure 55.** *Umbilicaria proboscidea*; section through cortex with irregular proliferations which scale off. (Magnification × 150.) **Figure 56.** *Parmelia saxatilis*; crystal layer on the thallus surface formed around pores of the cortex. (Magnification × 620.)

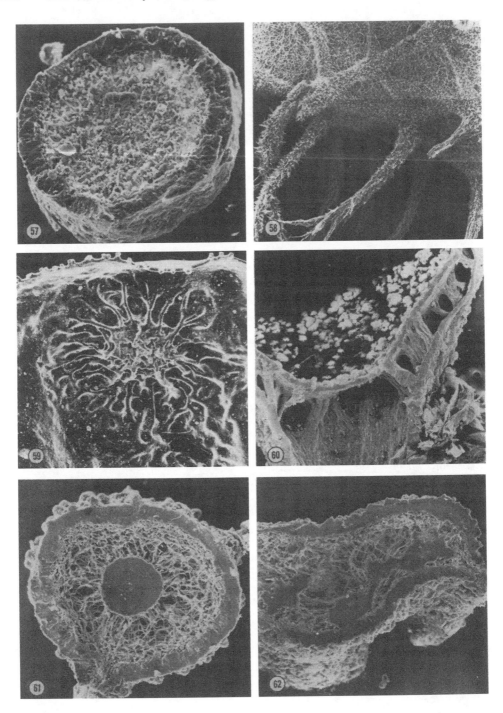

Figure 57. *Sphaerophorus fragilis*; cortex surrounding the medulla in the form of a cylinder. (Magnification × 60.) **Figure 58.** *Peltigera degenii*; veins and rhizinae on the lower surface. (Magnification × 13.) **Figure 59.** *Cetraria islandica*; ridges on the lower side of the thallus, giving stability to an apothecium on the upper side. (Magnification × 7.) **Figure 60.** *Cladonia chlorophaea*; diagonal supports in the hollow margin of a cup-shaped podetium. (Magnification × 8.) **Figure 61.** *Usnea filipendula*; section through the central strand, surrounded on all sides by a loose mesh of medullary hyphae. On the outside follow the annular algal zone and cortex. (Magnification × 26.) **Figure 62.** *Letharia vulpina*; transverse section through the thallus with a composite central strand. (Magnification × 27.)

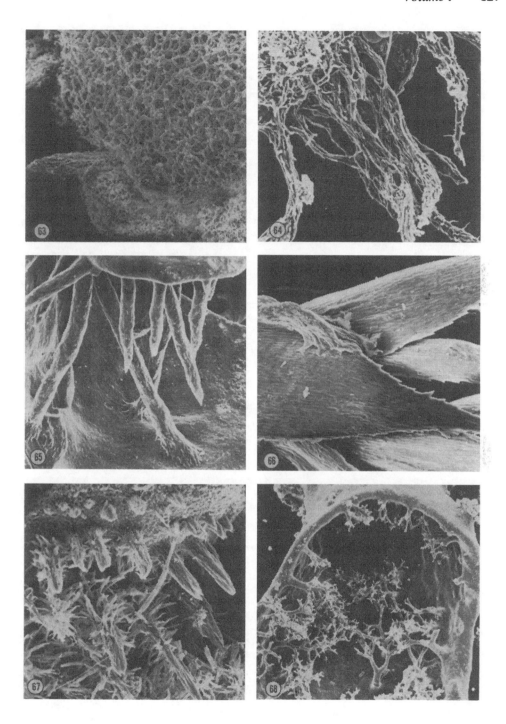

Figure 63. *Anzia ornata*; spongiostratum and rhizinae on the lower side of the thallus. (Magnification × 40.) **Figure 64.** *Cladonia foliacea*; rhizoptae. (Magnification × 80.) **Figure 65.** *Parmelia saxatilis*; simple pointed rhizinae without contact to the substrate and rhizinae fixed with a hold fast to the surface of another thallus. (Magnification × 55.) **Figure 66.** *Parmelia saxatilis*; holdfast of a rhizine on a moss leaf. (Magnification × 88.) **Figure 67.** *Physconia pulverulenta*; development of rhizinae at the margin of the thallus without contact to the substrate, older rhizinae covered with numerous small branches or hairs. (Magnification × 55.) **Figure 68.** *Parmelia perlata*; dichotomously forked rhizinae. (Magnification × 14.)

Figure 69. *Parmelia sulcata*; divaricately branched rhizinae. (Magnification × 50.) **Figure 70.** *Nephroma resupinatum*; hairs on the lower surface ending in numerous bulbously thickened cells. (Magnification × 265.) **Figure 71.** *Umbilicaria vellea*; rhizinomorphs. (Magnification × 15.) **Figure 72.** *Parmelia crinita*; cilia. (Magnification × 16.) **Figure 73.** *Anaptychia ciliaris*; black cilia covered with hyaline hairs. (Magnification × 120.) **Figure 74.** *Usnea filipendula*; branches covered with small fibrillae. (Magnification × 3.)

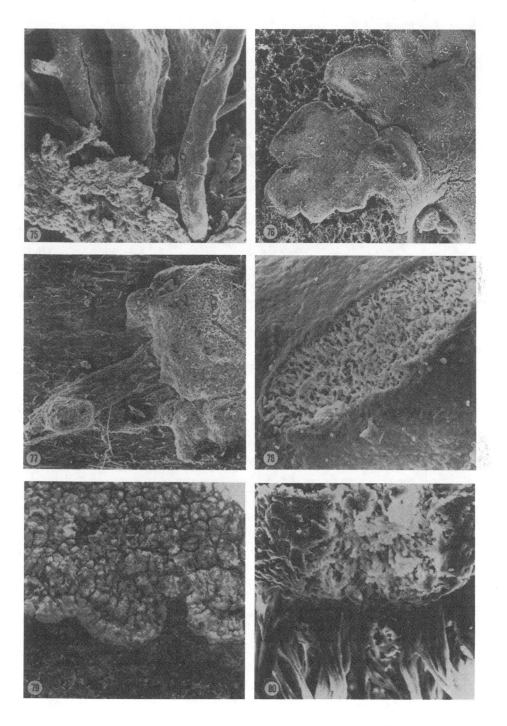

Figure 75. *Ramalina siliquosa*; holdfast with basal part of branches. (Magnification × 8.) **Figure 76.** *Parmeliella plumbea*; the foliose thallus is fringed by a dark felted prothallus. (Magnification × 10.) **Figure 77.** *Diploschistes scruposus*; thallus with outgrowing hyphal strand bearing a young thallus scale. (Magnification × 22.) **Figure 78.** *Cornicularia aculeata*; pseudocyphella. (Magnification × 140.) **Figure 79.** *Pertusaria leucosora*; prothallus with cartilaginous structure. (Magnification × 3.) **Figure 80.** *Pseuocyphellaria faveolata*; opening of a pseudocyphella filled with medullar hyphae. (Magnification × 150.)

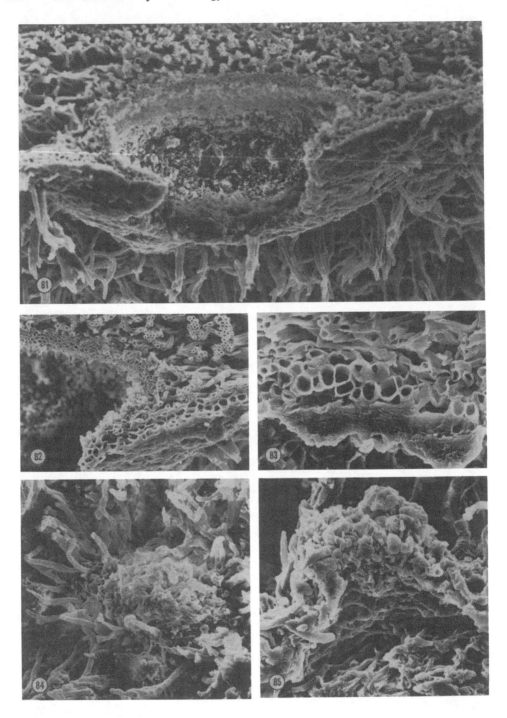

Figure 81. *Sticta canariensis*; section through a cyphella. (Magnification × 120.) **Figure 82.** *Sticta canariensis*; section through the inner and marginal part of a cyphella showing the spherical cells of its bottom. (Magnification × 170.) **Figure 83.** *Sticta sylvatica*; bottom of old cyphella conglutinated by mucilaginous substances. (Magnification × 480.) **Figure 84.** *Nephroma resupinatum*; tubercle between the hairs of the tomentum. (Magnification × 160.) **Figure 85.** *Nephroma resupinatum*; section through a tubercle showing an apical opening filled with short cells and a basal cavity. (Magnification × 350.)

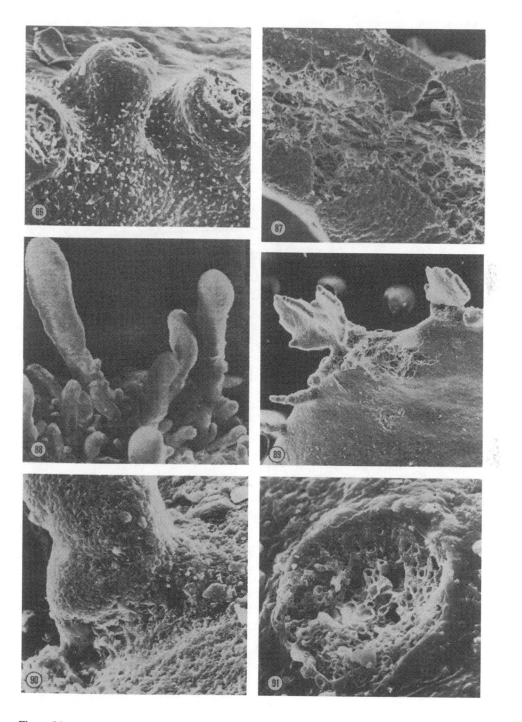

Figure 86. *Parmelia exasperata*; wart-like respiratory pores. (Magnification × 130.) **Figure 87.** *Ramalina siliquosa*; gaps in a thick cortex facilitating gaseous exchange. (Magnification × 80.) **Figure 88.** *Pseudevernia furfuracea*; isidia growing into lateral branches. (Magnification × 25.) **Figure 89.** *Cetraria islandica*; pycnidia on the thallus margin growing into new lobes. (Magnification × 16.) **Figure 90.** *Parmelia conspersa*; isidium with degeneration of its cortical base. (Magnification × 200.) **Figure 91.** *Parmelia conspersa*; respiratory opening formed by breaking off of an isidium. (Magnification × 350.)

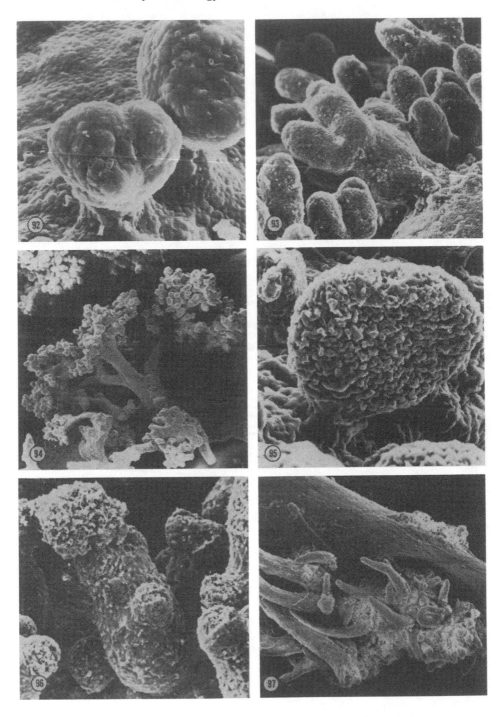

Figure 92. *Parmelia glabratula*; spherical isidia. (Magnification × 275.) **Figure 93.** *Parmelia tiliacea*; cylindrical and branched isidia. (Magnification × 80.) **Figure 94.** *Umbilicaria pustulata*; coralloid isidium. (Magnification × 20.) **Figure 95.** *Collema flaccidum*; isidium of a gelatinous lichen, surface irregularly shrunken due to drying. (Magnification × 250.) **Figure 96.** *Parmelia subaurifera*; sorediate isidium. (Magnification × 185.) (From Beltman, H. A., *Bibl. Lichenol.*, 11, 1, 1978. With permission.) **Figure 97.** *Alectoria nidulifera*; isidia growing in a soralium. (Magnification × 60.)

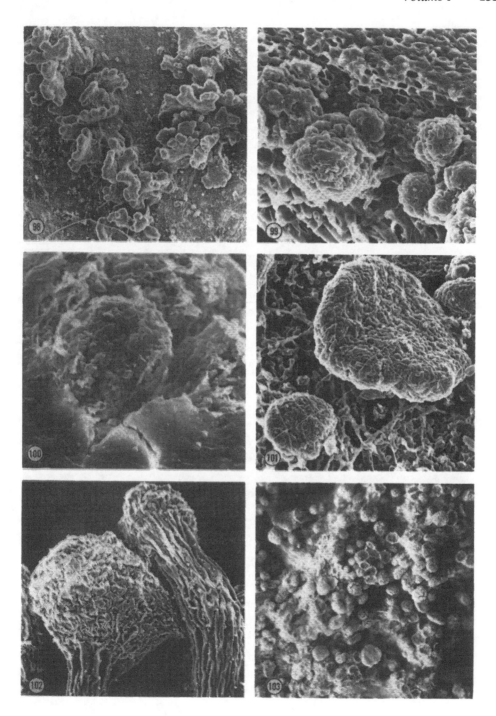

Figure 98. *Peltigera praetextata*; isidia formed along cracks due to injury. (Magnification × 16.) **Figure 99.** *Peltigera praetextata*; development of primordia of isidia in a crack. (Magnification × 200.) **Figure 100.** *Parmelia saxatilis*; young isidium growing out of a pseudocyphella. (Magnification × 385.) **Figure 101.** *Peltigera lepidophora*; development of isidia on the thallus from free-living algae trapped by the hairs of the cortex. (Magnification × 130.) **Figure 102.** *Lempholemma vesiculiferum*; globose hormocystangium, surface shrunken due to drying. (Magnification × 100.) **Figure 103.** *Baeomyces rufus*; schizidia. (Magnification × 20.)

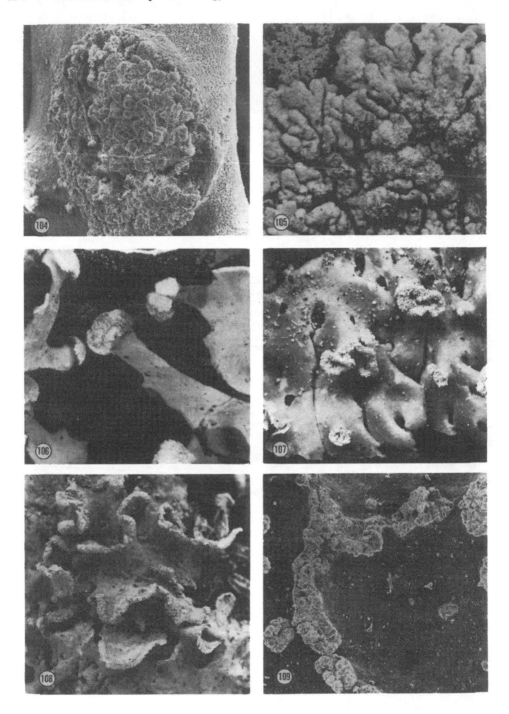

Figure 104. *Roccella fuciformis*; maculiform soralium. (Magnification × 27.) **Figure 105.** *Physcia caesia*; spherical soralia. (Magnification × 3.) **Figure 106.** *Parmelia perlata*; capitate soralia. (Magnification × 7.) **Figure 107.** *Menegazzia terebrata*; cuff-shaped soralia. (Magnification × 4.) **Figure 108.** *Cetraria pinastri*; marginal soralia. (Magnification × 7.) **Figure 109.** *Parmelia sulcata*; fissure soralia. (Magnification × 22.)

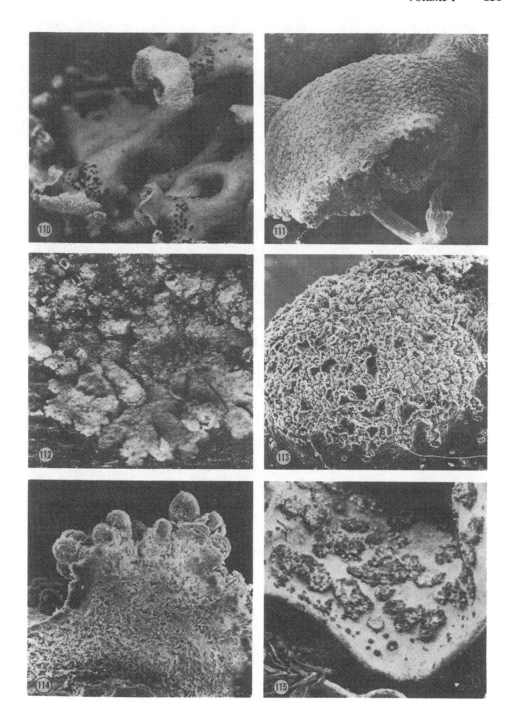

Figure 110. *Hypogymnia physodes*; labriform soralia. (Magnification × 4.) **Figure 111.** *Physcia ascendens*; vault soralium. (Magnification × 44.) **Figure 112.** *Parmeliopsis aleurites*; diffuse production of soredia on the surface of the thallus. (Magnification × 7.) **Figure 113.** *Hypogymnia tubulosa*; destruction of the cortex by the production of soredia in the algal layer. (Magnification × 35.) **Figure 114.** *Lobaria pulmonaria*; section through a soralium with soredia pushed upwards by medullar hyphae. (Magnification × 50.) **Figure 115.** *Peltigera spuria*; sterile thallus with soralia. (Magnification × 3.)

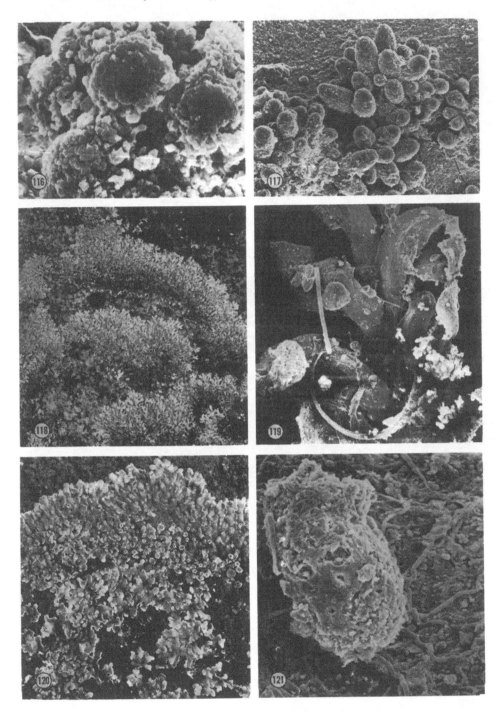

Figure 116. *Parmelia sulcata*; soredia with loose structure. (Magnification × 400.) **Figure 117.** *Lobaria pulmonaria*; soredia covered by a dense layer of hyphae. (Magnification × 27.) **Figure 118.** *Parmelia centrifuga*; concentric rings of thalli. (Magnification × .22.) **Figure 119.** *Baeomyces rufus*; soredia and thallus primordia on the moss *Diplophyllum albicans*. (Magnification × 20.) **Figure 120.** *Lecanora muralis*; regeneration of new thallus scales in the center of the lichen. **Figure 121.** *Pseudevernia furfuracea*; degenerating isidium with outgrowing hyphae. (Magnification × 170.)

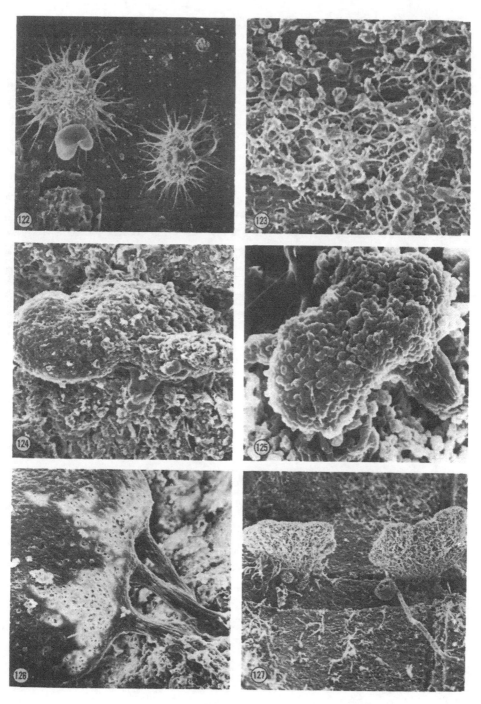

Figure 122. *Hypogymnia physodes*; soredia, arachnoidal stage of development. (Magnification × 75.) (From Schuster, G., *Bibl. Lichenol.*, 20, 1, 1985. With permission.) **Figure 123.** *Hypogymnia physodes*; hyphal net and free-living algae. (Magnification × 160.) (From Schuster, G., *Bibl. Lichenol.*, 20, 1, 1985. With permission.) **Figure 124.** *Hypogymnia physodes*; young thallus lobe growing from the basal tissue. (Magnification × 118.) (From Schuster, G., *Bibl. Lichenol.*, 20, 1, 1985. With permission.) **Figure 125.** *Physcia tenella*; primordium lifted above the surface of the substrate by rhizines. (Magnification × 190.) (From Schuster, G., Ott, S., and Jahns, H. M., *Lichenologist*, 17, , 1985. With permission.) **Figure 126.** *Parmelia sulcata*; young thallus with pores in the cortex and rhizinae. (Magnification × 100.) (From Ott, S., *Symbiosis*, 3, 57, 1987. With permission.) **Figure 127.** *Hypogymnia physodes*; thalli with ungelatinized cortex in wet, polluted environment. (Magnification × 33.) (From Schuster, G., *Bibl. Lichenol.*, 20, 1, 1985. With permission.)

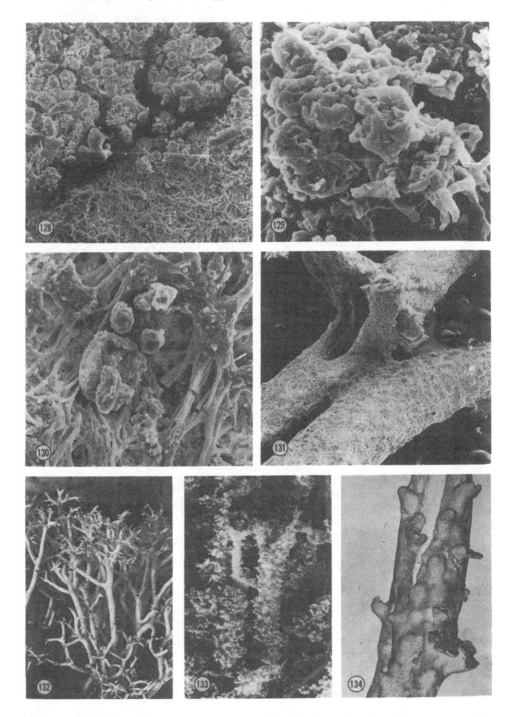

Figure 128. *Placynthium nigrum*; thallus with felty prothallus of fungal hyphae. (Magnification × 27.)
Figure 129. *Placynthium nigrum*; algae of the *Stigonema* type growing out of the thallus. (Magnification × 315.) **Figure 130.** *Placynthium nigrum*; gelatinized thallus primordia on the prothallus. (Magnification × 130.) **Figure 131.** *Cladonia rangiferina*; concrescent branches of different thalli. (Magnification × 12.) **Figure 132.** *Cladonia rangiferina*. **Figure 133.** *Cladonia squamosa*. (Magnification × .8.) **Figure 134.** Chimera of *Cladonia rangferina* and *Cladonia squamosa* resulting from fusion of two thalli. (Magnification × 16.)

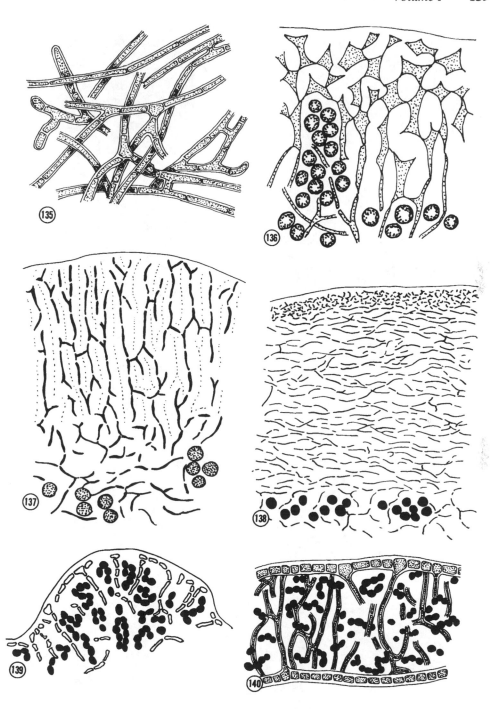

Figure 135. *Umbilicaria pustulata*; hyphae from the medulla with anastomoses. **Figure 136.** *Solorina crocea*; unevenly thickened cells with thin extensions of the lumina. **Figure 137.** *Sphaerophorus fragilis*; prosoplectenchymatous cortex. **Figure 138.** *Ramalina siliquosa*; double cortex. **Figure 139.** *Collema occultatum*; origin of the primitive cortex. **Figure 140.** *Leptogium sinuatum*; continuous cortex.

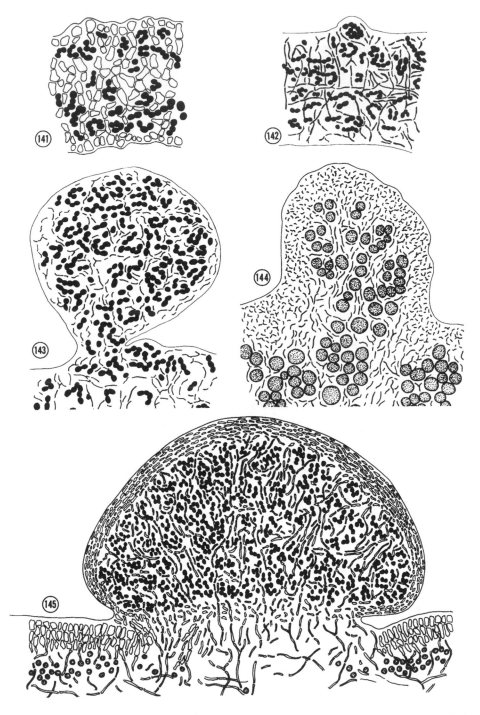

Figure 141. *Leptogium diffractum*; primitive cortex. **Figure 142.** *Collema flaccidum*; isidium developing from an agglomeration of algal cells. **Figure 143.** *Collema flaccidum*; isidium. **Figure 144.** *Parmelia tiliacea*; outgrowing isidium. **Figure 145.** *Peltigera aphthosa*; cephalodium containing cyanobacteria on a thallus containing green algae.

REFERENCES

1. **Henssen, A. and Jahns, H. M.,** *Lichenes,* Georg Thieme Verlag, Stuttgart, 1974.
2. **Scott, G. D.,** *Plant Symbiosis,* 2nd ed., Crane-Russak, New York, 1971.
3. **Schuster, G.,** Die Jugendentwicklung von Flechten, ein Indikator für Klimabedingungen und Umweltbelastungen, *Bibl. Lichenol.,* 20, 1, 1985.
4. **Poelt, J. and Wunder, H.,** Über biatorinische und lecanorinische Berandung von Flechtenapothecien, untersucht am Beispiel der *Caloplaca ferruginea*-Gruppe, *Bot. Jahrb. Syst. Pflanzengesch. Pflanzengeogr.,* 86, 256, 1967.
5. **Moser-Rohrhofer, M.,** Der vegetative Flechtenthallus — ein Derivat des Ascocarps?, *Anz. Oesterr. Akad. Wiss. Math. Naturwiss Kl.,* 6, 109, 1969.
6. **Jahns, H. M.,** Untersuchungen zur Entwicklungsgeschichte der Cladoniaceen unter besonderer Berücksichtigung des Podetium-Problems, *Nova Hedwigia,* 20, 1, 1970.
7. **James, P. W. and Henssen, A.,** The morphological and taxonomic significance of cephalodia, in *Lichenology: Progress and Problems,* Brown, D. H., Hawksworth, D. L., and Bailey, R. H., Eds., Academic Press, New York, 1976, 27.
8. **Brodo, I. M. and Richardson, D. H. S.,** Chimeroid associations in the genus Peltigera, *Lichenologist,* 10, 157, 1978.
9. **Reinke, J.,** Abhandlungen über Flechten I-IV, *Jb. Wiss. Bot.,* 26, 495, 1894; 28, 39, 1895; 29, 171, 1896.
10. **Hallbauer, D. K. and Jahns, H. M.,** Attack of lichens on quartzitic rock surfaces, *Lichenologist,* 9, 119, 1977.
11. **Doppelbaur, H. W.,** Studien zur Anatomie und Entwicklungsgeschichte einiger endolithischen pyrenocarpen Flechten, *Planta,* 53, 246, 1959.
12. **Kushnir, E., Tietz, A., and Galun, M.,** 'Oil Hyphae' of endolithic lichens and their fatty acid composition, *Protoplasma,* 97, 47, 1978.
13. **Poelt, J.,** Die lobaten Arten der Flechtengattung *Lecanora* Ach. sensu ampl. in der Holarktis, *Mitt. Bot. Staatss. Munchen,* 2, 411, 1958.
14. **Jahns, H. M., Pfeifer, K., and Schuster, G.,** Die Variabilität vegetativer und generativer Flechtenstrukturen und ihre Bedeutung für die Systematik, *Ber. Dtsch. Bot. Ges.,* 95, 313, 1982.
15. **Kershaw, K. A. and Harris, G. P.,** A technique for measuring the light profile in a lichen canopy, *Can. J. Bot.,* 49, 609, 1971.
16. **Kärenlampi, L.,** Distribution of chlorophyll in the lichen *Cladonia alpestris, Rep. Kevo Subarctic Res. Stat.,* 7, 40, 1971.
17. **Lange, O. L.,** Die funktionellen Anpassungen der Flechten und die ökologischen Bedingungen arider Gebiete, *Ber. Dtsch. Bot. Ges.,* 82, 3, 1969.
18. **Jahns, H. M. and Ott, S.,** Das Mikroklima dicht benachbarter Flechtenstandorte, *Flora,* 173, 183, 1983.
19. **Jahns, H. M.,** Morphology, reproduction and water relations — a system of morphogenetic interactions in *Parmelia saxatilis,* in *Beiträge zur Lichenologie, Festschrift J. Poelt,* Hertel, H. and Oberwinkler, F., Eds., J. Cramer, Vaduz, 1984, 715.
20. **Snelgar, W. P. and Green, T. G. A.,** Ecologically-linked variation in morphology, acetylene reduction, and water relations in *Pseudocyphellaria dissimilis, New Phytol.,* 87, 403, 1981.
21. **Ott, S.,** The juvenile development of lichen thalli from vegetative diaspores, *Symbiosis,* 3, 57, 1987.
22. **Larson, D. W.,** Differential wetting in some lichens and mosses: the role of morphology, *Bryologist,* 84, 1, 1981.
23. **Goyal, R. and Seaward, M. R. D.,** Metal uptake in terricolous lichens. I. Metal localization within the thallus, *New Phytol.,* 89, 631, 1981.
24. **Korf, R.,** Japanese discomycete notes I-VIII, *Sci. Rep. Yokohama Natl. Univ. Sec. 2,* 7, 7, 1958.
25. **Peveling, E.,** Fine structure, in *The Lichens,* Ahmadjian, V. and Hale, M. E., Jr., Eds., Academic Press, New York, 1973, 147.
26. **Hale, M. E., Jr.,** Lichen structure viewed with the scanning electron microscope, in *Lichenology: Progress and Problems,* Brown, D. H., Hawksworth, D. L., and Bailey, R. H., Eds., Academic Press, New York, 1976, 1.
27. **Lawrey, J. D.,** *Biology of Lichenized Fungi,* Praeger, New York, 1984.
28. **Ozenda, P.,** Lichens, in *Handbuch der Pflanzenanatomie,* Vol. 6, Linsbauer, K., Ed., Borntraeger, Berlin, 1963, 9.
29. **Poelt, J. and Romauch, E.,** Die Lagerstrukturen placodialer Küsten- und Inlandsflechten, ein Beitrag zur ökologischen Anatomie der Flechten, in *Beiträge zur Biologie der niederen Pflanzen. Systematik, Stammesgeschichte, Ökologie,* Frey, W., Hurka, H., and Oberwinkler, F., Eds., Gustav Fischer, Stuttgart, 1977, 141.
30. **Darbishire, O. V.,** Über das Wachstum der Cephalodien von *Peltigera aphthosa* L., *Ber. Dtsch. Bot. Ges.,* 45, 221, 1927.

31. **Jahns, H. M.,** Die Entwicklung von Flechten-Cephalodien aus *Stigonema*-Algen, *Ber. Dtsch. Bot. Ges.,* 85, 615, 1972.

32. **Ertl, L.,** Über die Lichtverhältnisse in Laubflechten, *Planta,* 39, 245, 1951.

33. **Richardson, D. H. S.,** The transplantation of lichen thalli to solve some taxonomic problems in *Xanthoria parietina* (L.) Th.Fr., *Lichenologist,* 3, 386, 1967.

34. **Rao, D. N. and Le Blanc, F.,** A possible role of atranorin in the lichen thallus, *Bryologist,* 68, 284, 1965.

35. **Snelgar, W. P., Green, T. G. A., and Wilkins, A. L.,** Carbon dioxide exchange in lichens: resistances to CO_2 uptake at different thallus water contents, *New Phytol.,* 88, 353, 1981.

36. **Lange, O. L. and Tenhunen, J. D.,** Moisture content and CO_2 exchange of Lichens. II, *Oecologia (Berlin),* 51, 426, 1981.

37. **Peveling, E.,** Die Darstellung von Oberflächenstrukturen von Flechten mit dem Raster-Elektronenmikroskop, *Ber. Dtsch. Bot. Ges. (Neue Folge),* 4, 89, 1970.

38. **Hale, M. E., Jr.,** Fine structure of the cortex in the lichen family Parmeliaceae viewed with the scanning-electron microscope, *Smithson. Contrib. Bot.,* 10, 1, 1973.

39. **Hale, M. E., Jr.,** Pseudocyphellae and pored epicortex in the Parmeliaceae: their delimitation and evolutionary significance, *Lichenologist,* 13, 1, 1981.

40. **Peveling, E. and Poelt, J.,** Glaszilien in der Flechtenfamilie Physciaceae: ihre Ultrastruktur und die Unterschiede gegenüber Rhizinen, *Nova Hedwigia,* 25, 639, 1974.

41. **Hannemann, B.,** Anhangsorgane der Flechten, ihre Strukturen und ihre systematische Verteilung, *Bibl. Lichenol.,* 1, 1, 1973.

42. **Poelt, J.,** Morphologie der Flechten, Fortschritte und Probleme, *Ber. Dtsch. Bot. Ges.,* 99, 3, 1986.

43. **Steiner, M. and Poelt, J.,** *Caloplaca* sect. Xanthoriella, sect. nov.: Untersuchungen über die ''*Xanthoria lobulata*-Gruppe'' (Lichenes, Teloschistaceae), *Plant Syst. Evol.,* 140, 151, 1982.

44. **Jørgensen, P. M. and Jahns, H. M.,** *Muhria,* a remarkable new crustose lichen genus from Scandinavia, *Notes R. Bot. Gard. Edinburgh,* 44, 1987.

44a. **Jordan, W. P.,** The internal cephalodia of the genus *Lobaria, Bryologist,* 73, 669, 1970.

44b. **Jordan, W. P. and Rickson, F. R.,** Cyanophyte cephalodia in the lichen genus *Nephroma, Am. J. Bot.,* 58, 562, 1971.

45. **Henssen, A.,** An interesting new species of *Lempholemma* from Canada, *Lichenologist,* 4, 99, 1969.

46. **Jahns, H. M. and Schuster, G.,** Morphogenetische Untersuchungen an *Cetraria islandica, Beitr. Biol. Pflanz.,* 55, 427, 1981.

47. **Jahns, H. M., Beltman, H. A., and v. d. Knaap, P.,** *Cladonia ecmocyna* Nyl., investigations on the ontogeny, *Acta Bot. Neerl.,* 18, 627, 1969.

48. **Groenhart, P.,** Two new Malaysian lichens, *Blumea Suppl.,* 4, 107, 1958.

49. **Beltman, H. A.,** Vegetative Strukturen der Parmeliaceae und ihre Entwicklung, *Bibl. Lichenol.,* 11, 1, 1978.

50. **Linkola, K.,** Über die Thallusschuppen bei *Peltigera lepidophora* (Nyl.), *Ber. Dtsch. Bot. Ges.,* 31, 52, 1913.

51. **Bitter, G.,** *Peltigera* — Studien I. Rückseitige Apothecien bei *Peltigera malacea.* II. Das Verhalten der oberseitigen Thallusaschuppen der *Peltigera lepidophora* (Nyl.), *Ber. Dtsch. Bot. Ges.,* 22, 248, 1904.

52. **Jahns, H. M. and Fritzler, E.,** Flechtenstandorte auf einer Blockhalde, *Herzogia,* 6, 243, 1982.

53. **Bitter, G.,** Zur Morphologie und Systematik von *Parmelia,* Untergattung *Hypogymnia, Hedwigia,* 40, 171, 1901.

54. **Du Rietz, G. E.,** Die Soredien und Isidien der Flechten, *Sven. Bot. Tidskr.,* 18, 371, 1924.

55. **Frey, E.,** Beiträge zu einer Lichenenflora der Schweiz, II. III. Die Familie Physciaceae, *Ber. Schweiz. Bot. Ges.,* 73, 389, 1963.

56. **Jahns, H. M., Tinz-Dubiel, A., and Blank, L.,** Hygroskopische Bewegungen der Sorale von *Hypogymnia physodes, Herzogia,* 4, 15, 1976.

57. **Wetmore, C. M.,** New type of soredium in the lichen family Heppiaceae, *Bryologist,* 77(2), 208, 1974.

58. **Pisut, I. and Jelinkova, E.,** Über die Arthberechtigung der Flechte *Lecanora conizaeoides* Nyl ex. Cromb., *Preslia,* 43, 254, 1971.

59. **Dahl, E.,** Studies in the macrolichen flora of South West Greenland, *Medd. Groenl.,* 150(2), 1, 1950.

60. **Poelt, J.,** Das Konzept der Artenpaare bei den Flechten, *Ber. Dtsch. Bot. Ges. (Neue Folge),* 4, 187, 1970.

61. **Poelt, J.,** *Bestimmungsschlüssel Europäischer Flechten,* Cramer, Lehre, 1969.

62. **Jahns, H. M. and Seelen, E. J. R.,** *Baeomyces* -Funde aus dem Himalaya. Untersuchungen zur Taxonomie der Gattung *Baeomyces* (III), *Khumbu Himal.,* 6(2), 101, 1974.

63. **Malone, C. P.,** Observations on *Endocarpon pusillum:* the role of rhizomorphs in asexual reproduction, *Mycologia,* 69, 1042, 1977.

64. **Jahns, H. M. and Smittenberg, J. C.,** *Baeomyces roseus* Pers. — Ontogenie und Regeneration der Fruchtkörper, *Herzogia,* 2, 79, 1970.

65. **Jahns, H. M.**, The cyclic development of mosses and the lichen *Baeomyces rufus* in an ecosystem, *Lichenologist*, 14, 261, 1982.
66. **Schuster, G., Ott, S., and Jahns, H. M.**, Artificial cultures of lichens in the natural environment, *Lichenologist*, 17, 247, 1985.
67. **Jahns, H. M., Mollenhauer, D., Jenninger, M., and Schönborn, D.**, Die Neubesiedlung von Baumrinde durch Flechten I, *Nat. Mus.*, 109, 40, 1979.
68. **Ahmadjian, V.**, *The Lichen Symbiosis*, Blaisdell, Waltham, Mass., 1967.
69. **Tschermak-Woes, E.**, *Myrmecia reticulata* as a phycobiont and free-living — free-living *Trebouxia* — The problem of *Stenocybe septata*, *Lichenologist*, 10, 69, 1978.
70. **Bubrick, P., Galun, M., and Frensdorff, A.**, Observations on free-living *Trebouxia* de Puymaly and *Pseudotrebouxia* Archibald, and evidence that both symbionts from *Xanthoria parientina* (L.) Th. Fr. can be found free-living in nature, *New Phytol.*, 97, 455, 1984.
70a. **Ott, S.**, Sexual reproduction and developmental adaptions in *Xanthoria parietina*, *Nordic J. Bot.*, 7, 219, 1987.
71. **Ahmadjian, V.**, Resynthesis of lichens, in *The Lichens*, Ahmadjian, V. and Hale, M. E., Jr., Eds., Academic Press, New York, 1973, 653.
72. **Ahmadjian, V., Jacobs, J. B., and Russell, L. A.**, Scanning electron microscope study of early lichen synthesis, *Science*, 200, 1062, 1978.
73. **Steiner, M.**, Wachstums- und Entwicklungsphysiologie der Flechten, in *Handbuch der Pflanzenphysiologie*, Vol. 1, Ruhland, W., Ed., Springer-Verlag, Berlin, 1965, 758.
74. **Hale, M. E., Jr.**, Single-lobe growth-rate patterns in the lichen *Parmelia caperata*, *Bryologist*, 73, 72, 1970.
75. **Nienburg, W.**, Anatomie der Flechten, in *Handbuch der Pflanzenanatomie*, Vol. 6, Linsbauer, K., Ed., Borntraeger, Berlin, 1926, 2.
76. **Geitler, L.**, Beiträge zur Kenntnis der Flechtensymbiose I-VII, *Arch. Protistenkd.*, 80, 378, 1933; 82, 51, 1934; 88, 161, 1937; 90, 489, 1938.
77. **Beschel, R.**, Individuum and Alter bei Flechten, *Phyton*, 6, 60, 1955.
78. **Jahns, H. M. and Frey, P.**, Thallus growth and the development of fruit bodies in *Peltigera canina*, *Nova Hedwigia*, 36, 485, 1982.
79. **Jahns, H. M., Herold, K., and Beltman, H. A.**, Chronological sequence, synchronization and induction of the development of fruit bodies in *Cladonia furcata* var. *furcata* (Huds.) Schrad., *Nova Hedwigia*, 30, 469, 1978.
80. **Jahns, H. M. and Ott, S.**, Flechtenentwicklung an dicht benachbarten Standorten, *Herzogia*, 6, 201, 1982.
81. **Jahns, H. M.**, Individualität und Variabilität in der Flechtengattung *Cladia* Nyl., *Herzogia*, 3, 277, 1972.

Section IV: The Fungus-Alga Relation

Chapter IV

THE FUNGUS-ALGA RELATION

Margalith Galun

I. INTRODUCTION

Ultrastructural studies on lichens started several years after the fine structure of other plant material had been examined almost routinely, mainly due to technical difficulties in adequately preparing the lichen tissue for electron microscope observations. Now, however, most of our knowledge on the physical relations between the fungus and the alga in the lichen thallus stems from electron microscopical investigations which at first substantiated and later extended previous light microscope observations.[1-3] It is still a problem, in many cases, to achieve equally good resolution of both the fungus and the alga in one and the same preparation. Each of the components would require different preparative treatments.

Details on the methods used for lichen ultrastructural studies can be found in References 4 to 7.

II. THE FUNGUS-ALGA RELATION IN SPECIES WITH GREEN PHYCOBIONTS

A. Discocarpous Species

Phycobionts with sporopollenin-containing cell walls (*Coccomyxa* and some *Myrmecia* species)[6] seem to be resistant to fungal penetration. In these cases the relation is by very tight adhesion of the mycobiont hyphae to the outermost cell wall layer of the phycobiont. A distinct rodlet pattern was discerned at the surface of the hyphae adhering to the *Coccomyxa* cells of some Peltigeralean species. Such a pattern was not observed on hyphae not in contact with these algal cells (Figure 1).[8,9]

In species with trebouxoid phycobionts (*Trebouxia* and *Pseudotrebouxia*, according to Archibald[10])* the contact ranges from intracellular penetration by fungal haustoria, through intermediate stages of incipient wall intrusions, to an association of wall-to-wall attachment or proximity.[9] The mode of contact is in correlation with the developmental differentiation of the lichen.[2,12-15]

Structurally simple crustose species feature intracellular invasion in which the haustorium pierces and grows through the algal cell wall (Figure 2). In this type, all the phycobiont cells from all phases of the normal life cycle of the alga (i.e., young, mature, senescing, and decaying cells[16]) could be intruded intracellularly. In more differentiated species (still crustose, but with a defined growth-form), fungal invasion in which the algal cell wall is not penetrated (Figure 3) but invaginated along the haustorium** appeared in cells at the mature or senescent phase. Finally, in lichens with a foliose or fruticose differentiation, haustoria were detected almost only in pre-existing decayed or entirely distorted algal cells.[1-3,12-20,30]

With the advancement of differentiation, the number of algal cells attacked by any of the above-mentioned ways also decreases. Peveling[21] suggested that algal cell degeneration may also be caused by fungal haustoria.

---

* *Trebouxia* and *Pseudotrebouxia* have recently been recombined into one genus — *Trebouxia*[11] (see also Chapter II.B).

** Originally described as intramembranous when examined with light microscope,[1] then changed to intrawall[12] or intraparietal[9] — all three describe invasion without piercing the phycobiont's cell wall.

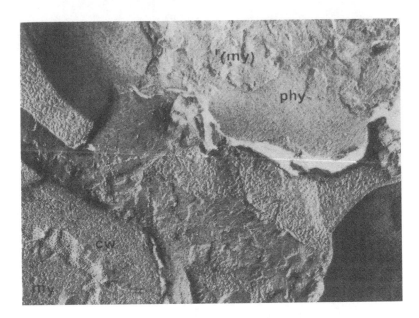

FIGURE 1.   Transmission electron microscopy (TEM) of "freeze-fracture preparation of the mycobiont (my)-*Coccomyxa*-phycobiont (phy) contact site in *Peltigera venosa*. The fracture plane follows the broken mycobiont (my) cell, its cell wall (cw) and the inner surface of the outermost cell wall layer; with a distinct rodlet (r) pattern, then the outer surface of the trilaminar, sporpollenin-containing outermost cell wall layer of the *Coccomyxa* phycobiont (phy). The outermost wall layer (rodlet layer) of the mycobiont tightly adheres to the phycobiont wall surface." (Magnification × 31,500.) (From Honegger, R., *Lichenologist*, 16, 118, 1984. With permission.)

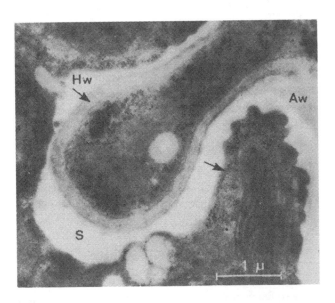

FIGURE 2.   Section through intracellular haustorium in algal cell of *Lecanora olea*. Note space (S) between haustorial wall (Hw) and algal plasmalemma (arrow); such space does not appear in decaying cells. Algal wall (Aw) disappeared at site of penetration (arrow). (From Galun, M. et al., *New Phytol.*, 69, 599, 1970. With permission.)

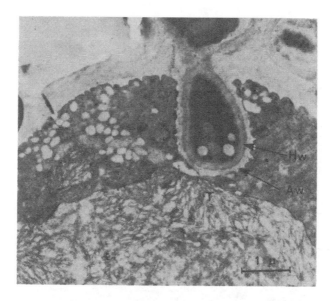

FIGURE 3. Section through intrawall haustorium in senescing algal cell of *Lecanora subplanata*. Algal wall (Aw) invaginated along the haustorial wall (Hw). T = thylakoids. (From Galun, M. et al., *New Phytol.*, 69, 599, 1970. With permission.)

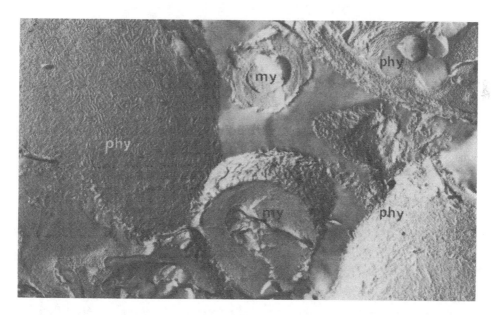

FIGURE 4. TEM of freeze-fracture preparation of the mycobiont (my)-phycobiont (phy) interface in *Cladonia macrophylla*. "The irregularly tessellated surface layers of myco- and phycobiont cell walls are continuous at the contact site." (Magnification × 15,000.) (From Honegger, R., *Lichenologist*, 16, 118, 1984. With permission.)

Mycobiont and phycobiont cells of some structurally complex Lecanoralean species exhibit at their outermost wall layer an irregularly tessellated pattern which merges at the contact site[9] (Figure 4).

The degree of intimacy between the symbionts is also affected by environmental conditions. While in juxtaposition in moderate climates, the symbionts of the same species demonstrate intracellular penetration when under extreme xeric conditions.[15,22,23]

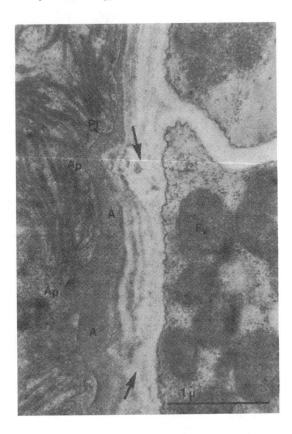

FIGURE 5.   Section through interaction zone between algal
cell and fungus in *Dermatocarpon hepaticum*; note translucent
area (arrow) at point of contact and deposited wall material
(?) (A). Ap = algal protoplast; Pl = plasmalemma; Fv =
fungal vacuole. (From Galun, M. et al., *Protoplasma*, 73,
457, 1971. With permission.)

The fungal-algal contact in species containing the filamentous green phycobiont *Trente-pohlia* was found to be either by fungal ensheathment of the algal cells[24] or by incipient haustoria and penetration into degenerating algal cells.[25] Only very few *Trentepohlia*-lichens have been examined.

In all discocarpous species the fungal haustoria retain their cell wall, albeit somewhat thinner than in the hyphae they derive from, and the algal plasmalemma remains intact even with the deepest fungal penetration (see Figure 2).

B. Pyrenocarpous Species

The relation between the symbionts in pyrenocarpous species is completely different from that of discocarpous species.[15,26-28] It is, however, in similar correlation with phyletic differentiation and is similarly affected by environmental conditions. At the site of interaction the fungal and algal cell walls disintegrate. A naked, membrane-bound, fungal protoplast emerges towards the algal cell, but does not succeed in attaching itself to the algal protoplast because the algal protoplast retreats and additional cell wall material (?) is deposited into the space formed which then merges with the original algal cell wall (Figures 5 and 6). Senescing and decaying phycobionts, when attacked, are penetrated by intracellular wall-bound haustoria in a manner similar to the process in the discocarpous type. Here also, only a small percentage of the algal cells are attacked.

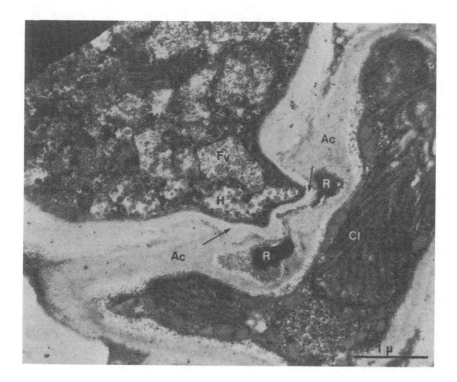

FIGURE 6.   Later stage of interaction than in Figure 5; note naked haustorium (H) and dissolved region in algal cell wall (Ac) (arrow). R = presumably cell wall remnants; Cl = chloroplast; Fv = fungal vacuole. (From Galun, M. et al., *Protoplasma*, 73, 457, 1971. With permission.)

Identical as well as similar phycobiont genera participate in both the discocarpous and the pyrenocarpous types. Hence, the mode of contact between the symbionts is determined by the fungus — independent of the species to which the algal companion belongs. This does not contradict with the fact that one fungus is capable of forming two distinct associations: one with a green alga and one with a cyanobacterium, such as is the case in cephalodia (see Chapter III, Section VI) or in "chimeroid associations".[29]

Thus, a mere proximity or wall-to-wall contact between the symbionts (Figure 7) is the most common relationship in green-algal lichens. It appears, therefore, that penetration of the fungus into the algal cell is not an obligatory condition for the mutual relationship (see Table 1). Consequently, any physiological interaction between the symbionts has to pass the barriers of two cell walls and often also intercellular matrix. This corresponds with the hypothesis of Collins and Farrar[30] on mass transfer by diffusion (see Chapter VI.A) and with the findings of Hessler and Peveling[31] on the transfer of ^{14}C-metabolites from the phycobiont to the mycobiont through the plasmalemma and cell walls.

The haustoria in primitively structured thalli, which gradually decrease and disappear with evolutionary development, may be a vestigal remnant of a parasitic fungus-alga relation.[32,33] It is not understood whether fungal invasions, wherever present, take place by mechanical means or by cell-wall digesting enzymes.[34] The dissolution of both the fungal and algal cell walls in pyrenocarpous lichens (see Figures 5 and 6) suggests an enzymatic involvement.

The hyphal web in the algal layer of a mature thallus is much looser than the tight hyphal envelopment of the algal cells during thallus development (compare Figure 8 here with Figure 5 in Chapter VIII.C of Volume II). The algal cells of soredia — free or intact — (see Chapter III, Section VII) are also very tightly entrapped by the fungal hyphae.

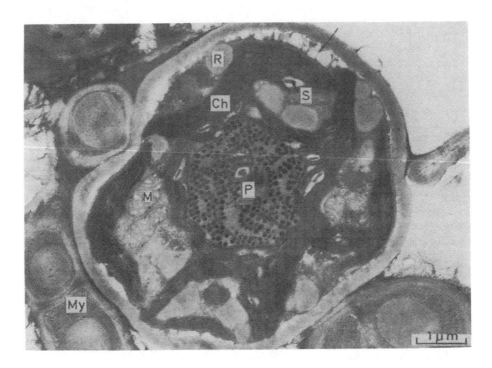

FIGURE 7. Section through a *Trebouxia* cell with closely attached mycobionts (My) in a thallus of *Cladonia incrassata*. Ch = chromatophor; M = mitochondria; P = pyrenoid; R = storage bodies; S = starch. (From Hessler, R. and Peveling, E., Z. *Pflanzenphysiol.*, 86, 287, 1978. With permission.)

Table 1
HAUSTORIA IN LICHENS

Algal genus	Lichen species	% algal cells invaded	Type of haustorium[a]	Ref.
	Haustoria Absent or Very Rare			
Nostoc	*Collema tenax*	0	—	48
	Leptogium hildenbrandii	0	—	35
	Peltigera canina	0	—	7,35
	P. rufescens	0	—	35
	P. polydactyla	0	—	49
Calothrix	*Lichina pygmaea*	0	—	37
Coccomyxa	*Icmadophila ericetorum*	0	—	6,9,50
	Peltigera aphthosa	0	—	9,50
	P. leucophlebia	0	—	6
	P. venosa	0	—	6
	Botrydina vulgaris	0	—	50
	Coriscium viride	0	—	50
	Solorina saccata	0	—	6
Myrmecia	*Baeomyces rufus*	0	—	6
	Sarcogyne pruinosa	0	—	15
Trebouxia	*Aspicilia* sp.	0	—	16
	Hypogymnia physodes	0	—	51
	Parmelia aurulenta	0	—	52
	Physcia aipolia	0	—	53

Table 1 (continued)
HAUSTORIA IN LICHENS

Algal genus	Lichen species	% algal cells invaded	Type of haustorium[a]	Ref.
	Usnea pruinosa	0	—	18
	U. rockii	0	—	18
	Xanthoria parietina	0	—	14,30
	Caloplaca aurantia	0	—	22
	C. flavovirens	0	—	14
	C. velana	0	—	14
	C. pyraceae	0	—	14
	C. erythrocarpa	0	—	14

Haustoria Found in Decaying and Degenerated Cells Only

Algal genus	Lichen species	% algal cells invaded	Type of haustorium[a]	Ref.
Trebouxia	*Squamarina crassa*	40	c,w	12,16
	Lecidea opaca	—	m,w	13
	Lecanora radiosa	—	m,w	23
	L. subplanata	20—25	m,w	12
	L. muralis	—	c,w	17,21
Myrmecia	*Lecidea decipiens*	—	—	13
	Dermatocarpon hepaticum	10—20	m,p,c,w	27
Stigonema	*Stereocaulon paschale*	—	—	43
Cephaleuros	*Strigula elegans*	10—20	—	55

Haustoria Found in Healthy and Degenerated Cells

Algal genus	Lichen species	% algal cells invaded	Type of haustorium[a]	Ref.
Trebouxia	*Lecanora olea*	50—60	c,w	12
	L. rubina	—	m	54
	Lecidea olivacea	—	c,w	13
	Catillaria reichertiana	10	c,w	15
	Parmelia sulcata	—	c,w	34
	Verrucaria spp.	10—15	m,p	15
	Protoblastenia metzleri	10—13	m,w	15
Scytonema	*Lichenothrix riddlei*	—	c,w	41
	Heppia echinulata	—	c,w	40
	Dictyonema irpicinum	—	c,w	45
	D. moorei	—	c,w	46
	Cora pavonia	—	c,w	47

[a] c = intracellular; m = intrawall[12] or intraparietal;[9] p = fungal plasmalemma naked; w = fungus wall bound; — = % not included.

Modified from Collins, C. R. and Farrar, J. F., *New Phytol.*, 81, 71, 1978.

III. THE FUNGUS-ALGA RELATION IN SPECIES WITH CYANOBACTERIAL SYMBIONTS

Among the cyanobacteria, *Nostoc* is the most common in lichens both in the thallus as well as the second "algal" partner in most cephalodia (see Chapter III, Section VI).

Many lichens with *Nostoc* have an unstratified, gelatinous thallus (see Chapter III, Section IV) in which both symbionts are embedded in a polysaccharidic matrix. In stratified *Nostoc*-containing species, the *Nostoc* is organized in a defined layer, ensheathed in a fibrillar matrix

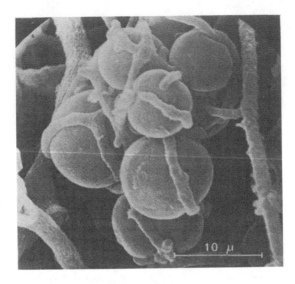

FIGURE 8.   Scanning electron micrograph of fractured preparation of the algal layer in *Ramalina duriaei* thallus. (Courtesy of Dr. Jacob Garty.)

FIGURE 9.   Schematic presentation of a section through a *Collema* thallus. Arrow = *Nostoc*; double arrow = fungal hypha. (From Ozenda, P. and Clauzade G., *Les Lichens,* Masson et Cie, Editeurs, Paris, 1970, 18. With permission.)

of acidic polysaccharide and proteins. The matrix material apparently belongs to the *Nostoc* sheath.[7] However, fungal intrusion is rare in both types.[35] The symbionts are either in physical contact with each other or, more often, particularly in the gelatinous type, at a distance up to several micrometers (Figure 9). An exception are the haustoria-intruding *Nostoc* of several *Lempholemma* species.[36]

Peveling[37] suggested that the vesicles that appear in the inner part of the gelatinous sheath of cyanobionts (Figure 10), not penetrated by haustoria, like *Nostoc* or *Calothrix* (of *Lichina*

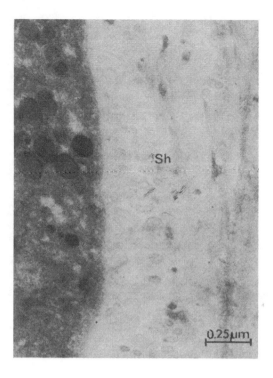

FIGURE 10.   Part of algal cell of *Calothrix* with its sheath (Sh) and many vesicles of different shape. (From Peveling, E., *New Phytol.*, 72, 343, 1973. With permission.)

pygmaea), play a role in nutrient transport from cyanobiont to mycobiont. Such vesicles were found also in free-living cyanobacteria and were suggested to be involved in sheath formation.[38,39]

Other cyanobionts encounter haustorial penetration: *Scytonema* of *Heppia echinulata*[40] and *Lichenothrix riddlei*,[41] *Gloeocapsa** of *Gonohymenia mesopotamica* and *G. sinaica*[43] (Figure 11), and degenerating *Stigonema* cells in the cephalodia of *Stereocaulon paschale*.[44] The physical myco-cyanobiont contact in these species is rather close but not intracellular. The external sheath of the algal cell wall disappears, but the penetrating haustorium, which retains its own cell wall, only pushes against the cyanobiont's cell wall and invaginates its four layers as well as the plasma membrane (Figure 12) without piercing through them. Heterocysts are usually not invaded by the fungus.

Intracellular haustoria piercing the cyanobiont's cell wall were observed in *Scytonema* cells of the basidio-lichens *Dictyonema irpicinum*,[45] *D. moorei*,[46] and *Cora pavonia*.[47]

It may be concluded that in cyanolichens, as in the green-algal lichens, haustoria are not essential for the biotrophic interaction between the symbionts (see Table 1).

*   On the taxonomic identity of *Gloeocapsa*, see Reference 42.

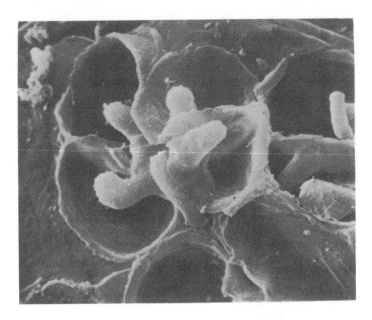

FIGURE 11.   Scanning electron micrograph of *Gonohymenia sinaica* cyanobiont cells with haustoria (thallus was broken by tearing, which removed cyanobacterial protoplast). (Magnification × 10,000.) (Courtesy of Dr. Paul Bubrick.)

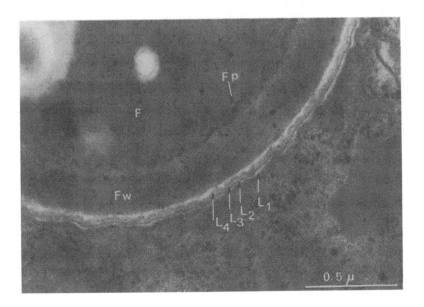

FIGURE 12.   Portion of penetrating hypha appressed against the invaginated four walls layers (L_1 to L_4) of the cyanobiont in *Gonohymenia sinaica*. F = fungus; Fp = fungal plasmalemma; Fw = fungal wall. (From Paran, N. et al., *Arch. Mikrobiol.*, 76, 103, 1971. With permission.)

REFERENCES

1. **Tschermak, E.,** Untersuchungen über die Beziehungen von Pilz und Alge im Flechtenthallus, *Ost. Bot. Z.,* 90, 233, 1941.

2. **Plessl, A.,** Über die Beziehungen von Haustorientypus und Organisationshöhe bei Flechten, *Ost. Bot. Z.,* 110, 194, 1963.

3. **Geitler, L.,** Über Haustorien bei Flechten und über *Myrmecia biatorella* in *Psora globifera, Ost. Bot. Z.,* 110, 270, 1963.

4. **Peveling, E.,** Elektronmikroskopische Untersuchungen an Flechten. II. Die Feinstruktur von *Trebouxia* — Phycobionten, *Planta,* 87, 69, 1969.

5. **Jacobs, J. B. and Ahmadjian, V.,** The ultrastructure of lichens. II. *Cladonia cristatella:* the lichen and its isolated symbionts, *J. Phycol.,* 7, 71, 1971.

6. **Honegger, R. and Brunner, U.,** Sporopollenin in the cell walls of *Coccomyxa* and *Myrmecia* phycobionts of various lichens: an ultrastructural and chemical investigation, *Can. J. Bot.,* 59, 2713, 1981.

7. **Boissiere, M.-C.,** Cytochemical ultrastructure of *Peltigera canina:* some features related to its symbiosis, *Lichenologist,* 14, 1, 1982.

8. **Honegger, R.,** Ascus structure and function, ascospore determination, and phycobiont cell wall types associated with the Lecanorales (lichenized Ascomycetes), *J. Hattori Bot. Lab.,* 52, 417, 1982.

9. **Honegger, R.,** Cytological aspects of the mycobiont-phycobiont relationship in lichens, *Lichenologist,* 16, 111, 1984.

10. **Archibald, P. A.,** *Trebouxia* de Puymaly (Chlorophyceae, Chlorococcales) and *Pseudotrebouxia* gen. nov. (Chlorophyceae, Chlorosarcinales), *Phycologia,* 14, 125, 1975.

11. **Gärtner, G.,** Taxonomische Probleme bei den Flechtenalgengattungen *Trebouxia* und *Pseudotrebouxia* (Chlorophyceae, Chlorellales), *Phyton (Horn, Austria),* 25, 101, 1985.

12. **Galun, M., Paran, N., and Ben-Shaul, Y.,** The fungus-alga association in the Lecanoraceae: an ultrastructural study, *New Phytol.,* 69, 599, 1970.

13. **Galun, M., Ben-Shaul, Y., and Paran, N.,** The fungus-alga association in the Lecideaceae: an ultrastructural study, *New Phytol.,* 70, 483, 1971.

14. **Galun, M., Ben-Shaul, Y., and Paran, N.,** Fungus-alga association in lichens of the Teloschistacae: an ultrastructural study, *New Phytol.,* 70, 837, 1971.

15. **Kushnir, E. and Galun, M.,** The fungus-alga association in endolithic lichens, *Lichenologist,* 9, 123, 1977.

16. **Galun, M., Paran, N., and Ben-Shaul, Y.,** Structural modifications of the phycobiont in the lichen thallus, *Protoplasma,* 69, 85, 1970.

17. **Peveling, E.,** Elektronenoptische Untersuchungen an Flechten. I. Strukturveränderungen der Algenzellen von *Lecanora muralis* (Schreb.) Rabenh. beim Eindringen von Pilzhyphen, *Z. Pflanzenphysiol.,* 59, 172, 1968.

18. **Chervin, R. E., Baker, G. E., and Hohl, H. R.,** The ultrastructure of phycobiont and mycobiont in two species of *Usnea, Can. J. Bot.,* 46, 241, 1968.

19. **Peveling, E.,** Elektronenoptische Untersuchungen an Flechten. III. Cytologische Differenzierungen der Pilzzellen im Zusammenhang mit ihrer symbiotischen Lebensweise, *Z. Pflanzenphysiol.,* 61, 151, 1969.

20. **Malachowski, J. A., Baker, K. K., and Hooper, G. K.,** Anatomy and alga-fungal interactions in the lichen *Usnea cavernosa, J. Phycol.,* 16, 346, 1980.

21. **Peveling, E.,** Fine structure, in *The Lichens,* Ahmadjian, V. and Hale, M. E., Eds., Academic Press, New York, 1973, 147.

22. **Ben-Shaul, Y., Paran, N., and Galun, M.,** The ultrastructure of the association between phycobiont and mycobiont in three ecotypes of the lichen *Caloplaca aurantia* var. *aurantia, J. Microsc. (Oxford),* 8, 415, 1969.

23. **Galun, M., Paran, N., and Ben-Shaul, Y.,** An ultrastructural study of the fungus alga association in *Lecanora radiosa* growing under different environmental conditions, *J. Microsc. (Oxford),* 9, 801, 1970.

24. **Meier, J. L. and Chapman, R. L.,** Ultrastructure of the lichen *Coenogonium interplexum* Nyl., *Am. J. Bot.,* 70, 400, 1983.

25. **Withrow, K. and Ahmadjian, V.,** The ultrastructure of lichens. II. *Chiodecton sanguineum, Mycologia,* 75, 337, 1983.

26. **Galun, M., Paran, Y., and Ben-Shaul, Y.,** Electron microscopic study on the lichen *Dermatocarpon hepaticum* (Ach.) Th. Fr., *Protoplasma,* 73, 457, 1971.

27. **Galun, M., Kushnir, E., Behr, L., and Ben-Shaul, Y.,** Ultrastructural investigation on the alga-fungus relation in pyrenocarpous lichen species, *Protoplasma,* 78, 187, 1973.

28. **Brunner, U.,** Ultrastrukturelle und chemische Zellwanduntersuchungen an Flechtenphycobionten aus Gattungen der Chlorophyceae (Chlorophytina) unter besonderer Berücksichtigung sporopolleninähnlicher Biopolymere, Ph.D. thesis, University of Zürich, Zürich, 1985.

29. **Brodo, M. J. and Richardson, D. M. S.**, Chimeroid associations in the genus *Peltigera, Lichenologist*, 10, 157, 1980.

30. **Collins, C. R. and Farrar, J. F.**, Structural resistance to mass transfer in the lichen *Xanthoria parietina, New Phytol.*, 81, 71, 1978.

31. **Hessler, R. and Peveling, E.**, Die Lokalisation von ^{14}C-Assimilaten in Flechtenthalli von *Cladonia incrassata* Floerke und *Hypogymnia physodes* (L.) Ach., *Z. Pflanzenphysiol.*, 86, 287, 1978.

32. **Ahmadjian, V. and Jacobs, J. B.**, Relationship between fungus and alga in the lichen *Cladonia cristatella* Tuck., *Nature (London)*, 289, 169, 1981.

33. **Galun, M.**, Symbiotic relationships in plants, in *Dynamic Aspects of Host-Parasite Relationship*, Vol. 3, Zuckermann, A., Ed., Israel University Press, Jerusalem, 1979.

34. **Webber, M. M. and Webber, P. J.**, Ultrastructure in lichen haustoria: symbiosis in *Parmelia sulcata, Can. J. Bot.*, 48, 1521, 1970.

35. **Peveling, E.**, Elektronische Untersuchungen an Flechten. IV. Die Feinstruktur einiger Flechten mit Cyanophyceen-Phycobionten, *Protoplasma*, 68, 209, 1969.

36. **Tschermak-Woess, E.**, Haustorienbefall und inäquale Teilung des *Nostoc* — phycobionten von *Lempholemma botryosum* (Lichinaceae), *Plant Syst. Evol.*, 137, 317, 1981.

37. **Peveling, E.**, Vesicles in the phycobiont sheath as possible transfer structures between the symbionts in the lichen *Lichina pygmaea, New Phytol.*, 72, 343, 1973.

38. **Boissiere, J.-C. and Boissiere, M.-C.**, Etude ultrastructural et cytochmique de la paroi de *Nostoc* libres et lichenisés, *C. R. Acad. Sci. Ser. D*, 278, 2767, 1974.

39. **Honegger, R.**, Cytological aspects of the triple symbiosis in *Peltigera aphthosa, J. Hattori Bot. Lab.*, 52, 379, 1982.

40. **Marton, K. and Galun, M.**, *In vitro* dissociation and reassociation of the symbionts of the lichen *Heppia echinulata, Protoplasma*, 87, 135, 1976.

41. **Tschermak-Woess, E., Bartlett, J., and Peveling, E.**, *Lichenothrix riddlei* is an ascolichen and also occurs in New Zealand — light and electron microscopical investigation, *Plant Syst. Evol.*, 143, 293, 1983.

42. **Bubrick, P. and Galun, M.**, Cyanobiont diversity in the Lichinaceae and Heppiaceae, *Lichenologist*, 16, 279, 1984.

43. **Paran, N., Ben-Shaul, Y., and Galun, M.**, Fine structure of the blue-green phycobiont and its relation to the mycobiont in two *Gonohymenia* lichens, *Arch. Mikrobiol.*, 76, 103, 1971.

44. **Bergman, B. and Huss-Danell, K.**, Ultrastructure of *Stigonema* in the cephalodia of *Stereocaulon paschale, Lichenologist*, 15, 181, 1983.

45. **Slocum, R. D.**, Light and electron microscopic investigations in the Dictyonemataceae (Basidiolichens). II. *Dictyonema irpicinum, Can. J. Bot.*, 58, 1005, 1980.

46. **Ahmadjian, V.**, Algal/fungal symbiosis, in *Progress in Phycological Research*, Vol. 1, Round, F. E. and Chapman, D. J., Eds., Elsevier, Amsterdam, 1982, 179.

47. **Roskin, P. A.**, Ultrastructure of the host-parasite interaction in the basidiolichen *Cora pavonia, Arch. Mikrobiol.*, 70, 176, 1970.

48. **Bousfield, J. and Peat, A.**, The ultrastructure of *Collema tenax*, with particular reference to microtubule-like inclusions and vesicle production by the phycobiont, *New Phytol.*, 76, 713, 1976.

49. **Peat, A.**, Fine structure of the vegetative thallus of the lichen *Peltigera polydactyla, Arch. Mikrobiol.*, 61, 212, 1968.

50. **Peveling, E. and Galun, M.**, Electron-microscopical studies on the phycobiont *Coccomyxa* Schmidle, *New Phytol.*, 77, 713, 1976.

51. **Durell, L. W.**, An electron microscope study of algal, hyphal contact in lichens, *Mycopathol. Mycol. Appl.*, 31, 273, 1966.

52. **Fahselt, D., Hayden, D. B., and Marinda, M.**, Structure of the lichen *Parmelia aurulenta, Can. J. Bot.*, 51, 2197, 1973.

53. **Brown, R. M. and Wilson, R.**, Electron microscopy of the lichen *Physcia aipolia, J. Phycol.*, 4, 230, 1968.

54. **Jacobs, J. R. and Ahmadjian, V.**, The ultrastructure of lichens. I. A general study, *J. Phycol.*, 5, 227, 1969.

55. **Chapman, R. L.**, Ultrastructural investigations on the foliicolous pyrenocarpous lichen, *Strigula elegans, Phycologia*, 15, 191, 1976.

Section V: Reproduction

Chapter V.A

ASCI, ASCOSPORES, AND ASCOMATA

André Bellemère and Marie Agnes Letrouit-Galinou

I. INTRODUCTION

Ascomata are structures in which asci are clustered. This term was recently introduced by some mycologists to replace the term "ascocarp" which is still commonly used. Asci (containing ascospores) are the sexual reproducing structures of the mycobiont.

In this chapter, structural features and the development, as well as similarities and dissimilarities, of asci and ascomata between lichenized and nonlichenized Ascomycetes will be considered.

Structural and ontogenetical data on asci and ascomata have been summarized in the past 20 years in several studies. Some are general surveys of the "lichen sexual reproduction" (which means that of the mycobiont),[1-6] others deal with the asci, especially their apical apparatus, as observed either by light microscopy[7] or by electron transmission microscopy.[8-10] Here, we shall be concerned mainly with new and well documented information on asci, ascospores and ascomata.

It is noteworthy that important features of the sexual development in Ascomyetes have first been observed in lichens: asci in *Pertusaria*,[11] hooks of ascogenous hyphae in *Parmelia*,[12] trichogynes in *Lecidea*[13] and *Collema*,[14] and the amyloid apical ring of asci in *Peltigera*.[15]

The ascogonial apparatus is known in lichens since the second half of the last century[13,14] and is demonstrated in nearly all studies dealing with ascomata development since then. It has more rarely been observed in nonlichenized Ascomycetes. This conspicuous apparatus has, in nearly all the species, the same structure (the *Collema* type).[16] It is (Figure 1) a filamentous structure, generally formed by several ascogonial hyphae with a dense cytoplasm. Its base is an enlarged glomerule of intertwined filaments, from which straight hyphae, the trichogynes, emerge. Each of the ascogonial hyphae is multicellular and uniseriate, and has a distorted base of cells which are larger than those of the surrounding vegetative hyphae. In contrast to nonlichenized Ascomycetes, there is no clearly differentiated foot. The trichogynes tips of the ascogonial hyphae are more or less erect, sometimes branched, and protrude outside the thallus. They are thin and have elongated cells. Besides this common type of the ascogonial apparatus, slightly variant ones, though also filamentous, are known. For instance, in Peltigeraceae there are large multinucleate ascogonial cells; in Baeomycetaceae, where isolated, coiled ascogonial hyphae are joined by thin connecting threads, trichogynes are rarely observed; in *Cladonia* the ascogonial hyphae are straight; and in *Lecidella elaeochroma* a large part of the bulky ascogonial complex is not functional and does not generate sporophytic hyphae.

Sometimes, several of those complicated ascogonial bulks are formed close together in clumps. In some cases, they appear directly among the thalline hyphae, generally in the algal layer. More often, they are surrounded by a differentiated area. When ascomata are lirellae, ascogonial hyphae develop continuously at the young undifferentiated growing ends. Sometimes ascogonial apparati are observed in old parts of degenerating ascomata.

The existence of protruding trichogynes has been confirmed by scanning electron microscopy[17,18] by which their progressive aging and final degeneration have also been discerned. Stahl's theory[14] assuming that trichogynes are female receptors for male spermatia, which was controversed,[16] has recently been confirmed by observations of *Cladonia furcata*[19]

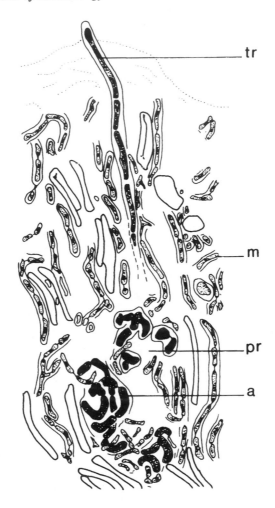

FIGURE 1.   Ascogonial apparatus. a = ascogonial hyphae; m = thallus medulla; pr = glomerule of intertwined filaments; tr = trichogyne.

where spermatia sunk into the trichogyne wall and the resulting rounded holes were seen (Figure 2). The big multinucleate cells observed at the margin of a *Peltigera* thallus[20] were perhaps ascogonial.

The sporophytic apparatus consists of all the filaments which originate from ascogonial cells and finally lead to ascus production. Their terminal parts (the ascogenous hyphae), just beneath the asci, have usually dicaryotic cells with lateral clumps.[21,22] Exceptionally, in *Baeomyces* and *Icmadophila*, the clumps are missing.[1,23] Relations between the ascogonial cells and the terminal ascogenous hyphae are not clear and several types of intermediary filaments have been described.[3]

II. ASCI AND ASCOSPORES

Asci are cells (sporocysts) in which, after caryogamy and meiosis, endocellular spores (ascospores) are produced, generally eight, more or less. The asci of lichens are usually small (about 50 μm in length), however, in some cases, they are larger (over 150 μm in *Stenhammarella* sp. and over 300 μm in some *Pertusaria*). Several are smaller (i.e., various

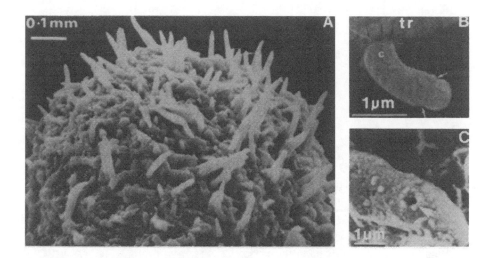

FIGURE 2.   (A) Emerging trichogynes; (B) trichogyne (tr) in contact (black arrows) with spermatia; (C) hole in trichogyne wall where spermatia entered.

Arthoniaceae). They are usually claviform, but some are shaped differently: subglobose in Arthoniaceae, flask-shaped in Thelocarpaceae, and cylindrical in Caliciales and other genera.

Nearly all lichen asci have a thick wall at the mature stage and a thin wall at the young stage. This wall is formed by main two layers, the exo- and the endoascus, as in other Ascomycetes.[24,25] Most pyrenolichens and a number of discolichens have "bitunicate" asci with a dehiscence of the fissitunicate (jack-in-the-box) type like those known in the Loculoascomycetes. In Lecanorales, asci are often covered with an amyloid fuzzy coat, and the endoascus is bistratified with the internal layer more or less developed and iodine reactive. There is an apical thickening, more or less amyloid, with internal differentiations and the dehiscence is generally of the rostrate type. Epiplasm of lichen asci has been little studied and no specific characteristics are known. Occasionally lipids and glycogen have been mentioned. Details about ascal cytology are given in the next paragraphs.

A. Origin of Asci

In lichens, as in all Ascomycetes, asci originate from proascal dicaryotic cells. Most often, these are the fertile cells of a dangeardium provided with a clamp at its basis. Several variants have been described in nonlichenized Ascomycetes,[1] but only a few cases are known in lichens (*Icmadophila ericetorum*,[1] *Baeomyces rufus*[22]). In *Pertusaria pertusa*[26] ascogenous hyphae have uninucleate cells; the dicaryotic cells producing asci are formed by fusion of two of those cells, there is no dangeardium formation.

As usual in Ascomycetes, asci originate only from terminal dangeardia of ascogenous hyphae. The nuclei of the median dicaryotic cell of the dangeardium fuse. Then an excrescence is formed at its top which elongates, building up the ascus. Asci are generally associated in cyma, because the foot cell of the dangeardium, beneath the ascus, can give rise laterally to a new fertile dangeardium. In *Chaenotheca trivialis*[27] not only the terminal clamped cell is sporogenous but also the cells underneath and, therefore, a chain of sporogenous cells is produced. In this case, caryogamy and meiosis take place in every dicaryotic cells of the ascogenous hyphae. Those cells act as asci. The uninucleate clamped cells in *Diploicia canescens* and other species[23] may be the result of a similar development, but this is restricted to caryogamy, without subsequent meiosis and spore formation. The development of an ascus inside an old empty one is not rare, since proascal cells may regain activity.

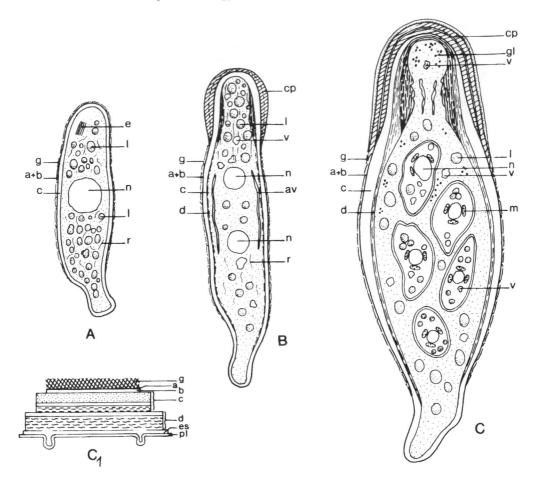

FIGURE 3.   Ascus development in *Caloplaca marina*. (A) After fusion of the two nuclei caryogamy; (B) after meiotic division; (C) after mitosis; (C₁) details of ascus wall at the median part of the ascus. a = a-layer; av = ascus vesicle; b = b-layer; c = c-layer; cp = cap; d = d-layer; e = endoplasmic reticulum; es = periplasmic space; g = gelified external fuzzy coat; gl = glycogen granules; l = lipid; m = mitochondria; n = nucleus; pl = plasmalemma; r = reticulum elements; v = vacuole.

B. Ascus Development: an Example, *Caloplaca marina*

During the development, important modifications occur in the ascus cytoplasm and the ascus wall. Caryogamy, meiosis, and later divisions are like those observed in nonlichenized Ascomycetes.[23,28-30] Recent electron microscopy studies revealed information on storage products and ascospore formation.[10] The complete development of the ascus has been studied in *Caloplaca marina*[21] and is presented in the following four sections.

1. The Young Uninucleate Ascus

Soon after the fusion of the two nuclei (Figure 3A), the proascal cell begins to elongate. The very young ascus, about 25 × 5 μm, has a narrow base, is slightly inflated in the center at the nucleus level, and somewhat narrowed at the top. A clear stratification appears in the epiplasm. Numerous lipid globules form clusters surrounded by mitochondria and elements of the endoplasmic reticulum. These are absent at the top where clusters of microvesicles are visible in a dense cytoplasm, similar to hyphal "Spitzenkörper". Underneath, above the lipids, piling up of reticulum is frequent. The ascus wall (Figure 3C) is thin, especially at the apex (i.e., 0.5 μm), but already clearly layered. In the median part of the ascus, the wall is already made of the four characteristic layers (Figure 3C$_1$) which are, from outside to inside:

1. The a-layer, thin and rich in Patag-reactive polysaccharids[31] with a gelified external fuzzy coat.
2. The b-layer, very thin without polysaccharids and difficult to discern.
3. The c-layer, the main part of the wall (about 0.5 μm), underlayered and slightly Patag-reactive.[31]
4. The d-layer, thinner than the c-layer (i.e., 0.1 to 0.2 μm), Patag-reactive,[31] undulated and stratified, and lacking at the top.

There is a thin electron-transparent periplasmic space between the wall and the epiplasm. The d-layer is considered as the internal part of the endoascus.[32] Consequently, the c-layer is the external part of the endoascus, and the a- and b-layers forming together the exoascus. However, some authors consider, on a functional basis, that the c-layer is part of the exoascus.[33]

At the end of this stage, the ascus reaches nearly its final length (Figure 3B). Lipid droplets, above the nucleus, are less crowded and reach the apex. Piling up of reticulum no longer exists. Reticular elements are irregularly dispersed in a specialized area without organelles, but near myelin-like structures. Other reticular elements lie more or less longitudinally. Microvesicles are numerous and fuse with the plasmalemma in relation with wall growth and differentiation. The enlarged nucleus is still median. At this level, there are clusters of fibrils close to both the wall and the nucleus; their role is unknown. Underneath, lipids are less numerous and no longer reach the ascus base; they are intermingled with vacuoles. The lateral wall slightly thickens and the d-layer is more apparent. At the top, a more prominent thickening of the wall with a large external Patag-reactive cap indicates the beginning of differentiation of the apical apparatus.

2. Apical Apparatus Formation

In lichens, as in others Ascomycetes, the apical apparatus (Figures 3B and 3C) at the ascus apex is a thickening of the wall in which various differentiations occur. In *Caloplaca marina* this apparatus consists of the amyloid cap (which is an amyloid differentiation of the a- and b-layers and part of the c-layer) and a subapical thickening of the d-layer with different modifications. The latter results from apposition of different wall materials discharged from two different kinds of cytoplasmic vesicles. Glycogen granules, hardly to be seen earlier, are now more abundant in the ascus, and vacuoles are distinct above the nucleus.

3. Ascosporogenesis

The formation of an ascus vesicle (Figure 3B) indicates in *Caloplaca marina*, as in other Ascomycetes, the beginning of ascosporogenesis. This vesicle, which develops close to the plasmalemma, is a cylinder open at both ends whose wall consists of two trilaminar membranes. It forms about when the division of the fusion nucleus has started, while the differentiation of the apical apparatus begins. Its origin is still unclear.

The delimitation of the ascospores (Figure 4A) results from indentation and then fragmentation of the ascus vesicle around each of the eight nuclei. The different stages of the ascospore development are easily defined according to the structure of the wall:

1. The primordial stage (Figure 4B): the space between the two layers of the delimiting membrane is very thin.
2. The primary stage (Figure 4C): a homogenous electron-transparent material is deposited between the two lamellae. At the beginning, the spore wall is thin and undulated, then it becomes thicker and rigid.

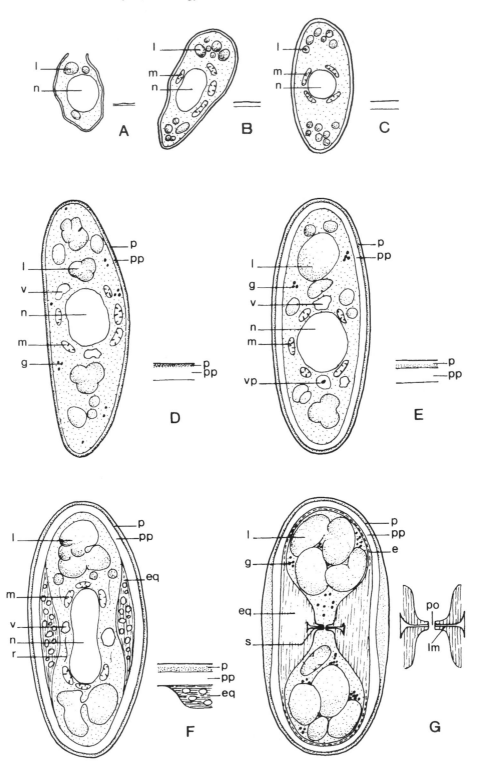

FIGURE 4.   Ascosporogenesis in *Caloplaca marina* (see text). e = endosporal layer; eq = equatorial thickening; g = glycogen; l = lipid; lm = clear median plate; m = mitochondria; n = nucleus; p = perispore; po = pore; pp = proper wall; r = reticular elements; s = septum; v = vesicle; vp = vacuolar precipitate.

3. The secondary stage (Figures 4D and 4E): the spore wall is now bilayered with an internal proper wall,[34] generally electron transparent, and an external electron-dense perispore.

4. The tertiary stage (Figures 4F and 4G): the above stages have been found similar in all the lichens studied. At the tertiary stage, special features appear according to the species. In *Caloplaca marina,* an equatorial thickening forms progressively by deposition of new material between the proper wall and the plasmalemma. The structure of this new layer is clearly different from that of the proper wall, therefore, it is named endospore. Later on, in old spores, a septum may develop at the level of the equatorial thickening.

During the development of the spores, important modifications occur in the sporoplasm (Figure 4). Up to the end of the primary stage, clusters of lipid droplets are observed at the poles, while the large median nucleus is surrounded with mitochondria. Then, the lipid droplets fuse and some small vacuoles and glycogen granules appear. When the endosporal thickening is well developed, the nucleus elongates and divides and each of the resulting nuclei reaches a polar locule. Glycogen is now more abundant and often close to the lipid droplets.

4. Ascus Dehiscence

Then all the layers of the apical apparatus lengthen upwards (Figure 8A) and reach the hymenium surface. The ascus tip ruptures and the ascospores are ejected.

C. Variations in Ascus Development

Many asci in lichens differ from those of *Caloplaca.* The origin of this diversity is better understood by a comparative study of the development of their structural elements.

1. Ascus Cytoplasm (Ascoplasm)

The nature of the storage products of asci (glycogen, lipids, vacuolar precipitates) differs according to the species.[10] Furthermore, in a given species, their quantity varies in the course of the development. The amount of glycogen is sometimes very small and can be overlooked. Different types of microvesicles are always present and are related with the wall modifications. A "Spitzenkörper" probably exists in the young elongating ascus. Growing asci are known to be remarkable for the abundance of endoplasmic reticulum. Concentric bodies (see Chapter VIII.B) are quite infrequent in asci and have only been mentioned in abortive asci.

2. Ascus Wall

An ascus wall structure of four layers, similar to that described in *Caloplaca marina,* exists in all lichens. A fuzzy coat is usually present; it is frequently amyloid, sometimes only partially. The variations of a- and b-layers are only slight. The c-layer generally has three underlayers; its thickness varies in different proportions according to the species, but stays more or less constant all along the ascus. The d-layer, which forms rather late, is generally well developed, often finely stratified, and differs from the c-layer by its structure and texture. In some cases, it is of a more or less uniform thickness; more frequently, it thickens only at the apical region where it forms the main part of the apical apparatus.

3. Apical Apparatus

The apical apparatus begins to develop when the ascus is at its full length and meiosis starts. It forms by apposition of various materials discharged from numerous epiplasmic microvesicles at the internal face of the d-layer. In the resulting thickening, infrastructural modifications appear in different regions. Usually, differentiation is optimal when ascospores are still young, at the stage between the primary to the secondary stage of their development. The resulting apices vary greatly according to the species.[9,10] Variations mostly occur in the d-layer of the ascus wall but not exclusively. These variations serve as characteristics used in systematics, for instance, genera in the Lecanoraceae are better defined according to the apical structure of asci[35] (see Chapter X).

Taking as a reference the apices of young *Rhizoplaca* and *Physcia* asci, several typological trends (described below) can be recognized, as shown in Figure 5.

In the young ascus of *Rhizoplaca*, the d-layer, which is only thickened laterally, delimits a large ocular chamber.

In the *Teloschistes* type (for example, *Caloplaca marina*), the ascus apex at the mature stage is very similar to that of the *Rhizoplaca* young ascus, but it is more differentiated, with an external cap and an internal noticeable polysaccharidic differentiation of the apical thickening.

The *Catillaria* type is characterized by a strong homogenous and Patag-reactive[31] thickening of the d-layer, but an ocular chamber is lacking.

In the *Lecanora* type and its variants (Figure 5, upper line), the apical thickening is characterized by an internal, Patag-negative region — the "axial body" — surrounding a partially developed ocular chamber. From *Lecidella* to *Lecanora* and *Candelariella*, the axial body becomes more and more prominent. In *Physcia*,[36] the differentiating apex is at first like the young *Rhizoplaca* (but the d-layer is thick above the ocular chamber), then it is like the *Lecidella* type, and, finally, at the mature stage it belongs to the *Lecanora* type.

In the *Psora* trend, the ocular chamber is reduced. A polyssacharidic differentiation develops in the apical thickening around the axial body. It protrudes downwards into the epiplasm forming a "pendant" in *Lecidea*. With the development of the axial body, this differentiation is reduced to a tube in *Psora*. In the *Collema* variant, the apical part of the thickening is relatively reduced, but the subapical part becomes well developed.

In the bitunicate asci (*Opegrapha*), the d-layer is thick not only at the apex but nearly all along the ascus wall. It is finely stratified and subdivided into two underlayers (d_1 and d_2), the inner one having frequently an accordion-like appearance.[37] It shows no polysaccharidic differentiations. It must be noticed that the term "bitunicate" is used here in a purely structural sense without any reference to the dehiscence. *Rhizocarpon* slightly differs by the presence of a small polysaccharidic differentiation at the extreme apex of the d_1-sublayer.

Peltigera type is somewhat intermediate between the bitunicate and *Collema* types. The d-layer is thick, polysaccharidic, but not sublayered. It lacks an apical thickening, axial body, and surrounding polysaccharidic tube; however, a well-developed and strongly Patag-reactive[31] pendant is present. Laterally, the d-layer forms a subapical thickening. This is longer than in *Collema* and extends downwards along about the upper fifth of the ascus length.

In *Phlyctis*, the c- and d-layers are present all along the ascus wall, but they are both thin, thus having the usual characters of very young asci. This may be either a primitive or a regressive trait.

4. Ascosporogenesis and Ascospores

The ascus vesicle has been observed in all the lichens studied with the transmission electron microscope.[10] Its origin is still in question. The spore delimitation generally occurs as in *Caloplaca*. Cases where ascospores are less or more than eight have been rarely studied.[26] At the primary stage, differences in spore shape are already visible. At the secondary stage, differences in sporoplasm have been insufficiently studied, but the diversity of the spore wall is better known. Concerning the proper wall, the differentiation of an endospore is not common and rarely reaches the importance observed in *Caloplaca*. The perispore is rather

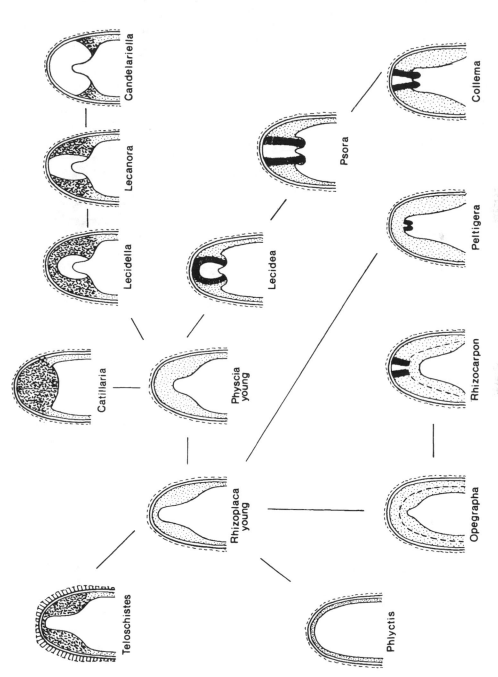

FIGURE 5.   Different types of ascus apical apparati (see text).

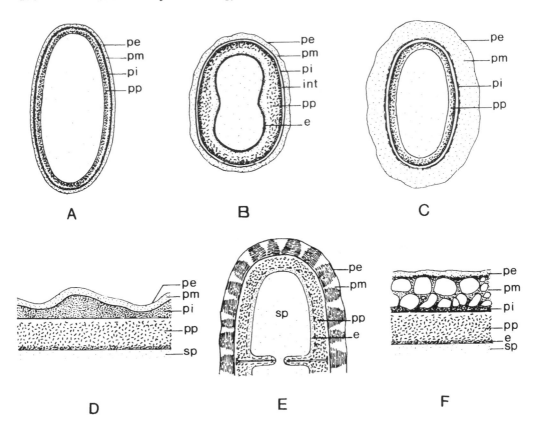

FIGURE 6.   Various perispore developments in different spore types (see text). pe = perispore; pm = middle perispore layer; pi = internal perispore layer; pp = proper wall; int = intermediary wall; e = endosporal layer; sp = sporoplasm.

variable. It is either reduced, as in *Lecanora* (Figure 6A)[9] and *Orphniospora* (Figure 6B), or well developed, especially in halonate spores where the perispore is swollen and more or less gelified, as in *Porpidia* (Figure 6C). It may be of uniform thickness (i.e., *Lecanora*) or irregular, as in *Pannaria* (Figure 6D). The perispore is usually subdivided into three layers; the middle and the internal ones are generally more Patag-reactive.[31] The characteristics of the middle layer play a prominent part in spore diversity (i.e., *Opegrapha*, Figure 6F). Spore ornamentation usually results from its heterogenity (i.e., *Calicium salicenum*,[38] Figure 6E).

Septation — Most of the common lichens (*Lecanora, Lecidea, Parmelia, Cladonia*) have unicellular spores. Pluricellular, either uniseptate (*Ramalina, Physcia*), pluriseptate (*Graphis*), or muriform (*Rhizocarpon*) spores are also frequent. Whatever the final septation type is, the first septum is always transverse. All the septa, the first and following ones, form similarly. The very young septum is a thin annular protrusion at the internal face of the wall. It grows centripetally and is progressively covered by new wall materials, as the rest of the spore wall. A median plate, thin and clear, nonpolysaccharidic, more or less swollen at its basis (ampulla-like), remains distinct in the septum in the course of its development. In the fully developed septa, a central pore always persists. According to the genera, septa are more or less thick and are pigmented or not. The lenticular appearance of cells in spores of the *Graphis* type is due to an enlargement of the septal base at the ampulla level, where no special increase of wall deposit is noticeable. The time of septum formation differs with genera. As previously seen, it develops late in *Caloplaca*, after the endospore equatorial

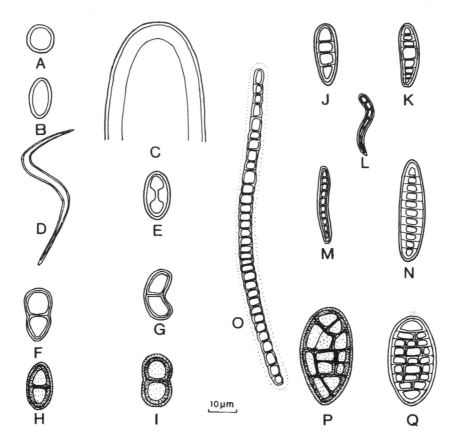

FIGURE 7. Different types of mature spores. (A) *Sphaerophorus* sp.; (B) *Tephromela atra;* (C) *Pertusaria corallina*; (D) *Sarrameana paradoxa*; (E) *Caloplaca marina*; (F) *Arthonia didyma*; (G) *Ramalina fraxinea*; (H) *Diploicia canescens*; (I) *Tholurna* sp.; (J) *Opegrapha atra*; (K) *Opegrapha vulgata*; (L) *Scoliciosporum umbrinum*; (M) *Bacidia muscorum*; (N) *Graphis scripta*; (O) *Conotrema urceolatum*; (P) *Rhizocarpon geographicum*; (Q) *Graphina anguina*.

deposit, whereas in *Physcia*, it is formed earlier before the endospore is distinct. Despite this difference in development, morphologically, mature bicellular spores can be similar in both genera. Rarely, in aging spores, cells may separate along the median plate of the septum as in *Tylophorella*. Sometimes, splitting is incomplete and occurs only at the external wall layers and a distinct furrow results as in some Caliciales.

Pigmentation — Pigmented spores are frequent in lichens. Pigmentation usually appears when the spore wall is already well differentiated and sometimes later. In some species, pigmentation is not uniform, such as colored bands in some *Rinodina* species.[39] Ultrastructural studies have shown that pigmentation results from the formation of granules containing micrograins of pigment, probably melanines. The granules may first appear in undefined zones. Then, they fuse and form a continuous sheath which is independent of the preexisting structural layers of the wall.[40]

Mature spores — There is a great diversity of mature spores in lichens, as shown in Figure 7. There are differences in dimension, shape, septation, and pigmentation. Spore content was rarely studied. However, it could provide distinctive characteristics, for instance, the features of lipids may greatly differ according to the species. The presence of germ pores, frequently used as a systematic criterion in nonlichenized Ascomycetes, has rarely been mentioned in lichens, perhaps because of their small size, such as in *Rhizocarpon*.[41] Germinating spores were often observed in asci. With the aid of electron microscopy, it has

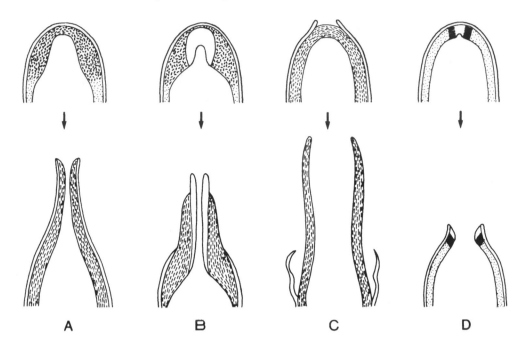

FIGURE 8.   Ascus dehiscence. (A) *Teloschistes* type; (B) rostrum type; (C) fissitunicate type; (D) eversion type.

been established that the germ tube wall originates from the periplasmic space of the spores. Ascosporal conidia are rare in lichens and are known only in species with a single muriform spore per ascus.[6,42,43]

5. Ascus Dehiscence

Several dehiscence types are known in lichens. Some occur only in lichens, such as the *Teloschistes* and rostrum types, whereas some types, common in nonlichenized Ascomycetes, have not been observed in lichens, such as the operculate type. Recent improvement of our knowledge on dehiscence stems from electron microscope studies.[9,10]

In the *Teloschistes* type (*Caloplaca marina*, Figure 8A), all the wall layers lengthen equally and reach the hymenial surface where the wall ruptures. The *Teloschistes* type has been observed in all the Teloschistaceae species and the related genus *Fuscidea*. It is also known in the genus *Letrouitia*[44] and seems to be present also in the genus *Pertusaria*.

The rostrum type (Figure 8B) is very common in lichens (e.g., Lecanoraceae, Parmeli-aceae, Physciaceae, Cladoniaceae). First, the external layers a-, b-, and c- rupture at the top. Then the apical thickening (d-layer) lengthens into a rostrum which reaches the top of the hymenium where it breaks out. Concomitantly, the axial body also lengthens and gelifies, allowing the spores to be ejected.

The fissitunicate (jack-in-the-box) type (Figure 8C) is regularly found in the asci of the bitunicate type defined above. First, as in the rostrum type, there is a rupture of the outer layers of the wall (a-, b-, c-). Then there is a lengthening of the inner spore wall not only at the top of the ascus but also of a large portion of the lateral wall. Moreover, the evaginating part not only reaches the hymenium surface but often emerges above it. During the fissi-tunicate dehiscence, the d-layer is strongly modified. Nearly all the pyrenolichens, the Pyrenulales, as well as the Verrucariales, have fissitunicate asci.[45] This type is also common in discolichens, such as the Arthoniales and the Opegraphales. The dehiscence in *Rhizocarpon* asci differs slightly, particularly by a reduced evagination. In *Peltigera*, with an amyloid pendant of the apex, dehiscence is similar to the fissitunicate type, but elongation only occurs at the level of the subapical thickening.

Other types of dehiscence are rare in lichens:

1. The eversion type (Figure 8D), common in nonlichenized Ascomycetes, is rare in lichens (*Saccomorpha*). In this type there is a very reduced elongation of the whole wall at the top, then an opening by a pore and a slight eversion of the apical thickening.

2. In the Caliciales type, the ascospores become free as the result of the disintegration of the ascus wall and a "mazaedium" is formed from spores and remnants of paraphyses.

III. ASCOMATA

Ascomata are found on the upper surface of the thallus. In some exceptional cases, such as in some *Rhizocarpon* species, the apothecia are directly connected to the fungal hypothallus. They are either scattered all over the thallus, concentrated in the center of the thallus, or, as in *Cetraria* and *Peltigera*, marginally located. In *Nephroma*, they are at first upright marginal and finally turned downwards. In *Cladonia*, *Stereocaulon*, *Baeomyces*, and some other genera, they are formed on special outgrowths of the thallus which are called podetia.

There are two major types of ascomata:

1. Apothecia, with an open and usually circular hymenial surface, often about 1 mm in diameter, sometimes smaller or larger. They are brightly colored or dark, prominent on the thallus or immersed. Lirellae, which are elongated, Y- or star-shaped, and have a longitudinal split as an opening, are an exception.

2. Perithecia, which appear as black dots on the thallus, but are in fact flask-shaped, immersed fruiting bodies with an apical pore as an opening.

Ascomata with intermediate characters, frequently named "perithecioid apothecia", are known in *Lichina*, some Pertusariaceae, and Thelotremaceae. The components of the ascomata are the hymenium, the hypothecium, and the excipulum.

1. The hymenium consists of asci and interascal filaments of different ontogenetical value (and part of the hamathecium[46]). When the apical part of the hymenium appears differentiated, it is known as epithecium.

2. The hypothecium is the part beneath the hymenium which contains the sporophytic apparatus. Often a distinct layer, the subhymenium, containing the ascogenous hyphae, exists at its top below the hymenium.

3. The excipulum, which envelops the hymenium and the hypothecium is devoid of sporophytic elements. It is either thallus-like in color and structure or entirely fungal, dark and bright colored. Ontogenetical terms as "amphithecium", "thalline margin", "parathecium", or "excipulum proprium" are frequently applied to define parts of the excipulum.

A. Ascoma Development: Two Examples

The development of ascoma is diverse and complex. Only two examples, apothecial and perithecial, are described in detail in the following sections.

1. Diploicia (Buellia) canescens[23] (Figure 9I)

The first sign of ascoma development (Figure 9A) is the appearance of knobs of ascogonial filaments scattered in the thallus at the lower part of the algal layer, then a net-like structure, the primordium, is formed around them (Figure 9B). The primordium enlarges and converts into a differentiated "ébauche"* (primary corpus) (Figure 9C). Its central part, around the fertile elements, is the carpocentrum. It is surrounded by the envelope whose basal part becomes a thin floor, whereas the upper part becomes the roof, made of coherent, thick-walled hyphae through which numerous trichogynes emerge to the upper thallus surface.

At the next stage (Figure 9D), the ébauche broadens and the carpocentrum shows two

* "Ébauche" is a french word for a sculpture (or any work) at its beginning when its shape is still rough, unachieved.

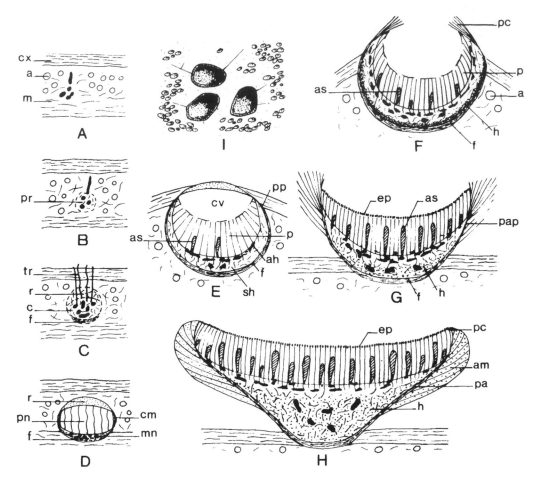

FIGURE 9. Ascoma development in *Diploicia (Buellia) canescens* (see text). a = algae; ah = ascogenous hyphae; am = amphithecium; as = asci; c = carpocentrum; cm = circumcentral muff; cv = cavity of the ascoma; cx = cortex; ep = epithecium; f = floor; h = hypothecium; m = medulla; mn = meniscus; p = paraphyses; pa = parathecium; pap = parathecial apparatus; pc = parathecial crown; pn = paraphysoid net; pp = proparathecium; pr = primordium; r = roof; sh = subhymenium; tr = trichogyne.

superposed regions. The thin basal one, or basal meniscus, keeps a network structure and contains the sporophytic apparatus. In the thick upper region, the meshes of the hyphal net (the paraphysoid net) are more or less stretched vertically. The envelope structure is hardly modified; around the paraphysoid net, it is completed by a muff (circumcentral muff) whose parallel hyphae are tightened and subvertical.

Later (Figure 9E), the carpocentrum becomes modified. At first, the filaments of the paraphysoid net rupture and the ascomatal cavity is formed. Then new filaments, the paraphyses, arise from the basal meniscus; they are arranged parallel in a well-defined layer. Among their bases, ascogenous hyphae are distinct. Moreover, a proparathecium, extension of the margins of the meniscus and the floor, develops laterally around the carpocentral region. It produces new paraphyses which add laterally to the carpocentral hymenium; ascogenous hyphae develop at their bases.

At the next stage, the roof ruptures under the thallus cortex (Figure 9F) and at the margin of the opening a parathecial crown organizes. It is formed by short diverging hyphae whose disposition results from the initiation of a sympodial growth. The hypothecium thickens by adjunction of the bases of the paraphyses to its upper part, whereas its basal part turns brown.

Along the sides of the ascoma, the crown develops basipetally and generates a parathecial apparatus (Figure 9G). The internal branches are new paraphyses (parathecial paraphyses) and the external ones are filaments with brown walls. Those add laterally to the preexisting basal excipulum, and the final structure of the apothecium is accomplished (Figure 9H). The hymenium now contains numerous asci with mature ascospores. The differentiated apex of the paraphyses form a brown epithecium. The pigmentation of the hypothecium progresses upwards and now reaches its secondary part formed by the paraphyses bases.

2. *Verrucaria cazzae*[45] *(Figure 10F)*

A small net-form primordium, located near the basis of the algal layer and lacking ascogonial elements, is the first stage observed in the course of the development (Figure 10A). Then, the primordium enlarges (Figure 10B). At its top, an epicentral cone is generated. Trichogynes originating from the ascogonial apparatus, now present in the lower part of the ébauche, the carpocentrum, grow through this cone.

The broadened and cork-shaped ébauche (Figure 10C) grows both downwards and upwards without reaching the thallus surface. In the modified carpocentrum, three parts become distinct. The lower part (basal meniscus), contains the fertile elements. The median part is an ascomatal cavity which is covered by the upper part of the carpocentrum. This is bell-shaped (subhymenial bell) and its margin is connected to that of the meniscus. The lower face of this bell generates filaments with free tips downwards (descending filaments), in a basipetal process. Its upper part develops into an apical point which penetrates the axial part of the enlarged epicentral cone.

By subsequent growth (Figure 10D), the top of the ascoma reaches the thallus surface while, below its base, a thin floor differentiates from the thallus. The structure of the subhymenial meniscus remains unchanged while young asci arise from it in the enlarged ascomatal cavity. The subhymenial bell ruptures at its axial part as well as the lower part of the overtopping apical point, so the ostiolar canal, still closed at the top, is forming. Along this canal, periphyses are produced by a distinct clear sheath of parallel hyphae rearranged from the apical point. Those periphyses follow the numerous descending filaments upwards. The epicentral cone, laterally pushed and curved, becomes pigmented, and gives rise to an involucrum.

The enlarged ascoma grows down into the thallus because a lateral muff, devoid of lateral filaments, develops between the subhymenial meniscus and the subhymenial bell; consequently, the descending filaments and the periphyses get located at the upper part of the ascoma. The ostiolar canal opens up at its top where no periphyses develop. The adult ascoma, where mature asci are numerous, is a perithecium. The external and basal parts of its wall which is of thalline origin, has become dark (Figure 10E). Descending filaments become less conspicuous and even disappear in old ascomata.

B. Comparative Remarks and Conclusion

In the above mentioned examples, three structural stages can be identified: those of the primordium, the ébauche, and the parathecial apparatus. In both species, the primordium is formed by a net of filaments surrounding the ascogonial apparatus, but, in *Diploicia canescens*, this net develops after the ascogonial apparatus while it is the contrary in *Verrucaria cazzae*.

Then, in both cases, the primordium turns into an ébauche with two different parts, the carpocentrum and the surrounding envelope, which themselves quickly subdivide. So, in *Diploicia*, the carpocentrum is formed by a basal meniscus, which generates paraphyses, and by a paraphysoid net which finally ruptures giving placc to the ascomatal cavity. In *Verrucaria*, the meniscus does not produce any paraphyses, the paraphysoid net is missing, and the ascomatal cavity appears early. Moreover, a subhymenial bell, not present in *Di-*

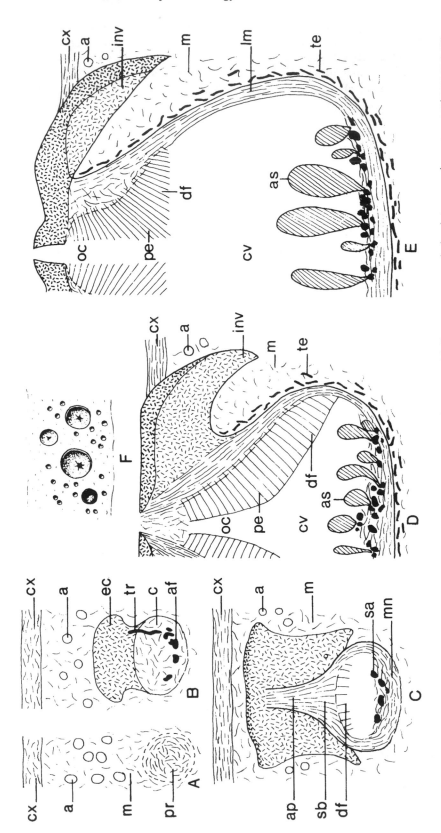

FIGURE 10.   Ascomata development in *Verrucaria cazzae* (see text).   a = algae; af = ascogonial filaments;   ap = apical point; as = asci; c = carpocentrum; cv = cavity of the ascomata; cx = cortex; df = descending filaments; ec = epicentral cone; inv = involucrum; lm = lateral muff; m = medulla; mn = meniscus; oc = ostiolar canal; pe = periphyses; pr = primordium; sa = sporophytic apparatus; sb = subhymenial bell; te = thalline envelope; tr = trichogyne.

ploicia, exists here and generates descending filaments. In both species, the envelope is composed by a floor and a roof. In *Diploicia,* the floor is thin at first, then becomes thicker and turns brown while the roof, though relatively thick, disappears early. In *Verrucaria,* the roof develops considerably; it turns brown and persists at the mature stage. The floor is at first relatively thin, but later it enlarges by adjunction of thalline filaments.

The parathecial apparatus, at the third structural stage, is a new element with a sympodial growth added to the ébauche. It is only present in *Diploicia, Verrucaria* being devoid of any parathecial formations.

With these two examples, it is clear that ascomata are formed from several structural elements and that each of them develops diversely. Moreover, a functional diversity interferes with this structural complexity. So the two studied examples show that the growth characteristics of the whole ascoma and of each element differ and that the opening modalities are not the same.

In both studied species, the primordium enlargement is reduced but equal in all directions (that is not the case in lichens with lirellae[3,47]). The widening of the ébauche also is roughly equivalent, but its modalities differ. In *Verrucaria,* there is a long and important development of the roof and a vertical growth due to the muff, while in *Diploicia,* there is a centrifugal extension of the floor and the meniscus resulting in the formation of a proparathecium, and then the elongation of the parathecial elements provokes an important enlargement.

The opening of the ascoma appears early in *Diploicia* where it widens as the hymenium enlarges (apothecial type of ascoma). In *Verrucaria,* it forms late and remains narrow (perithecial type), with concomitant periphyses development.

The comparison of the development of *Diploicia* and *Verrucaria* shows how functional differences of diverse structural components can give rise to a large diversity in the development of ascomata in lichens. As in nonlichenized Ascomycetes,[1,62] various ontogenetical types have been described in lichens.[1-6,12,23,45-61] Their systematical significance can be in accordance with data obtained from ascus studies as in *Ramalina,*[56] but sometimes they may differ (for instance *Parmelia* compared with *Lecanora*[4,35,63]). Many genera still need to be investigated.

REFERENCES

1. **Chadefaud, M.,** Végétaux non vasculaires (crytogamie), in *Traité de Botanique Systématique,* Vol. 1, Chadefaud, M. and Emberger, L., Eds., Masson, Paris, 1960, 429.
2. **Letrouit-Galinou, M. A.,** The apothecia of the Discolichens, *Bryologist,* 71, 297, 1968.
3. **Letrouit-Galinou, M. A.,** Sexual reproduction, in *The Lichens,* Ahmadjian, V. and Hale, M. E., Eds., Academic Press, New York, 1973, chap. 2.
4. **Henssen, A. and Jahns, H. M.,** *Lichenes, eine Einführung in die Flechtenkunde,* Georg Thieme Verlag, Stuttgart, 1974.
5. **Henssen, A.,** Studies in the developmental morphology of lichenized Ascomycetes, in *Lichenology: Progress and Problems,* Brown, D. H., Hawksworth, D. L., and Bailey, R. H., Eds., Academic Press, New York, 107, 1976.
6. **Henssen, A.,** The Lecanoralean centrum, in *Ascomycete Systematics,* Don Reynolds, R., Ed., Springer-Verlag, Berlin, 1981, chap. 10.
7. **Letrouit-Galinou, M. A.,** Les asques des Lichens et le type archaeascé, *Bryologist,* 30, 76, 1973.
8. **Bellemère, A. and Letrouit-Galinou, M. A.,** The Lecanoralean ascus. An ultrastructural preliminary study, in *Ascomycete Systematics,* Don Reynolds, R., Ed., Springer-Verlag, Berlin, 1981, chap. 5.
9. **Bellemère, A. and Letrouit-Galinou, M. A.,** Development and differentiation of lichen asci including dehiscence and sporogenesis, in *Progress and Problems in Lichenology in the Eighties,* Peveling, E., Ed., J. Cramer, Vaduz, 1987, 137.
10. **Honegger, R.,** Ascus structure and function, ascospore delimitation and phycobiont cell wall types associated with the Lecanorales (lichenised Ascomycetes), *J. Hattori Bot. Lab.,* 52, 417, 1982.

11. **Micheli, P. A.,** *Nova Plantarium Genera*, 1729.

12. **Moreau, F. and Moreau, F.,** Recherches sur quelques Lichens des genres *Parmelia, Physcia* et *Anaptychia*, *Rev. Gen. Bot.*, 37, 385, 1925.

13. **Fuisting, W.,** Beiträge zur Entwicklungsgeschichte der Lichenen, *Ostmaerk. Bot. Z.*, 26, 641, 1868.

14. **Stahl, E.,** *Beiträge zur Entwicklungsgeschichte der Flechten*, Felix, Leipzig, 1877.

15. **Nylander, W.,** *Synopsis Methodica Lichenum Omnum Hucusque Cognitorum*, Vol. 1, Martinet, Paris, 1858—1860.

16. **Moreau, F. and Moreau, F.,** Les phénomènes cytologiques de la reproduction chez les Champignons des Lichens, *Botaniste*, 20, 1, 1928.

17. **Jahns, H. M.,** The trichogynes of *Pilophorus strumaticus, Bryologist*, 76, 414, 1973.

18. **Jahns, H. M., Herold, K., and Beltman, H. A.,** Chronological sequence, synchronization and induction of the development of fruit bodies in *Cladonia furcata* var. *furcata* (Huds.) Schrad., *Nova Hedwigia*, 30, 427, 1978.

19. **Honegger, R.,** Scanning electron microscopy of the contact site of conidia and trichogynes in *Cladonia furcata, Lichenologist*, 11, 16, 1984.

20. **Boissiere, M. C.,** Cytologie du *Peltigera canina* (L.) Willd, en microscopie électronique. I. Premières observations, *Rev. Gen. Bot.*, 79, 167, 1972.

21. **Bellemère, A. and Letrouit-Galinou, M. A.,** Le dévelopment des asques et ascospores chez le *Caloplaca marina* Wedd. et chez quelques Lichens de la famille des Téloschistaceae *(Caloplaca, Fulgensia, Xanthoria)*: étude ultranstructurale, *Crytog. Bryol. Lichenol.*, 3, 95, 1982.

22. **Rudolph, E. D. and Giesy, R. M.,** Electron microscope study of lichen reproductive structures in *Physcia aipolia, Mycologia*, 58, 786, 1966.

23. **Letrouit-Galinou, M. A.,** Recherches sur l'ontogénie et l'anatomie comparées des apothécies de quelques Discolichens, *Rev. Bryol. Lichenol.*, 34, 413, 1966.

24. **Mayne, F.,** Anatomie et morphologie comparées des asques de quelques lichens, *Rev. Bryol. Lichenol.*, 15, 203, 1946.

25. **Chadefaud, M., Letrouit-Galinou, M. A., and Janex-Favre, M. C.,** Sur l'origine phylogénétique et l'évolution des Ascomycètes des Lichens, Colloque sur les Lichens, *Bull. Soc. Bot. Fr.*, 79, 1968.

26. **Erbisch, F. H.,** Ascus and ascospore development of five species of the lichen-forming genus *Pertusaria, Bryologist*, 72, 178, 1969.

27. **Schmidt, A.,** Ascus Typen in der Familie Caliciaceae, *Ber. Dtsch. Bot. Ges. (Neue Folge)*, 4, 127, 1970.

28. **Maire, R.,** Recherches cytologiques sur quelques Ascomycètes, *Ann. Mycol. Berlin*, 3, 123, 1905.

29. **Moreau, F. and Moreau, F.,** Recherches sur les Lichens de la famille des Peltigéracées, *Ann. Sci. Natur.: Bot. Biol. Veg.*, 1, 29, 1919.

30. **Stevens, R. B.,** Morphology and ontogeny of *Dermatocarpon aquaticum, Am. J. Bot.*, 28, 59, 1941.

31. **Thiery, J. P.,** Mise en évidence des polysaccharides sur coupes fines en microscopie electronique, *J. Microsc. (Oxford)*, 6, 987, 1967.

32. **Chadefaud, M., Letrouit-Galinou, M. A., and Janex-Favre, M. C.,** Sur l'évolution des asques et du type archaeascé chez les Discolichens de l'ordre des Lécanorales, *C. R. Acad. Sci.*, 257, 4003, 1963.

33. **Parguey-Leduc, A., and Janex-Favre, M. C.,** La paroi des asques chez les Pyrénomycètes: étude ultra-structural. I. Les asques bituniques typiqués, *Can. J. Bot.*, 60, 1222, 1982.

34. **Bellemère, A., Melendex-Howell, L., Nicholas, A., and Rossignol, J. L.,** Etude ultrastructurale comparative due développement des ascospores chez la lignée sauvage et chex des mutants à ascospores "ceinturées" ou "albinos" de l' *Ascobolus immersus* Pers. ex Fr., *Cryptog. Mucol.*, 2, 299, 1981.

35. **Hafellner, J.,** Studie in Richtung einer natürlischen Gliederung der Sammel-Familien Lecanoraceae and Lecideaceae, *Beiträge zur Lichenologie, Festschrift J. Poelt (Beiheft 79 sur Nova Hedwigia)*, Hertel, H. and Oberwinkler, F., Eds., J. Cramer, Vaduz, 1984, 241—371.

36. **Honegger, R.,** The ascus apex in lichenized fungi. I. The *Lecanora-, Peltigera-* and *Teloschistes-* types, *Lichenologist*, 10, 47, 1978.

37. **Reynolds, D. R.,** Wall structure of a bitunicate ascus, *Planta*, 98, 244, 1971.

38. **Tibbel, L.,** The Caliciales of boreal North America, *Symb. Bot. Ups.*, 21, 1—128, 1975.

39. **Mayrhofer, H.,** Ascosporen and Evolution der Flechtenfamilie Physciaceae, *J. Hattori Bot. Lab.*, 52, 313, 1982.

40. **Bellemère, A. and Hafellner, J.,** L'appareil apical des asques et la paroi des ascospores du *Catolechia wahlenbergii* (Ach.) Flotow ex Koerber et de l'*Epilichen scabrosus* (Ach.) Clem. ex Haf. (Lichens, Lécanorales): étude ultrastructurale, *Cryptog. Bryol. Lichenol.*, 1, 4, 1983.

41. **Honegger, R.,** The ascus apex in lichenized Fungi. II. The *Rhizocarpon* -type, *Lichenologist*, 12, 157, 1980.

42. **Henssen, A.,** A corticolous species of *Gyalectidium* from Costa Rica, *Lichenologist*, 13, 155, 1981.

43. **Hafellner, J. and Bellemère, A.,** Über die Bildung phialidischer Konidien in den mauerförmigen, einzeln im Ascus liegenden Sporen von *Brigantiaea leucoxantha* (lichenizierte Ascomycetes, Lecanorales), *Nova Hedwigia*, 38, 169, 1983.

44. **Hafellner, J. and Bellemère, A.**, Elektronenoptische Untersuchunger an Arten der Flechtengattung *Letrouitia* gen. nov., *Nova Hedwigia*, 35, 263, 1982.

45. **Janex-Favre, M. C.**, Recherches sur l'ontogénie, l'organisation et les asques de quelques Pyrénolichens, *Rev. Bryol. Lichenol.*, 37, 421, 1970.

46. **Eriksson, O.**, The families of bitunicate Ascomycetes, *Opera Bot.*, 60, 1, 1981.

47. **Janex-Favre, M. C.**, Sur les ascocarpes, les asques et al position systématique des lichens du genre *Graphis*, *Rev. Bryol. Lichenol.*, 33, 242, 1964.

48. **Doppelbaur, H. W.**, Studien zur Anatomie und Entwicklungsgeschichte einiger endolithischen pyrenocarpen Flechten, *Planta*, 53, 246, 1959.

49. **Henssen, A.**, Eine revision der Flechtenfamilien Lichinaceae and Ephebeceae, *Symb. Bot. Ups.*, 18, 1, 1963.

50. **Janex-Favre, M. C.**, L'ontogénie et l'organisation des ascocarpes des *Lichina* et la position systématique de ces Lichens, *Bull. Soc. Bot. Fr.*, 114, 145, 1967.

51. **Henssen, A.**, Die Apothecienentwicklung bei *Umbilicaria* Hoffm. emend. Frey, *Ber. Dtsch. Bot. Ges.*, 4, 103, 1970.

52. **Jahns, H. M.**, Untersuchungen zur Entwicklungsgeschichte der Cladoniaceen unter besonderer Berüksichtigung des Podetien-Problems, *Nova Hedwigia*, 20, 1, 1970.

53. **Janex-Favre, M. C.**, L'ontogénie et al structure des apothécies de l' *Umbilicaria cylindrica*, *Rev. Bryol. Lichenol.*, 40, 59, 1974.

54. **Letrouit-Galinou, M. A.**, Le développement des apothécies du *Gyalecta carneolutea* (Turn.) Oliv. (Discolichen, Gyalectacée), *Bull. Soc. Mycol. Fr.*, 90, 23, 1974.

55. **Keuck, G.**, Ontogenetisch-systematische Studie über *Erioderma* im Vergleich mit anderen cyanophilen Flechtengattungen, *Bibliotheca Lichenologica*, J. Cramer, Vaduz, 1977, 6.

56. **Keuck, G.**, Die systematische Stellung der Ramalinaceae, *Ber. Dtsch. Bot. Ges.*, 92, 507, 1980.

57. **Sipman, H. J. M.**, A monograph of the lichen family Megalosporaceae, *Bibliotheca Lichenologica*, J. Cramer, Vaduz, 1983, 18.

58. **Wagner, J.**, Etude du thalle et des périthèces due Pyrénolichen *Endocarpon pusillum* Hedw., *Thèse de Doctorat de Bème cycle*, Cytologie et Morphogénèse Végétales, Univ. P. et M. Curie, Paris, 1984.

59. **Janex-Favre, M. C.**, Développement et structure des apothécies de l' *Aspicilia calcarea* (Discolichen), *Cryptog. Bryol. Lichenol.*, 6, 25, 1985.

60. **Janex-Favre, M. C. and Ibrahim-Ghaleb, M.**, L'ontogénie et la structure des apothécies du *Xanthoria parietina* (L.) Beltr. (Discolichens), *Cryptog. Bryol. Lichenol.*, 7, 457, 1986.

61. **Letrouit-Galinou, M. A. and Bellemère, A.**, Ascomatal development in lichens: a review, *Cryptog. Myc.*, in press.

62. **Bellemère, A.**, Contribution à l'étude du développement de l'apothécie chez les Discomycètes inoperculés, *Bull. Soc. Mycol. Fr.*, 83, 393 and 753, 1967.

63. **Letrouit-Galinou, M. A.**, Les apothécies et les asques du *Parmelia conspersa* (Discolichen, Parméliacée), *Bryologist*, 73, 39, 1970.

Chapter V.B

CONIDIOMATA, CONIDIOGENESIS, AND CONIDIA

David L. Hawksworth

I. INTRODUCTION

Conidium-producing structures, or "conidiomata", occur in perhaps as many as 8,000 (59%) of the 13,500 species of lichen-forming fungi recognized (see Chapter II.A).[1] Despite this frequency, they have received remarkably little critical study, and a few key works from the last century[2-4] remained the only substantial contributions on the topic until 1980.[5]

Conidiomata, and the conidia themselves, show a wide degree of variation in structure, the extent of which is only now beginning to be appreciated. It is important to stress that the terms conidiomata and conidia are descriptive of their morphology and that their use here does not imply that they have any specific biological role.[6]

The role of conidia in lichens has been the subject of much debate and uncertainty for many years.[7,8] In some instances the conidia function as spermatia as a part of the sexual reproductive process of the lichens (Chapter V.A),[7-10] while in others they act as diaspores which can give rise to new lichenized thalli on contact with appropriate photobionts. In this last case, the conidia can often germinate and produce mycelial colonies or even young pycnidial conidiomata in pure culture.[11] However, these roles cannot be assumed to be mutually exclusive as species in which the conidia are able to act in either spermatial or dispersal capacities are known in the nonlichenized fungi.[12] In the majority of species there is no firm evidence as to the biological role of the conidia, and in some instances they may even be functionless relics.[13] In the case of lichenized members of the Deuteromycotina (see Chapter II.A), in which no sexual stage is known, the conidia can be assumed to act as diaspores.

In certain genera more than one type of conidioma or conidia may be produced by a single species.[3] This phenomenon has been particularly well documented in foliicolous lichens, for example *Arthonia macrosperma* and species of *Mazosia, Porina, Raciborskiella,* and *Strigula.*[14] In some *Mazosia* species both types of conidia can arise in the same conidioma.[14] The smaller conidia are often referred to as "microconidia" and the larger as "macroconidia". The production of micro- and macroconidia is not, by any means, confined to foliicolous groups but also occurs in, for instance, *Anisomeridium, Cryptolechia, Lecanactis, Micarea,* and *Opegrapha.* In such cases, the microconidia almost certainly function as spermatia as they are almost invariably thin-walled, bacilliform, or globose, only 0.5 to 1.5 μm wide, and contain so little protoplasm that they must be short-lived; the macroconidia are always much larger with more substantial protoplasm and so the larger reserves to be expected in a diaspore. Several species of *Micarea,* for example *M. denigrata,* have three conidial types which have been termed "micro-", "meso-", and "macroconidia"[15] (Figure 7); here again the microconidia probably serve as spermatia.

Micro- and macroconidia, and also the ascomata, almost always arise on the same thallus, although they often tend to be near the edges of the growing lobes in *Parmelia* subgen. *Amphigymnia,* and on younger branches in *Alectoria* and *Ramalina.* In *Cladonia,* however, they may be restricted to basal squamules or appear at the apices of podetia and persist through all growth stages,[16] whereas in *Ephebe* the ascomata arise from former conidiomata ("pycnoascocarps").[17]

Conidiomata, especially if they have a spermatial function, appear before the onset of ascoma formation and can become less frequent in older thalli, as in *Strigula elegans.*[14,18]

Dioecism, where conidiomata ("spermogonia") occur on one thallus, and trichogynes and, subsequently, ascomata occur on another, appears to occur in a few rare cases, notably *Tylophoron crassiusculum*[19] and *Lecidea verruca*.[20]

II. NOMENCLATURE

In contrast to the nonlichenized fungi, states of lichen-forming fungi characterized by the production of conidia ("anamorphs") cannot be given independent scientific names to states of the same species producing ascomata or basidiomata ("teleomorphs").[21,22] There are a few instances where mycologists have published new names for the conidial states of lichens, for example *Discosiella* for that of *Strigula*,[23] but these are not permitted under the rules of nomenclature currently in force.[21] An extension of the rules for naming such states to all fungi would be logical but there would be considerable difficulties in applying it to lichenized groups.[24]

It has been argued that spermatial states of lichens are not anamorphs, as they do not function as diaspores, and so should not be referred to by that name.[13] However, this criterion is difficult to apply in practice where the role of the conidia is often uncertain and where several types of conidia are produced in different conidiomata in the same species (see above). Spermatial states are, in any case, frequently regarded as anamorphs and sometimes given independent scientific names in the nonlichenized fungi.[25]

III. CONIDIOMATA

Knowledge of the variation of these multihyphal conidium-bearing structures in the lichen-forming fungi is still very imperfect. The conidiomata may be globose or flask-shaped ("pycnidia"), cupuliform ("acervular"), cushion-like ("sporodochia"), hooded or peltate ("campyiidia"), or erect ("synnemata", "hyphophores"). In a few cases, conidia arise directly from single hyphae rather than aggregated hyphae, as in cultures of the mycobiont of *Pertusaria pertusa*[26] and the anamorphs of *Coniocybe furfuracea*,[27] *Mycocalicium schefflerae*,[28] and, perhaps, *Scoliciosporum chlorococcum*.[29]

Globose to flask-shaped conidiomata opening by a single pore are the most commonly occurring type in the lichen-forming fungi; these can also be referred to as "pycnidial", and in the older literature are often termed "spermogonia" when presumed to have a sexual role[30] (Table 1). In the macrolichens, pycnidial condiomata are usually immersed in the thallus with only their brown to black, or rarely white, ostioles visible; they are particularly conspicuous in *Cladonia* where they may be pigmented and exude conidia in a mucilage which is reddish in some species (e.g., *C. ciliata*). In crustose lichens the pycnidial conidiomata are often raised and conspicuous (e.g., *Anisomeridium*, *Arthopyrenia* s.s., *Lecanora*, *Opegrapha*) and in some *Micarea* species (e.g., *M. stipitata*) are stalked.[15] Five types of pycnidial conidiomata have been recognized by recent workers (Figure 1).[1,5] Of these five types, all but the *Lecanactis*-type are clearly pycnidial in form; in that genus a more open disk-like structure, or "acervulum", is produced. This is apparently a rare type of conidioma among the lichenized fungi, but is also seen in the exclusively conidial *Lichingoldia gyalectiformis*.[31] However, so few species have been critically studied that many other types are to be expected. In *Arthopyrenia*, *Eopyrenula*, and *Mycomicrothelia*, the pycnidial conidiomata can be very similar to the ascomata in structure and may be clypeate in some *Mycomicrothelia* species.[32] Pycnidial conidiomata in lichenized fungi are almost exclusively unilocular in structure, exceptions being seen only in the Teloschistales and *Dermatocarpon*[1,5] (Figure 1E).

Badimia, *Loflammia*, *Sporopodium*, and certain other mainly (but not exclusively) folicolous species (Figure 2A) had been regarded as lichenicolous fungi by many authors and

Table 1
**CURRENT DESCRIPTIVE TERMINOLOGY OF
CONIDIOMATA, CONIDIOPHORES,
CONIDIOGENOUS CELLS, AND CONIDIA
COMPARED WITH NOW OBSOLETE TERMS
BASED ON POSTULATED FUNCTIONS**

	Obsolete	
Current	**Asexual function**	**Sexual function**
Pycnidial conidioma	Pycnide	Spermogone
	Pycnidium	Spermogonium
Conidiophores and	Basidium	Fulcrum
conidiogenous cells		Spermatiophore
		Sterigma
Conidiogenesis	Progemmation	Spiculation
Conidia	Conidium	Spermatium
	Pycnidiospore	
	Pycnoconidum	
	Pycnospore	
	Stylospore	

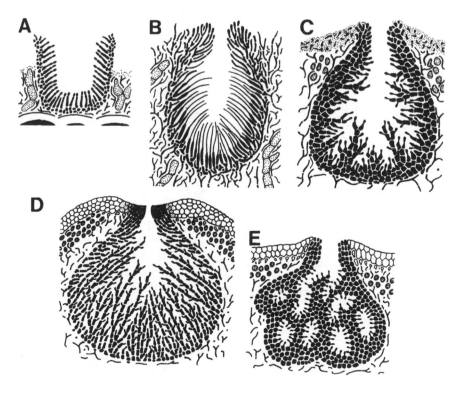

FIGURE 1.   Diagram of vertical sections to illustrate selected types of conidiomata in lichen-forming fungi.[5] (A) *Lecanactis*-type (also in *Arthonia* and *Byssoloma*); (B) *Roccella*-type (in all Roccellaceae); (C) *Umbilicaria*-type (also in *Acroscyphus, Cetraria, Hypogymnia,* and *Parmelia*); (D) *Lobaria*-type; (E) *Xanthoria*-type (also in *Dermatocarpon* and *Endocarpon*).

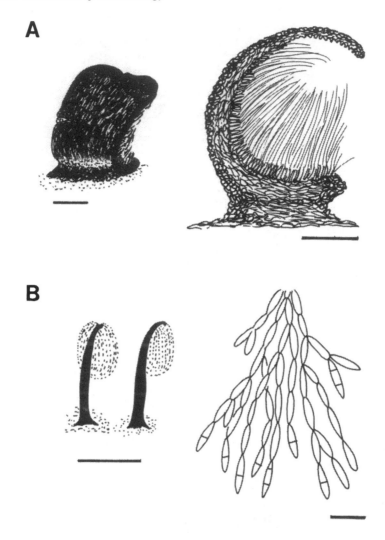

FIGURE 2.   (A) Campylidium of *Pyrenotrichum splitgerberi* (scales = 250 μm); (B) hyphophores in *Aulaxina epiphylla*[35] (left-hand scale = 100 μm; right-hand scale = 10 μm).

assigned to a genus *Pyrenotrichum*.[14] However, these structures, termed "campylidia", have now been conclusively shown to be conidiomata.[33] They originate from ascoma-like initials which become extended upwards into the scale- or hood-like campylidia bearing conidia on their inner surfaces.[33]

A further special type of conidiomata, seen in the family Gomphillaceae (Graphidales), is formed from elongated hyphae to produce a synnematous structure which bears chains of conidia below a somewhat rounded apex (Figure 2B). These structures, termed "hyphophores",[34] are unknown outside this family. The role of conidia produced from hyphophores requires detailed study, but, in at least some species, ball-like agglomerations including photobiont cells are formed;[35] in such cases they can be presumed to act as vegetative propagules.

Sporodochia, conidiomata in which the conidiophores arise from a cushion-like tuft, are seen in two of the monotypic genera of exclusively conidial lichen-forming fungi, *Blarneya hibernica*[36] and *Cheiromycina flabelliformis*,[31,37] and also in the macroconidial phase of *Micarea adnata*.[15]

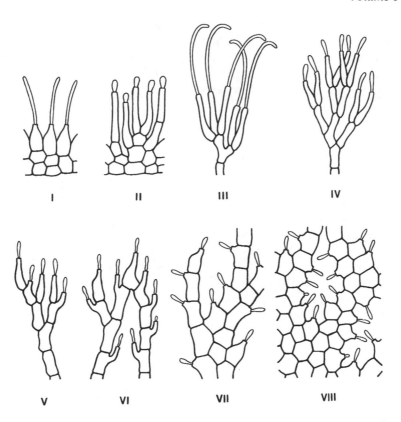

FIGURE 3.   Conidiophores and the arrangement of conidiogenous cells in selected lichen-forming fungi.[5] (I) *Arthonia*; (II) *Byssoloma, Lecanactis, Peltigera,* and *Thermutis*; (III) *Combea, Dirina, Dolichocarpus, Roccella,* and *Roccellina*; (IV) *Cladonia* and *Ramalina*; (V) *Alectoria, Cetraria, Parmelia acetabulum,* and *Omphalodium*; (VI) *Acroscyphus, Hypogymnia, Parmelia* p.p., *Phaeophyscia, Physconia,* and *Umbilicaria*; (VII) *Anaptychia, Lobaria, Nephroma,* and *Psoroma*; (VIII) *Dermatocarpon, Teloschistes,* and *Xanthoria.*

IV. CONIDIOPHORES AND CONIDIOGENOUS CELLS

The hyphae or specialized cells which support the conidiogenous cells are now termed "conidiophores", but were formerly inappropriately termed "basidia", "fulcra", or "spermatiophores" in early lichenological literature according to their presumed function (Table 1). The shape and branching of the conidiophores and of the conidiogenous cells themselves vary considerably in different genera of lichen-forming fungi. Various schemes to clarify these have been proposed by early workers of which the separation between "exobasidial fulcra" (where the conidiogenous cells are terminal with apically produced conidia) and "endobasidial fulcra" (where the conidiogenous cells are intercalary and bear conidia laterally) was widely used.[7] Such a classification oversimplifies the position and can mask important similarities, as conidiogenesis is then not considered. Eight types of arrangement (Figure 3) have been recognized by some recent authors,[1,5] but it must be stressed that such schemes are based on restricted surveys and many other types are to be expected.

Arrangements similar to those of types I to IV (Figure 3) are not dissimilar from types seen in plant pathogenic and saprobic Coelomycetes,[25] but types VII and VIII are scarcely known outside lichenized groups.

Studies of conidiophores and of the conidiogenous cells can be difficult due to their small size. However, the use of biologically active washing powders prior to fixing and critical point drying and examination in the scanning electron microscope (SEM) has recently been found to permit these minute structures to be examined in much more detail than previously possible.[13] Further application of this technique can be expected to increase dramatically our knowledge of these structures in the future.

V. CONIDIOGENESIS

The process by which conidia are produced from conidiogenous cells is termed "conidiogenesis". The variety of types of conidiogenesis in nonlichenized fungi has received detailed attention from mycologists over the last 30 years,[38,39] and the advent of electron microscopy has contributed significantly to its understanding in more recent years.[40] In all lichen-forming fungi so far studied conidiogenesis is blastic, that is, a recognizable conidium initial is formed before a septum delimits it. In the case of the first conidium to be produced, development is invariably holoblastic, involving both outer and inner wall layers of the conidiogenous cells (Figure 4). Subsequent conidia continue to develop holoblastically in exclusively conidial, lichenized genera such as *Blarneya*[36] and *Cheiromycina*,[31,37] but this is unusual. In most lichenized fungi subsequent conidia arise by enteroblastic conidiogenesis where the wall of the conidium is not produced from the original wall layers but from newly laid down, internal wall layers. The enteroblastic, or "phialidic", type of development has not been documented by transmission electron microscopy (TEM) in a selection of lichen conidiomata.[1,5,13,41] Extensions ("proliferations") may or may not occur prior to the formation of each conidium.[1,5] Where proliferations occur, ridge-like scars, or "annellations", remain on the conidiogenous cell (Figure 5B).

Phialidic and annellidic conidiogenesis have often been regarded as mutually exclusive categories by mycologists, but recent reappraisals of the terminology used show that it is merely the timing of similar events which causes the differences[42] (Figure 5). Indeed, mixtures of proliferating and nonproliferating phialides can be found in the same pycnidial conidiomata in lichenized[1,5] as well as certain lichenicolous Coelomycetes.[43,44]

In many species, because of the minute size of the area to be resolved, the method of conidiogenesis cannot be determined satisfactorily by light microscopy and requires transmission electron microscopy for an unequivocal elucidation. A substantial volume of work is still required before we can be regarded as having data on this aspect from a representative sample of the lichen-forming fungi.

VI. CONIDIA

The conidia produced from conidiomata or directly from hyphae in the lichenized fungi have received a variety of names according to their supposed function (Table 1). In adopting the use of the neutral term "conidium" here, it is recognized that the roles may be as spermatia, diaspores, or both.

The range of form found among the conidia produced by lichen-forming fungi is considerable. They may be subglobose, bacilliform, ellipsoid, falcate, sigmoid, thread-like, branched, simple- or nonseptate, colorless, or pigmented (Figure 6). The size varies from *circa* 1 × 0.5 μm in certain spermatial conidia to 100 × 1.5 to 2 μm in the macroconidia of *Arthonia macrosperma*.[14]

The minute conidia which contain scarcely any protoplasm almost certainly have a spermatial function (see above and Chapter V.A), whereas those which are much larger and/or septate and/or pigmented contain much larger nutritional reserves and most probably serve as diaspores (see above). In the exclusively conidial lichenized fungi[1,31] (see Chapter II.A) conidia are the only form of propagation (Figure 7). In the aquatic *Lichingoldia gyalectiformis*, their sigmoid shape (Figure 7A) is particularly suited to dispersal in flowing water.[31]

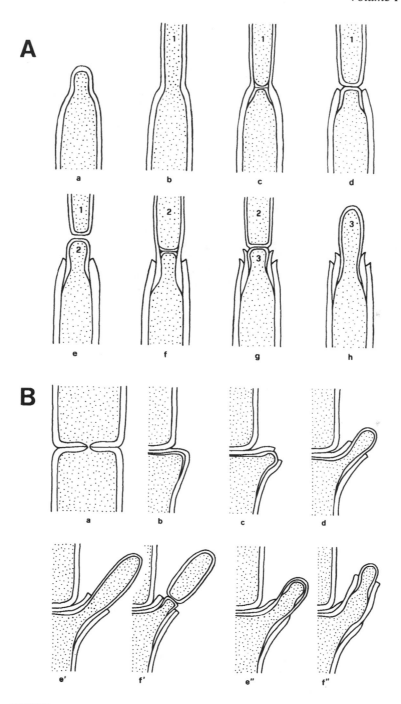

FIGURE 4. Examples of methods of conidiogenesis in selected lichen-forming fungi based on transmission electron microscopy.[5] (A) Arthoniales and Opegraphales; and (B) Lecanorales, Peltigerales, and Teloschistales. a to h = stages; e' f' and e'' f'' = alternatives; 1 to 3 = successively produced conidia. The first formed conidium is produced holoblastically and successive conidia form enteroblastically from phialidic conidiogenous cells (see Figure 5) which may proliferate (B, e'', and f'').

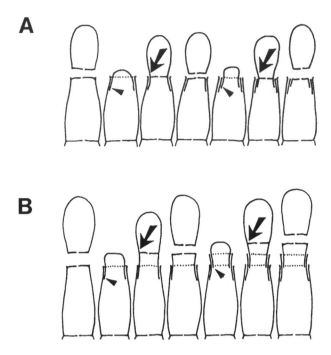

FIGURE 5.   Development of conidiogenous cells.[42] (A) Phialidic; (B) annellidic.[42] In both cases, proliferation (arrowheads) is enteroblastic and conidial initiation (arrows) is holoblastic.

Conidia arising from muriform ascospores while still enclosed within asci are seen in a variety of nonlichenized ascomycetes, and this phenomenon is encountered in several genera of tropical lichens, for example, certain species of *Lopadium, Myxodictyon, Sporopodium,* and *Tricharia.*[14]

Specialized conidium-like structures, termed "brood grains" ("Brutkörner"), have been found arising from the undersurfaces of lobes and rhizines in *Umbilicaria* species.[45] These germinate on damp surfaces and clearly serve as propagules. Such structures are not generally searched for by students of the macrolichens and may well be more widespread than is at present supposed.

Where two, or even three, types of conidia are produced by a single species (see above), these can different substantially in size, shape, and also even color (Figure 8). In these instances the "microconidia" are presumed to act as spermatia, and the "macroconidia" as diaspores. Experimental evidence for this is almost entirely lacking, a key exception being the discovery that macroconidia of *Lecanactis abietina* grow to produce hyphal mats in pure culture.[11] In most cases conidia are discharged in a mucilagenous mass which is exuded as a drop from the ostioles of pycnidial conidiomata. The drop appears to be excuded forcibly on drying in the lichenized coelomycete *Woessia.*[31] Dry conidia, in contrast, are usual in hyphomycete states.[27,31,36]

VII. SYSTEMATIC AND EVOLUTIONARY CONSIDERATIONS

Characteristics derived from the conidiomata of lichens have been largely ignored by taxonomists during the present century. In the last 5 years they have been found to merit more critical study at all levels of the taxonomic hierarchy. At the ordinal level the conidiomata are one of the features distinguishing the Teloschistales (Figure 1E) from other

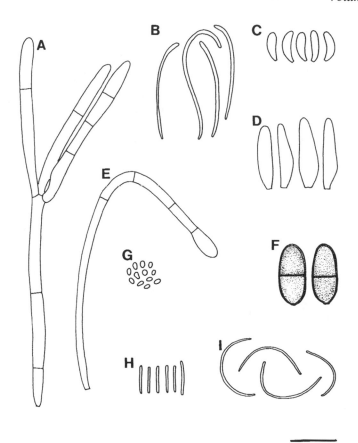

FIGURE 6. Conidia from selected lichen-forming fungi in which ascomata are also known, to illustrate the range of types produced. (A) *"Pyrenotrichum" staurosporum*; (B) *Lecanora chlarona*; (C) *L. saligna*; (D) *Lecanactis abietina* (macroconidia); (E) *"Pyrenotrichum" splitgerberi*; (F) *Mycomicrothelia melanospora*; (G) *Anisomeridium biforme* (microconidia); (H) *Parmelia acetabulum*; (I) *Opegrapha vulgata*. Scale − 10 μm.

lecanoralean types; at the rank of family it characterizes the Roccellaceae (Figure 1B), while distinctive chains of conidiogenous cells are so far only known from the Lobariaceae (Figure 1D). Nevertheless, too much importance should not be attached to correlations of this type while so many genera remain unstudied from this standpoint.

Conidiomatal features are especially valuable in some generic separations. Conidium shape is one of the principle characteristics separating *Foraminella* with falcate conidia from *Parmeliopsis* with minute bifusiform conidia with one to two subapical swellings.[46] The presence of unciform conidia in *Punctelia* supports its separation from *Parmelia*,[47] and minute ellipsoid conidia that of *Phaeophyscia* from both *Physcia* and *Physconia* which have longer subcylindrical conidia.[48]

At the species level conidium shape and size rather than the structure of the conidiomata or method of conidiogenesis assume major importance and have been found to be of value in a large number of genera, including *Aspicilia*,[49] *Heterodea*,[50] *Micarea*,[15] *Opegrapha*,[51] *Parmelia*,[52] and various foliicolous genera.[14] In the separation of *Ramalina cuspidata* from *R. siliquosa* it is the color of the pycnidial conidiomata which is of prime importance.[53]

The type of conidiomata has been used to refer certain species which have no known ascomatal stage to genera characterized by such states, as in *Lecanactis subabietina*.[54] That

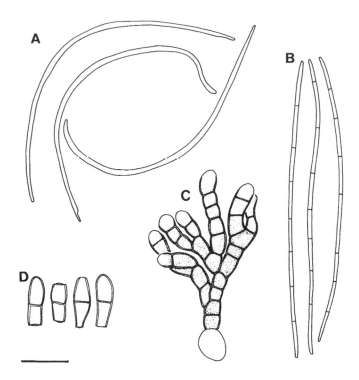

FIGURE 7.   Conidia from selected lichen-forming fungi in which ascomata are unknown, to illustrate the range of types produced. (A) *Lichingoldia gyalectiformis*; (B) *Hastifera tenuispora*; (C) *Cheiromycina flabelliformis*; (D) *Blarneya hibernica*. Scale = 10 μm.

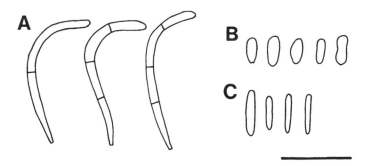

FIGURE 8.   Conidium types produced by *Micarea denigrata*.[15] (A) Macroconidia; (B) mesoconidia; and (C) microconidia. Scale = 10 μm.

such cases occur is not surprising in view of the considerable number of species from a wide range of families which are regularly found only with conidiomata, for example, *Anisomeridium nyssaegenum, Cliostomum griffithii, Lecanactis abietina, Micarea hedlundii,* and *Opegrapha vermicellifera.*

   In cases where no ascomatal stage has been observed and the conidiomata themselves are not easily correlated with an ascomatal genus, genera are delimited on the basis of characteristics derived from the conidiomatal state alone.[1,31,36] Many of these fungi have distinctive lichenized thalli which may be almost placodioid (e.g., *Nigropuncta*[31,44]), minutely isidiate (e.g., *Woessia*[31]), or irregularly areolate (e.g., *Lichingoldia*,[31] (Figure 7A). In *Blar-*

neya (Figure 7D), the characteristic lichen products, lecanoric and schizopeltic acids, are formed.[36] As with the plant pathogenic and saprobic conidial fungi, there can be little doubt that such conidial lichens have been derived from lichenized ascomatal genera, which in many cases may no longer be extant. The switch to dependence on conidia for propagation must be viewed as one of the evolutionary strategies open to lichenized fungi.

Finally, it is important to draw attention to the danger of confusion between conidiomata on lichen thalli which are a part of the mycobionts and those which belong to one of the many conidial lichenicolous fungi which can be encountered.[44,55] Where the invading fungus is parasitic, galls, necrotic patches, or other discolorations can occur, but difficulty arises when the fungus causes little or no damage or is even a parasymbiont.[56] The possibility of confusion is exacerbated in that certain lichenicolous fungi, for example, *Lichenosticta*,[44] have unusual arrangements of the conidiogenous cells outside the lichen-forming fungi; such fungi perhaps evolved from some lichenicolous lichen ancestor.

Conversely, cases are known where new names for conidial fungi have been based on the normal conidiomata of the host, as in *Pyrenotrichum* campylidia on a variety of mainly foliicolous lichens,[33] and *Pleurosticta* on the conspicuous pycnidial conidiomata of *Parmelia acetabulum*[45,57] (Figure 6H). The distinctive conidia of *Lecanora saligna* (Figure 6C) have been described as lichenicolous fungi on at least three occasions.[45] Considerable caution is, therefore, warranted when embarking on the study of conidial structures on lichen thalli.

REFERENCES

1. **Vobis, G. and Hawksworth, D. L.,** Conidial lichen-forming fungi, in *Biology of Conidial Fungi,* Vol. 1, Cole, G. T. and Kendrick, B., Eds., Academic Press, New York, 1981, 245.
2. **Lindsay, W. L.,** Memoir on the spermogones and pycnides of filamentous, fructiculose, and foliaceous lichens, *Trans. R. Soc. Edinburgh,* 22, 101, 1859.
3. **Lindsay, W. L.,** Memoir on the spermogones and pycnides of crustaceous lichens, *Trans. Linn. Soc. London,* 28, 189, 1872.
4. **Glück, H.,** Entwurf zu einer vergleichenden Morphologie der Flechten-Spermogonien, *Verh. Naturhist. – Med. Ver. Heidelberg,* n.f., 6(2), 81, 1899.
5. **Vobis, G.,** Bau und Entwicklung der Flechten-Pycnidien und ihrer Conidien, *Bibl. Lich. Vaduz,* 14, 1, 1980.
6. **Kendrick, W. B. and Nag Raj, T. R.,** Morphological terms in Fungi Imperfecti, in *The Whole Fungus,* Vol. 1, Kendrick, B., Ed., National Museums of Canada, Ottawa, 1979, 43.
7. **Steiner, J.,** Über- die Funktion und den systematischen Wert der Pycnoconidien der Flechten, *Wien Festschr. Feier Zweihundertj. Bestch. K.K. Gymnasium,* 119, 1901 [not seen].
8. **Smith, A. L.,** *Lichens,* Cambridge University Press, London, 1921.
9. **Jahns, H. M.,** Untersuchungen zur Entwicklungsgeschichte der Cladoniaceen unter besonderer Berücksichtigung des Podetiums-Problems, *Nova Hedwigia,* 20, 1, 1970.
10. **Honegger, R.,** Scanning electron microscopy of the contact site of conidia and trichogynes in *Cladonia furcata, Lichenologist,* 16, 11, 1984.
11. **Vobis, G.,** Studies on the germination of lichen conidia, *Lichenologist,* 9, 131, 1977.
12. **Esser, K. and Kuenen, R.,** *Genetik der Pilze,* Springer-Verlag, Berlin, 1965.
13. **Honegger, R.,** Ultrastructural studies on conidiomata, conidiophores, and conidiogenous cells in six lichen-forming ascomycetes, *Can. J. Bot.,* 62, 2081, 1984.
14. **Santesson, R.,** Foliicolous lichens. I. A revision of the taxonomy of the obligately foliicolous lichenized fungi, *Symb. Bot. Ups.,* 12(1), 1, 1952.
15. **Coppins, B. J.,** A taxonomic study of the lichen genus *Micarea* in Europe, *Bull. Br. Mus. Nat. Hist. Bot.,* 11, 17, 1983.
16. **Jahns, H. M., Herold, K., and Beltman, H. A.,** Chronological sequence, synchronization and induction of the development of fruit bodies in *Cladonia furcata* var. *furcata* (Huds.) Schrad., *Nova Hedwigia,* 30, 469, 1978.
17. **Henssen, A.,** Eine Revision der Flechtenfamilien Lichinaceae und Ephebaceae, *Symb. Bot. Ups.,* 18(1), 1, 1963.

18. **Ward, M.**, On the structure, development and life history of the tropical epiphyllous lichen *Strigula complanata, Trans. Linn. Soc. London Ser. 2*, 2, 87, 1884.
19. **Tibell, L. and Hawksworth, D. L.**, unpublished data, 1983.
20. **Poelt, J.**, Eine diözische Flechte, Plant Syst. Evol., 135, 81, 1980.
21. **Voss, E. G. et al., Eds.**, International Code of Botanical Nomenclature adopted by the Thirteenth International Botanical Congress, Sydney, August 1981, *Regnum Veg.*, 111, i, 1983.
22. **Hawksworth, D. L.**, Recent changes in the international rules affecting the nomenclature of fungi, *Microbiol. Sci.*, 1, 18, 1984.
23. **Nag Raj, T. R.**, Genera coelomycetum. XIX. *Discosiella*, a lichenized mycobiont, *Can. J. Bot.*, 59, 2519, 1981.
24. **Hawksworth, D. L.**, The taxonomy of lichen-forming fungi: reflections on some fundamental problems, in *Essays in Plant Taxonomy*, Street, H. E., Ed., Academic Press, New York, 1978, 211.
25. **Sutton, B. C.**, *The Coelomycetes*, Commonwealth Mycological Institute, Kew, 1980.
26. **Lallemant, R.**, Recherches sur le développement en cultures pures in vitro du mycobionte du discolichen *Pertusaria pertusa* (L.) Tuck., *Rev. Bryol. Lichenol.*, 43, 255, 1977.
27. **Honegger, R.**, The hyphomycetous anamorph of *Coniocybe furfuracea, Lichenologist*, 17, 273, 1985.
28. **Samuels, G. J. and Buchanan, D. E.**, Ascomycetes of New Zealand. V. *Mycocalicium schefflerae* sp. nov., its ascal ultrastructure and *Phialophora* anamorph, *N. Z. J. Bot.*, 21, 163, 1983.
29. **Riedl, H.**, Die Flechte *Basidia chlorococca* (Stenh.) Lettau und ihre Beziehungen zu Formgattungen der Fungi Imperfecti, *Phyton (Horn, Austria)*, 17, 357, 1976.
30. **Hawksworth, D. L., Sutton, B. C., and Ainsworth, G. C.**, *Ainsworth and Bisby's Dictionary of the Fungi*, 7th ed., Commonwealth Mycological Institute, Kew, 1983.
31. **Hawksworth, D. L. and Poelt, J.**, Five additional genera of conidial lichen-forming fungi from Europe, *Plant Syst. Evol.*, 154, 195, 1986.
32. **Hawksworth, D. L.**, A redisposition of the species referred to the ascomycete genus *Microthelia, Bull. Br. Mus. Nat. Hist. Bot.*, 14, 43, 1985.
33. **Sérusiaux, E.**, The nature and origin of campylidia in lichenized fungi, *Lichenologist*, 18, 1, 1986.
34. **Vězda, A.**, Foliicole Flechten aus der Republik Guinea (W. Afrika). I, *Cas. Slezskeho Muz. Vedy Prir. (Acta Mus. Silesiae Ser. A Sci. Nat.)*, 22, 67, 1973.
35. **Vězda, A.**, Flechtensystematische Studien. XI. Beiträge zur Kenntnis der Familie Asterothyriaceae (Discolichenes), *Folia Geobot. Phytotaxon Bohemoslovakae*, 14, 43, 1979.
36. **Hawksworth, D. L., Coppins, B. J., and James, P. W.**, *Blarneya*, a lichenized hyphomycete from southern Ireland, *Bot. J. Linn. Soc.*, 79, 357, 1980.
37. **Sutton, B. C. and Muhr, L.-E.**, *Cheiromycina flabelliformis* gen. et sp. nov. on *Picea* from Sweden, *Nord. J. Bot.*, 6, 831, 1986.
38. **Hughes, S. J.**, Conidiophores, conidia and classification, *Can. J. Bot.*, 31, 577, 1953.
39. **Kendrick, W. B., Ed.**, *Taxonomy of Fungi Imperfecti*, University of Toronto Press, Toronto, 1971.
40. **Cole, G. T. and Samson, R. A.**, *Patterns of Development in Conidial Fungi*, Pitman Publ., Marshfield, Mass., 1979.
41. **Honegger, R.**, Ascocarpontogenie, Ascusstruktur und -funktion bei Vertretern der Gattung *Rhizocarpon, Ber. Dtsch. Bot. Ges.*, 91, 579, 1978.
42. **Minter, D. W.**, New concepts in the interpretation of conidiogenesis in deuteromycetes, *Microbiol. Sci.*, 1, 86, 1984.
43. **Hawksworth, D. L.**, Taxonomic and biological observations on the genus *Lichenoconium* (Sphaeropsidales), *Persoonia*, 9, 159, 1977.
44. **Hawksworth, D. L.**, The lichenicolous Coelomycetes, *Bull. Br. Mus. Nat. Hist. Bot.*, 9, 1, 1981.
45. **Hasenhüttl, G. and Poelt, J.**, Über die Brutkörner bei der Flechtengattung *Umbilicaria, Ber. Dtsch. Bot. Ges.*, 91, 275, 1978.
46. **Meyer, S. L. F.**, Segregation of the new lichen genus *Foraminella* from *Parmeliopsis, Mycologia*, 74, 592, 1982.
47. **Krog, H.**, *Punctelia*, a new lichen genus in the Parmeliaceae, *Nord. J. Bot.*, 2, 287, 1982.
48. **Moberg, R.**, The lichen genus *Physcia* and allied genera in Fennoscandia, *Symb. Bot. Ups.*, 22(1), 1, 1977.
49. **Magnusson, A. H.**, Studies in species of *Lecanora*, mainly the *Aspicilia gibbosa* group, *K. Svenska Ventensk. Akad. Handl. Ser. 3*, 17(5), 1, 1939.
50. **Filson, R. B.**, A revision of the genus *Heterodea* Nyl., *Lichenologist*, 10, 13, 1978.
51. **Schaeur, T.**, Ozeanische Flechten in Nordalpenraum, *Port. Acta Biol.*, 8, 17, 1965.
52. **Culberson, W. L. and Culberson, C. F.**, Microconidial dimorphism in the lichen genus *Parmelia, Mycologia*, 72, 127, 1980.
53. **Sheard, J. W.**, The taxonomy of the *Ramalina siliquosa* species aggregate (lichenized ascomycetes), *Can. J. Bot.*, 56, 915, 1978.

54. **Coppins, B. J. and James, P. W.,** New or interesting British lichens. IV, *Lichenologist,* 11, 139, 1979.
55. **Hawksworth, D. L.,** The lichenicolous Hyphomycetes, *Bull. Br. Mus. Nat. Hist. Bot.,* 6, 183, 1979.
56. **Hawksworth, D. L.,** Secondary fungi in lichen symbioses: parasites, saprophytes and parasymbionts, *J. Hattori Bot. Lab.,* 52, 357, 1982.
57. **Santesson, R.,** Svampar som leva pa lavar, *Sven. Bot. Tidskr.,* 43, 141, 1949.

Section VI: Lichen Physiology

Chapter VI.A

CARBON METABOLISM

Margalith Galun

I. INTRODUCTION

Lichens excel in the wide range of their distribution and in their tolerance to extreme environmental conditions. The effects of many climatic and other environmental factors on physiological processes in lichens, including those related to their carbon metabolism (e.g., CO_2 gas exchange, photosynthesis, respiration) have, therefore, been investigated and are dealt with in several chapters of this book (i.e., Chapters VI.C, VII.B, XI).

Also, a wide variety of carbohydrates occurs in lichens, including polyols (sugar alcohols), mono-, oligo-, and polysaccharides. These are discussed in detail in Chapter IX.B.

As in all symbiotic associations between an autotroph and a heterotroph, the lichen mycobiont (the heterotroph) derives its carbohydrates from the photobiont (the autotroph). Again, comparable to the situation in other symbioses of this kind, the carbohydrate transferred is usually a single and simple molecule which is rapidly converted in an irreversible manner by the recipient to a carbohydrate unavailable to the donor.

II. CARBOHYDRATE EFFLUX AND TRANSFER

The type of carbohydrate released by the alga and supplied to the fungus is determined by the alga. In lichens containing cyanobacteria, the carbohydrate released and transferred to the fungus is glucose. In lichens containing green algae, the carbohydrate released and transferred to the fungus is a polyol — ribitol, erythritol, or sorbitol (Table 1).

Most of the pioneering work in identifying the nature of the photosynthates released and transferred was done by Smith and his co-workers,[1-4] with the aid of the "inhibition technique". This technique is based on the competition between labeled and unlabeled compounds. Lichen disks are allowed to photosynthesize in a solution of $NaH^{14}CO_3$ with a high concentration of the unlabeled counterpart of the transferred compound. The unlabeled compound is taken up by the fungus, whereas the radioactive compound produced by photosynthesis enters the medium and can then be measured by chromatography or radioautography. A high proportion of the carbon fixed by the alga passes to the fungus. This massive and selective carbohydrate efflux from algal cells is characteristic of the symbiotic situation and occurs only in the intact lichen. The carbohydrate metabolism of the "free" algae is strikingly different from that of the photobiont in the intact lichen: free-living cyanobacteria do not excrete glucose, and free-living green algal cells do not produce polyols in culture or, in some cases, produce them in small amounts. Moreover, when the photobionts are isolated from the thallus and grown in culture, they cease gradually to "lose" carbohydrates until the release declines to very small amounts or completely stops, with a subsequent increase of ^{14}C incorporation into the ethanol-insoluble fraction (Table 2).[5]

The transfer of carbohydrate is from living photobionts and does not depend on the mycobiont penetrating photobiont cells (see Chapter IV) nor on close contact between large surface areas of the symbionts. There is also no evidence to suggest that fungal factors stimulate efflux from the algae. It was, therefore, proposed that transfer from the alga is accomplished by a carrier-mediated facilitated diffusion process,[6-8] and occurs across inert material.

The mechanism(s) regulating the changes in carbohydrate metabolism of the photobionts when separated from the fungus is not yet known.

Table 1
NATURE OF
CARBOHYDRATES
TRANSFERRED

Photobiont	Mobile carbohydrate
Coccomyxa	Ribitol — $C_5H_7(OH)_5$
Myrmecia	Ribotol
Trebouxia	Ribitol
Trentepohlia	Erythritol — $C_4H_6(OH)_4$
Hyalococcus	Sorbitol — $C_6H_8(OH)_6$
Stichococcus	Sorbitol
Nostoc	Glucose — $C_6H_6(OH)_6$
Dichothrix	Glucose
Calothrix	Glucose
Scytonema	Glucose

Table 2
ALGAE OR LICHENS INCUBATED IN DISTILLED WATER CONTAINING
NaH^{14}CO$_3$ FOR 3 HR AT ROOM TEMPERATURE

Condition of alga	*Nostoc* from *Peltigera canina*	*Hyalococcus* from *Dermatocarpon miniatum*	*Coccomyxa* from *Peltigera aphthosa*	*Trebouxia* from *Xanthoria aureola*
% Fixed ^{14}C released from alga				
In lichen	60	55	65	40
0 hr isolate	15.3	26.1	23.1	8.0
24 hr isolate	7.0	6.2	3.6	1.0
Culture	4.0	1.3	1.4	2.5
% Fixed ^{14}C incorporated into ethanol-insoluble				
In lichen	9.0	2.0	21.0	1.0
0 hr isolate	27.8	29.6	56.2	35.3
24 hr isolate	50.7	40.2	62.8	53.2
Culture	46.0	50.3	50.8—72.3	57.9

	As % ^{14}C							
	Released	In cells	Released	In cells	Released	In cells	Released	In cells
"Mobile" carbohydrate								
0 hr isolate	22.0	Trace	91.0	73.0	85.0	49.0	85.0	83.5
24 hr isolate	0	0	15.0	75.7	0.9	5.1	20.0	58.5
Culture	0	0	0	Trace	0	Trace	17.1	32.8
Identity of carbohydrate								
	Glucose		Sorbitol		Ribitol		Ribitol	

Note: Algae took 30 min to 1 hr to isolate from lichens. Release from algae in lichens calculated from ^{14}C accumulation in fungal polyols.

From Smith, D. C., Transport from symbiotic algae and symbiotic chloroplasts to host cells, *Symp. Soc. Exp. Biol.*, 28, 485, 1974. With permission.

From electron microscope observations of serial sections of the algal layer in several species, it became clear that the photosynthetic products released by the photobiont and transferred to the mycobiont must cross the plasma membranes and cell walls of both symbionts and the extracellular matrices between them,[16] except for rare intracellular haustoria mainly into senescent cells. Furthermore, according to Collins and Farrar,[6] only 20% of the algal cell wall is in direct contact with fungal cell wall and a further 30 to 35% is in contact with the extracellular matrix between them. These authors postulate that, the extracellular matrix renders excreted sugar available to a larger fungal surface area than would otherwise be possible.

According to another hypothesis (see Chapter VI.C), lichen substances are involved in regulating carbohydrate transfer: urease, an algal product, breaks down urea accumulated in the fungus and the thus produced ammonia enhances the permeability of the algal membrane, permitting carbohydrate release in excess. Lichen substances, such as usnic acid, have the capacity of inactivating the enzyme urease. Where an excess of carbohydrates appears in the fungal pool, usnic acid is mobilized to inhibit urease. One of the obvious questions is, what accounts for transfer in lichens lacking lichen substances, such as of the Peltigeraceae.

III. UPTAKE AND CONVERSION OF TRANSFERRED CARBOHYDRATE BY THE MYCOBIONT

A large proportion of the sugar that reaches the fungus, whether glucose or ribitol, is immediately converted into mannitol. When disks of the *Nostoc*-containing thallus of *Peltigera polydactyla* were incubated in the light in $NaH^{14}CO_3$ solutions, about 70% of ^{14}C was incorporated into mannitol and about 10% into insoluble compounds during the first hr of the experiment.[5] Similarly to the glucose released by *Nostoc*, ribitol released by *Trebouxia* and by *Coccomyxa* are converted in the mycobiont into mannitol (and also arabitol, probably an intermediate product). Erythritol released by *Trentepohlia* accumulates partially as erythritol; sorbitol released by *Hyalococcus* accumulates as sorbitol or volemitol.[9]

The supply of carbohydrate to the fungus is much larger than that required for the slow growth of lichens (see Chapter VII.A). The large pools of polyols accumulated in the mycobiont provide the energy required for the rewetting process during the wetting and drying cycle of the lichen thalli (see Chapter VII.B.1) and as osmoregulators (see Chapter VII.B.3).

The conversion into mannitol is a one-way process; mannitol that accumulates in the fungus is not available to the alga. An assumptive pathway suggested by Feige[10] is the following: ribitol is at first converted into arabitol by epimerization, arabitol is then oxidized to arabinose by pentitol dehydrogenase: phosphorylation of arabinose yields arabinosephosphate which is isomerized to ribose-5-phosphate. From the latter, fructose-6-phosphate is produced in the pentosephosphate cycle, which is converted into mannitolphosphate and then yields mannitol via dephosphorylation (see also Chapter VI.C).

Exogenously supplied glucose is also incorporated into mannitol by the mycobiont of green algal lichen species,[10,11] whereas mannitol, mannose, and ribitol, supplied exogenously, are not incorporated by free-living fungi.[12]

Temperature has apparently very little effect on the amount of glucose transported and incorporated as mannitol in the mycobiont of the *Nostoc*-containing species: *Peltigera praetextata, P. polydactyla, P. rufescens,* and *Collema furfuraceum.* The level of thallus hydration, however, is of prime importance and correlates with the habitat occupied by each species. In *C. furfuraceum* and *P. rufescens* from relatively xeric habitats, glucose transfer and incorporation into mannitol occurs down to very low levels of thallus hydration, whereas in *P. polydactyla,* a species of mesic habitats, the amount of sugar incorporated into mannitol declines rapidly below 300% thallus moisture (by weight).[13,14]

The alternating wetting and drying cycles in the lichens in nature are assumed to control the adequate carbon supply to each one of the symbionts. At low hydration levels, photosynthates are retained by the algal partner, while the requirements of the fungal partner are satisfied at the higher hydration levels.[13,14]

In earlier experiments, when the samples were incubated in $NaH^{14}CO_3$ solutions and transfer examined by means of the "inhibition technique", no particular attention was directed to the percentage of thallus hydration. These experiments were, however, of great qualitative value in determining the nature of the sugars transferred. Recent techniques employed, e.g., thin layer chromatography, ion exchange chromatography, gas-liquid chromatography, and high-pressure liquid chromatography (summarized in Reference 15), allow more accurate quantitative estimations.

REFERENCES

1. **Drew, E. A. and Smith, D. C.,** Studies on the physiology of lichens. VIII. Movement of glucose from alga to fungus during photosynthesis in the thallus of *Peltigera polydactyla, New Phytol.,* 66, 389, 1967.
2. **Richardson, D. H. S., Smith, D. C., and Lewis, D. H.,** Carbohydrate movement between the symbionts of lichens, *Nature (London),* 214, 877, 1967.
3. **Richardson, D. H. S. and Smith, D. C.,** Lichen physiology. IX. Carbohydrate movement from the *Trebouxia* symbiont of *Xanthoria aureola* to the fungus, *New Phytol.,* 67, 61, 1968.
4. **Hill, D. J. and Smith, D. C.,** Lichen physiology. XII. The "inhibition technique", *New Phytol.,* 71, 15, 1972.
5. **Smith, D. C.,** Transport from symbiotic algae and symbiotic chloroplasts to host cells, *Symp. Soc. Exp. Biol.,* 28, 485, 1974.
6. **Collins, C. R. and Farrar, J. F.,** Structural resistance to mass transfer in the lichen *Xanthoria parietina, New Phytol.,* 81, 71, 1978.
7. **Smith, D. C.,** Symbiosis and the biology of lichenized fungi, *Symp. Soc. Exp. Biol.,* 29, 373, 1975.
8. **Smith, D. C.,** Mechanisms of nutrient movement between lichen symbionts, in *Cellular Interactions in Symbiosis and Parasitism,* Cook, C. B., Pappas, P. W., and Rudolph, E. D., Eds., Ohio State University Press, Columbus, 1980, 197—227.
9. **Feige, G. B.,** Untersuchungen zur Ökologie und Physiologie der marinen Blaualgenflechte *Lichina pygmaea* Ag. III. Einige Aspekte der photosynthetischen C-Fixierung unter osmoregulatorischen Bedingungen, *Z. Pflanzenphysiol.,* 77, 1, 1975.
10. **Feige, G. B.,** Probleme der Flechtenphysiologie, *Nova Hedwiga,* 30, 725, 1978.
11. **Feige, B.,** Zur Verwertung uniform ^{14}C-markierter Glucose und uniform ^{14}C-markierten Glycerins durch die Flechte *Cladonia convoluta, Z. Pflanzenphysiol.,* 63, 211, 1970.
12. **Galun, M., Braun, A., Frensdorff, A., and Galun, E.,** Hyphal walls of isolated lichen fungi-autoradiographic localization of precursor incorporation and binding of fluorescein-conjugated lectins, *Arch. Microbiol.,* 108, 9, 1976.
13. **Tysiaczny, M. J. and Kershaw, K. A.,** Physiological environmental interactions in lichens. VII. The environmental control of glucose movement from alga to fungus in *Peltigera canina* var. *praetextata* Hue., *New Phytol.,* 83, 137, 1979.
14. **MacFarlane, J. D. and Kershaw, K. A.,** Physiological environmental interaction in lichens. XIV. The environmental control of glucose movement from alga to fungus in *Peltigera polydactyla, P. rufescens* and *Collema furfuraceum, New Phytol.,* 91, 93, 1982.
15. **Richardson, D. H. S.,** The surface physiology of lichens with particular reference to carbohydrate transfer between symbionts, in *Surface Physiology of Lichens,* Vicente, C., Brown, D. H., and Legaz, M. E., Eds., Universidad Complutense de Madrid, Madrid, 1984, 24—55.
16. **Galun, M.,** unpublished observations.

Chapter VI.B

NITROGEN METABOLISM

Amar Nath Rai

INTRODUCTION

This Chapter deals with the physiological and biochemical aspects of nitrogen metabolism in lichens. Much of the information has become available recently, especially from studies on nitrogen-fixing lichens, including nitrogen fixation, regulation of nitrogenase and pathways of nitrogen metabolism.[1-6] Although, I have covered here aspects of nitrogen metabolism both in nitrogen-fixing and nonnitrogen-fixing lichens, one may find that this chapter contains relatively more information on nitrogen-fixing lichens. This is merely due to the fact that relatively less information is available on aspects of nitrogen metabolism in nonnitrogen-fixing lichens.

Nitrogen-fixing lichens almost always involve a nitrogen-fixing cyanobacterium as cyanobiont, and in some cases, in tripartite (three-membered) lichens, an additional phycobiont — a green alga.

NITROGEN CONTENT OF LICHEN THALLI

Total Nitrogen

The total nitrogen content of many lichen species has been investigated and the information is presented in Table 1. In general, cyanophilic lichens have a higher nitrogen content than lichens with only green algae as phycobionts. However, in the case of *Candelariella coralliza, Lecanora muralis,* and *Physcia tribacia* substantially higher nitrogen content has been reported[15] probably because these lichens were collected from nitrogen-rich substrates.

In *Peltigera aphthosa, Nostoc, Coccomyxa,* and the mycobiont constitute 6, 14.4, and 79.6% of the total thallus protein, respectively,[11] and *Nostoc* constitutes nearly 80% of the soluble protein in cephalodia[16] (see Chapters II.B and II.C). In *Peltigera canina* the *Nostoc* constitutes nearly 36% of the total thallus protein.[16] The above values were obtained by measuring the protein/chlorophyll ratios in cyanobiont and phycobiont cells isolated directly from the lichen thallus. Using this ratio the amount of cyanobiont or phycobiont protein in a lichen thallus was estimated by just measuring the chlorophyll content of the whole thallus. Such a procedure eliminates the problem of quantitative isolation of cyanobiont or phycobiont cells from the thallus where a large error is inherent due to loss of cells during isolation. This is the reason why early reports of cyanobiont or phycobiont contents as % of total thallus protein are very low.[9,17]

The nitrogen content of different parts of the *P. aphthosa* thallus and its phycobiont and cyanobiont are given in Table 2. The nitrogen content of both the cultured and symbiotic *Coccomyxa* is quite similar indicating that the phycobiont is not nitrogen starved in the symbiotic state. The higher nitrogen content in cephalodia is a reflection of the fact that *Nostoc* is the main nitrogen source.[11] Sampling away from cephalodia shows small but yet noticeable decrease in the nirtogen content of the thallus.[10]

Forms of Nitrogen

The total nitrogen pool in *Peltigera polydactyla* is reported to consist of 75% insoluble nitrogen and 25% soluble nitrogen. Of the latter nearly 50% is unidentified nitrogen and the rest is made up of ammonia, amino nitrogen, and amide nitrogen whereas nitrate and

Table 1
TOTAL NITROGEN CONTENT OF VARIOUS LICHEN THALLI

Lichen	Nitrogen content[a]	Ref.
Lichens with cyanobionts		
Lichina confinis	3.73	7
Lobaria laetevirens	2.20	8
L. pulmonaria	2.70	8
Peltigera aphthosa var. *leucophlebia*	3.00	9
P. aphthosa Willd.	3.20, 3.30	10,11
P. canina	3.30	12
P. polydactyla	3.60—4.50	13
P. praetextata	4.70	14
P. rufescens	2.22	7
Placopsis gelida	0.93	7
Sticta sylvatica	4.00	8
Lichens with eukaryotic phycobionts		
Anaptychia fusca	1.92	7
Candelariella corallizza	4.20—5.00	15
Cladonia foliacea	0.65	7
C. impexa	0.33	7
Cornicularia aculeata	0.38	7
Evernia prunastri	0.84	7
Lecanora atra	0.69	7
L. muralis	6.20—9.24	15
Ochrolechia parella	0.65	7
Parmelia physodes	0.49	7
P. sulcata	0.96	7
Physcia ascendens	1.15	7
P. tribacia	3.90—4.70	15
Ramalina siliquosa	0.93	7
Usnea subfloridana	0.37	7
Xanthoria candelaria	4.20—4.40	15
X. parietina	1.41	7

[a]   Expressed as percent of dry weight.

Table 2
CARBON AND NITROGEN CONTENT AND C:N RATIOS IN THE *PELTIGERA APHTHOSA* THALLUS AND ITS SYMBIONTS

Sample	Nitrogen (% dry wt)	Carbon (% dry wt)	Carbon:Nitrogen
Whole thallus	3.20	44.40	13.5:1
Excised cephalodia	4.07	38.83	9.5:1
Thallus without cephalodia	2.89	44.00	15.2:1
Mycobiont	2.15	41.86	19.5:1
Symbiotic *Coccomyxa*[a]	5.21	55.71	10.7:1
Free-living *Coccomyxa*[b]	5.30	51.80	9.7:1
Free-living *Nostoc*[b]	6.10	41.52	6.8:1

[a]   Fresh isolates.
[b]   Cultured isolates.

Table 3
AMMONIA AND AMINO ACID POOLS OF *PLATIMATIA GLAUCA*,
***HYPOGYMNIA PHYSODES* AND *PSEUDEVERNIA FURFURACEA*[a][b]**

Ammonia/Amino acids	*Platimatia glauca*	*Hypogymnia physodes*	*Pseudevernia furfuracea*
Alanine	3.0	0.9	2.8
Aminobutyric acid	1.2	0.3	2.4
Ammonia	4.2	2.3	3.8
Arginine	6.0	0.5	5.6
Asparagine	4.1	1.2	0.0[c]
Aspartic acid	1.2	1.8	4.2
Glutamic acid	5.9	4.8	27.0
Glutamine	33.2	7.5	14.2
Proline	0.8	0.6	3.3
Taurine	2.2	1.6	4.9

[a] Values expressed as nanomole per milligram of dry weight.
[b] For further information see Chapter VI.C of this volume and Reference 25.
[c] Actual value 0.02.

nitrite are negligible.[13] Information regarding amino acid composition is also available for various other lichens. Some of the earliest qualitative investigations of amino acids have been done with *Cladonia rangiferina*, *C. gracilis*, *Dermatocarpon moulinsii*, *Lobaria isidiosa*, *L. subisidiosa*, *Parmelia nepalensis*, *P. tinctorum*, *Peltigera canina*, *Ramelina sinensis*, *Roccella montangnei*, *Umbilicaria pustulata*, *Usnea flexilis*, *U. orientalis*, and *U. venosa*.[18-22] Glutamic acid and glutamine were found to be the most abundant free amino acids present. Similar studies have been conducted by Solberg on Norwegian lichens including *Anaptychia fusca*, *P. canina*, and *Omphalodiscus spodochrous*.[23,24]

An analysis of the soluble nitrogen fraction of *Sticta sylvatica*, *Lobaria laetevirens*, and *L. pulmonaria* has been performed by Goas and Bernard.[8] Glutamic acid was found to be the most abundant of all amino acids. A detailed composition of amino acid pools in *Platimatia glauca*, *Hypogymnia physodes*, and *Pseudevernia furfuracea* is given in Table 3. It is seen that the most abundant amides and amino acids are glutamine, asparagine, glutamic acid, aspartic acid, arginine, γ-amino butyric acid, alanine, and taurine. In these lichens, asparagine and arginine have been suggested to serve as the main nitrogen storage compounds.[25] In *P. furfuracea* the concentration of glutamic acid and proline are much higher than the other amino acids. Proline probably functions as an energy storage compound in this lichen.[26]

Analysis of the ammonia and amino acid pool in cephalodia of *Peltigera aphthosa* has shown that glutamic acid, glutamine, aspartic acid, and alanine are the major amino acids present. Glutamate was most abundant (50 μmol g^{-1} dry weight), followed by alanine (29 μmol g^{-1} dry weight), aspartic acid (11 μmol g^{-1} dry weight), and glutamine (10 μmol g^{-1} dry weight). Ammonia was 18 μmol g^{-1} dry weight (Table 4). On detachment of cephalodia from the main thallus the alanine pool increases due to the absence of a sink provided by the main thallus.[27] Similar studies on *P. canina* indicate that the largest amount of amino acid in the pool was alanine, followed by glutamine, glutamic acid, aspartic acid, and ammonia (Table 4).[28] The pools present in the cyanobiont were assayed by treating the thalli with digitonin (0.01%, w/v) to release the fungal amino acid pool after which the thalli were washed, and the remaining amino acid pool measured. The only major amino acids present in such thalli (cyanobiont) were glutamine, alanine, glutamic acid, aspartic acid, and ammonia. As seen in Table 4, a third of the total ammonia, glutamic acid, and glutamine pool of the thallus is present in the cyanobiont which is as expected because the cyanobiont constitutes nearly a third of the total thallus protein. Because the pools of aspartic

Table 4
POOLS OF AMMONIA AND MAJOR AMINO ACIDS IN THE
THALLUS AND CYANOBIONT OF *PELTIGERA CANINA* AND IN
THE CEPHALODIA OF *P. APHTHOSA*[27,28]

Ammonia/amino acids	*Peltigera canina* (μmol $\cdot$ g^{-1} thallus dry wt)		*Peltigera aphthosa* (μmol $\cdot$ g^{-1} cephalodial dry wt)
	Whole thallus	Cyanobiont[a]	Cephalodia
Alanine	145.20	29.10	29.00
Ammonia	33.80	10.80	18.00
Asparagine	3.80	trace[b]	trace
Aspartate	58.70	6.10	11.00
Glutamate	76.00	28.80	50.00
Glutamine	93.00	34.00	10.00
Glycine	14.40	4.00	trace
Serine	14.60	trace	trace
Threonine	11.00	trace	trace

[a] Values obtained from digitonin-treated thalli.
[b] Values below 1 nmol mg^{-1} dry weight are referred to as trace.

acid and alanine were selectively lost by digitonin treatment it has been suggested that these amino acids are located mainly in the mycobiont.[28]

Synthesis of several amino acids during photosynthesis in *C. rangiferina, Pseudevernia furfuracea, Collema* sp., *Peltigera horizontalis, Usnea florida, P. canina,* and *P. aphthosa* has been studied using ^{14}C-labeled sodium bicarbonate and carbon dioxide.[4,29-31] In *Collema* sp., *Pseudevernia furfuracea,* and *Cladonia rangiferina,* aspartic acid, glutamic acid, and alanine show ^{14}C-labeling within a 90-min exposure to ^{14}C-carbon dioxide. In *U. florida,* aspartic acid and alanine were labeled within a 1 min exposure to ^{14}C carbon dioxide. In *Peltigera horizontalis* and *P. canina,* however, only aspartic acid was labeled after 1- and 5-min exposures to ^{14}C carbon dioxide and ^{14}C sodium bicarbonate, respectively.

Apart from amino acids, various amines have also been found in lichens, particularly in members of the Stictaceae. Methylamine, dimethylamine, and trimethylamine are abundant in *L. laetevirens.*[32,33] A scheme summarizing the interrelationships between glycine, methylamines, sarcosine, and choline has been proposed in this lichen.[33] Bound di- and polyamines putrescine, cadavarine, agmatine, spermidine, and spermine are common in *Platimatia glauca, H. physodes,* and *Pseudevernia furfuracea.*[25] The concentrations of these, except spermine, are of the same order of magnitude or sometimes even higher than those of the free amino acids. Free di- and polyamines are reported to be involved in a number of physiological processes and a close relationship between polyamines, nucleic acids, and membranes has been suggested.[34] Polyamines protect nucleic acids, membranes, and ribosomes against thermal and enzymic destruction by binding and stabilizing them under extreme environmental conditions.[35,36] High polyamine contents could help such drought-resistant organisms as lichens to overcome drought stress conditions.[37,38]

A soluble nitrogenous compound, sticticin, has been isolated from the lichen *L. laetevirens* and it has been reported that at a thallus water content of 10 to 12%, the concentration of sticticin exceeds 1 *M* in this lichen.[39] Sticticin may play a role in osmoregulation. Using ^{14}C tyrosine, ^{14}C dihydroxyphenylalanine, and ^{14}C methionine, the route of sticticin biosynthesis has been investigated. It involves *N*-methylation of tyrosine followed by hydroxylation of the ring and esterification of the acid function.[39]

A group of nitrogenous compounds commonly found in cyanobacteria are the phycobiliproteins. Until recent years, very little information was available about these in symbiotic cyanobacteria.[40] In a preliminary study it has been reported that the *Nostoc* cyanobiont in the cephalodia of *Peltigera aphthosa* contains only phycoerythrin and no phycocyanin, while the free-living isolate contains both.[4] (For recent update see Chapter IX.C.)

Lichens may survive long periods of nitrogen starvation because of the very slow rate of protein breakdown as shown in the case of *P. polydactyla* where only 5 to 10% of the protein was degraded after a 6-day starvation.[41]

UTILIZATION OF EXOGENOUS COMBINED NITROGEN

Lichens have been shown to utilize exogenous urea, ammonia, nitrate, and amino acids. Urea causes an increase in the polyol/glucose ratio in lichens.[42,43] The carbon dioxide released during hydrolysis of urea is refixed, at a slow rate, into phenolics, amino acids, organic acids, and sugars.[42,43] Urease activity in lichens with green algae as phycobionts is induced on addition of urea[44-46] (see Chapter VI.C). Certain epiphytic lichens, e.g., *Evernia prunastri*, may hydrolyze urea exogenously by secreting urease into the surrounding medium. This exogenous production of carbon dioxide from urea may explain the slow rate of incorporation of the urea-derived carbon dioxide in this lichen.[47] Exogenous urea also causes an increase in the production of ribitol and mannitol (the translocatable and accumulation form of carbon, respectively) in *Cladonia sandstedei*.[48] Urea also enhances the synthesis of atranorin and its precursor, methyl β-orcinol carboxylate. It is believed that these effects may be due to the ammonia release from the urea.[42] In *Chlorella*, urea-derived ammonia has been shown to accelerate the breakdown of carbohydrates and a similar effect in lichens is not unlikely.[49]

An interaction among urea, urease, and secondary compounds may affect the phycobiont (such as increased respiration and carbohydrate breakdown, caused by ammonia, and the stimulation of photosynthesis, caused by carbon dioxide) by increasing the flow of nutrients from the phycobiont. Also, the production of urease has been shown to be controlled by lichen compounds such as usnic acid[43-46] (see also Chapter VI.C).

Many lichens have been shown to absorb nitrate and ammonium ions from the culture media.[57] Lichens with green algae as phycobionts take up much more nitrate, on a per unit thallus weight basis, than those with cyanobionts, though both types appear more or less equal in their capacity to absorb ammonia. In *P. polydactyla*, a cyanophilic lichen, the nitrate absorption is only 20% of the ammonia absorption from a solution of similar concentration (5 mM).[13] Furthermore, virtually all the nitrate absorbed from the medium remained as nitrate in the thallus.[13]

Ammonia absorption in *P. polydactyla*, *P. canina*, and *P. aphthosa* has been studied in some detail. In *P. polydactyla* the rate of ammonia absorption has been reported to be about 400 ng ammonium-nitrogen hr^{-1} mg^{-1} dry thallus weight.[13] This decreased over a period of time and ceased altogether after 12 hr. Addition of glutamic acid increased the rate of ammonia absorption as well as prolonged it beyond 12 hr.[13] Such results are similar to those shown by Macmillan for the fungus *Scopulariopsis brevicaulis*.[58] Ammonia absorption caused increase in ammonia and amino-nitrogen but not in the amounts of amide-nitrogen leading to the suggestion that amides do not have an important quantitative role in the nitrogen metabolism of *P. polydactyla*. Protein synthesis was very slow and in small amount, if any, and increased with the addition of glutamic acid.[13]

In *P. aphthosa* the rate of ammonia uptake from a 5 mM ammonium chloride solution is reported to be 59 and 53 nmol mg^{-1} fresh weight 24 hr^{-1} for the whole thallus and isolated cephalodia (approximately 3.6 and 3.12 μg ammonium-nitrogen taken up mg^{-1} dry weight 24 hr^{-1}), respectively.[59] Ammonia uptake rates have also been estimated in *P. aphthosa* and *P. canina* using ^{15}N-labeled ammonium chloride. Rates of about 2 and 1.5 μg ammonium-nitrogen mg^{-1} thallus dry weight hr^{-1} were found for *P. aphthosa* and *P. canina*, respectively (Figure 1). When [15]NH_4Cl uptake by cultured and symbiotic cyanobionts of *P*.

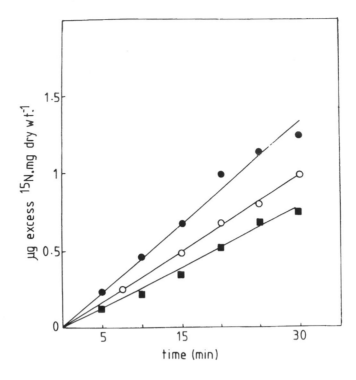

FIGURE 1.   Uptake and incorporation of exogenous ^{15}N-labeled ammonium chloride into *Peltigera aphthosa* and *P. canina*. Experimental lichen material was suspended in a 5 m*M* ammonium chloride solution labeled with 35 atom % ^{15}N. At time intervals, samples were processed to monitor ^{15}N label in the lichen material using a mass spectrometer. For details of methodology see Reference 11, Symbols: ○ = *P. aphthosa*, whole thallus; ● = *P. aphthosa*, cephalodia; ■ = *P. canina*, thallus.

aphthosa and *P. canina* was compared, uptake by symbiotic cyanobionts was found to be extremely slow (Figure 2). The cultured *Nostoc* CAN (the cultured cyanobacterial isolated from *P. canina*) and symbiotic cyanobionts of *P. canina* (digitonin-treated thalli) showed uptake rates of 500 and 20 μg ammonium-nitrogen mg^{-1} chlorophyll *a* hr^{-1}, respectively. The rates of uptake by cultured and symbiotic cyanobionts of *P. aphthosa* (*Nostoc* APH — the cultured cyanobacterial isolate of *P. aphthosa* — and digitonin-treated cephalodia, respectively) were 550 and 30 μg ammonium-nitrogen mg^{-1} chlorophyll *a* hr^{-1}. The extremely low rate of ammonium uptake in symbiotic cyanobionts may be due to the rate of ammonia assimilation in the symbiotic cyanobiont where the activity of glutamine synthetase (GS), the primary ammonia assimilating enzyme,[60-62] is extremely low (see below). It has also been suggested that the symbiotic cyanobiont in lichens may lack an ammonium transport system.[63] Although, such studies on symbiotic cyanobionts of lichens have not been done, due to difficulty in getting enough clean cells to do the experiments, studies on the symbiotic cyanobiont from the *Azolla-Anabaena* symbiosis suggest that the symbiotic cyanobionts do have an energy-dependent Δψ-driven ammonium transport system.[64]

Uptake and utilization of asparagine, glutamine, aspartic acid, and glutamic acid have been studied in *P. polydactyla*.[65] Relatively large amounts of asparagine actively accumulate in the tissues very rapidly. When the disappearance of asparagine from a 5 m*M* solution in which *P. polydactyla* disks were incubated, was followed by monitoring total nitrogen, amino-nitrogen, amide-nitrogen, and ammonium-nitrogen, it was found that over a 24 hr period ammonium-nitrogen increased continuously, suggesting deamidation of asparagine.

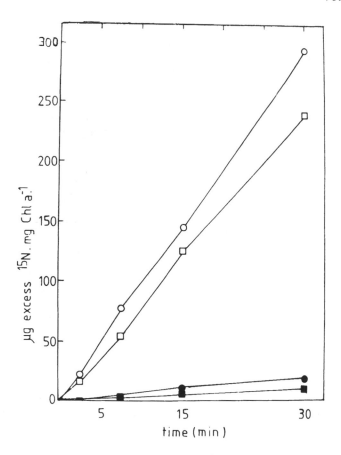

FIGURE 2.   Uptake and incorporation of exogenous ammonium chloride, labeled with ^{15}N, into *Peltigera aphthosa* and *P. canina*. Experimental details as in the caption to Figure 1. Symbols: ○ = *Nostoc* APH, the isolated cyanobiont of *P. aphthosa* cultured; ● = *Nostoc* cyanobiont in the cephalodia of *P. aphthosa* (digitonin-treated cephalodia); □ = *Nostoc* CAN, isolated cyanobiont of *P. canina* cultured; ■ = *Nostoc* cyanobiont in *P. canina* thallus (digitonin-treated thalli).

The finding that amide- and amino-nitrogen decline equally suggests that asparagine enters the tissues as intact molecules. The utilization of absorbed asparagine is very slow and involves deamidation of asparagine to ammonia as suggested by the release of ammonia into the medium. The asparagine uptake is partially inhibited by glucose.

In comparison to the rate of asparagine uptake (15.6 to 22.5 μg nitrogen mg^{-1} dry weight 24 hr^{-1}), the rates of glutamine, aspartic acid, and glutamic acid were slower being 9.07, 9.27, and 4.6 μg nitrogen mg^{-1} dry weight 24 hr^{-1}, respectively. Absorption of all these compounds is accompanied by a release of ammonia into the medium and by significant increases in the amounts of ammonia and amino-nitrogen in disks. Within the thallus, the algal region absorbs more asparagine and ammonia than the medulla.[66] After 22.5 hr, the ammonium-nitrogen absorbed mg^{-1} dry weight is reported to be 880 and 610 ng, respectively.[66] Over a 24 hr period, the absorption of aparagine is reported to be 17.4 and 14.5 μg nitrogen mg^{-1} dry weight by the algal and medulla regions, respectively. The algal region thus seems to be more metabolically active.

This high absorptive ability may enable accumulation of nutrients into the tissues during occasional periods of abundance which could then be used during periods of scarcity.

Table 5
ALANINE DEHYDROGENASE (EC 1.4.1.1)
ACTIVITIES IN *PELTIGERA CANINA* AND *P.*
APHTHOSA

Lichen	Activity (nmol product formed min^{-1} mg^{-1} protein)
Peltigera canina[28,72]	
Thallus	3.80
Cyanobiont	5.50
Free-living *Nostoc* (*Nostoc* CAN)[a]	22.90
P. aphthosa[30,59]	
Whole thallus	9.70
Thallus without cephalodia	10.00
Excised cephalodia	8.50
Mycobiont	15.00
Free-living *Nostoc* (*Nostoc* APH)[a]	11.00
Free-living *Coccomyxa*[a]	3.00

[a] Cultured isolates.

ENZYMES OF NITROGEN METABOLISM

Many of the enzymes of the nitrogen metabolism (see also Chapter VI.C) are liberated extracellularly. These include asparaginase, allantoicase, allantoinase, lichenase, and uricase, all having nitrogenous substrates.[67,68] Uricase activity is generally found to be absent in cyanophilic lichens, however, in other lichens variable levels of uricase activity have been reported.[69] Considerably high uricase activity was found in *Candelariella* sp.[69]

Many lichens have urease activity and the available data suggest that in cyanophilic lichens urease is a constitutive enzyme and in lichens containing green algae as phycobionts urease is urea-inducible.[44-46,70]

Carbamoylphosphate synthase (CPS) activity has been investigated in the cephalodia and main thallus of the lichen *Peltigera aphthosa*.[71] Cephalodia contain a much higher CPS activity than the main thallus. The ammonium-dependent CPS activity of the cephalodia and main thallus has been reported to be 1.7 and 0.1 nmol product formed min^{-1} mg^{-1} protein, respectively. The CPS activity is lower when glutamine is used as a substrate rather than ammonia. The glutamine-dependent CPS activity of cephalodia and main thallus is reported to be 0.25 and 0.05 nmol product formed min^{-1} and mg^{-1} protein, respectively.[71]

Alanine dehydrogenase (ADH) activity has been investigated in *P. aphthosa* and *P. canina* (Table 5).[27,28,59,72] In *P. canina* thalli ADH activity is found to be 3.8 nmol product formed min^{-1} mg^{-1} protein. A similar activity was observed in the cyanobiont cells directly isolated from the lichen thalli. In contrast, the activity in cultured *Nostoc* CAN cells is 22.9 nmol product formed min^{-1} mg^{-1} protein. In the case of *P. aphthosa*, ADH activity in different components shows the following trend: the activity in the whole thallus and in the thallus without cephalodia is similar; excised cephalodia show a slightly lower activity while the mycobiont shows significantly higher activity; and the cultured *Coccomyxa* cells contain very low activity of ADH.

Glutamate dehydrogenase (GDH) activity has been studied in the lichens *P. canina*, *P. aphthosa*, *Pseudevernia furfuracea*, and *L. laetevirens*.[25-28,59,73] In *P. furfuracea* an NADPH-dependent GDH activity of 58×10^{-3} units g^{-1} air-dried lichen material has been found.[25] The NADH-dependent activity is much lower (7×10^{-3} units g^{-1} air-dried lichen material). In *L. laetevirens*, two NADP-dependent GDH activities have been reported, of which the

Table 6
**NADPH-DEPENDENT GLUTAMATE DEHYDROGENASE (EC
1.4.1.4) ACTIVITY IN SOME LICHENS[25-28,59,72]**

Lichens	Activity (nmol product formed min^{-1} mg^{-1} protein)
Peltigera aphthosa[27,59]	
Whole thallus	90.00
Thallus without cephalodia	35.00
Excised cephalodia	398.00
Digitonin-treated cephalodia	0.00
Mycobiont (medullary hyphae)	24.00
Free-living *Nostoc* (*Nostoc* APH)[a]	2.00
Free-living *Coccomyxa*[a]	22.00
P. canina[28,72]	
Whole thallus	195.80[28]
	201.70[72]
Digitonin-treated thallus	4.15
Cyanobiont	2.40
Free-living *Nostoc* (*Nostoc* CAN)[a]	0.00
Pseudevernia furfuracea[25]	58.00[b]

[a]   Cultured isolates.
[b]   Value expressed as units $\times$ 10^{-3} g^{-1} air dried lichen material.

major one is located in the mycobiont. The enzyme has been isolated, purified, and its kinetics studied.[73] This enzyme is more active in glutamate biosynthesis rather than in glutamate oxidation and its activity is dependent on ammonia concentration, on ionic strength, and on pH. In *Peltigera canina*, higher NADPH-dependent activity has been reported in the mycobiont.[28,72] That this activity is associated with the mycobiont has been suggested by the observation that the cyanobiont in the lichen thallus shows only negligible activity (2.4 nmol product formed min^{-1} mg^{-1} protein in the cyanobiont as compared to 195 to 200 nmol product formed min^{-1} mg^{-1} protein in the whole *P. canina* thallus) (see Table 6). Furthermore, on treatment of the thallus with 0.01% (w/v) digitonin to selectively disrupt the fungal membranes,[74] virtually all the GDH activity is washed out, indicating its localization in the mycobiont. Similar studies on *P. aphthosa* indicate that the cephalodia have extremely high NADPH-dependent GDH activity and that all this activity is localized in the mycobiont (digitonin-treated cephalodia show no GDH activity).[27,59] The mycobiont hyphae in the main thallus, however, show much lower activity than the mycobiont hyphae of cephalodia which are in close contact with the nitrogen-fixing *Nostoc* cells.

Glutamate synthase (GOGAT) activity of various constituents of *P. canina* and *P. aphthosa* has been examined (Table 7).[75-77] Ferredoxin-dependent GOGAT activity in *P. canina* is localized in the cyanobiont and is less than 3% of the activity found in *Nostoc* CAN. The fungal hyphae do not contain any GOGAT, be it ferredoxin-dependent or NADPH-dependent. A similar situation is found in *P. aphthosa*. In cephalodia, the GOGAT activity is located in the cyanobiont and it is only about 3% of that in *Nostoc* APH. Both the cephalodia and medullary fungal hyphae do not show any GOGAT activity. The *Coccomyxa* on the other hand, has similar activity in symbiosis and in cultured cells. It is not known at present how the GOGAT activity of the cyanobiont is reduced to such a low level in symbiosis.

The earliest reports of aminotransferase activities in lichens are those of Bernard[78] and Bernard and Goas[79] who studied the activities of glutamate oxaloacetate transaminase (GOT), glutamate pyruvate transaminase (GPT), and glutamate decarboxylase (GDC) in members of the Stictaceae. GOT was found to be the major transaminating enzyme and its activities

Table 7

FERREDOXIN-DEPENDENT GLUTAMATE SYNTHASE (EC 1.4.7.1) ACTIVITY IN *PELTIGERA CANINA* AND *P. APHTHOSA*[75,76]

Lichen material	Activity (nmol product formed min^{-1} mg^{-1} protein)
P. canina thallus	0.28
Digitonin-treated *P. canina* thallus	0.84
Free-living *Nostoc* (*Nostoc* CAN)[a]	33.50
Excised cephalodia of *P. aphthosa*	0.58
Digitonin-treated cephalodia	0.72
Free-living *Nostoc* (*Nostoc* APH)[a]	31.29
Free-living *Coccomyxa*[a]	23.10
Symbiotic *Coccomyxa* from *P. aphthosa*[b]	20.94
Mycobionts of *P. canina* and *P. aphthosa*	not detectable

[a]   Cultured isolates.
[b]   Fresh isolates.

Table 8

ACTIVITIES OF SERINE TRANSHYDROXYMETHYLASE AND SOME AMINOTRANSFERASES IN *PELTIGERA APHTHOSA* AND *P. CANINA*[a]

Lichen material	Activity (nmol product formed min^{-1} mg^{-1} protein)				
	Serine transhydro-xymethylase (EC 2.1.2.1)	Glutamate glyoxylate transaminase (EC 2.6.1.4)	Glutamate oxaloacetate transaminase (EC 2.6.1.1)	Glutamate pyruvate transaminase (EC 2.6.1.2)	Aspartate pyruvate transaminase (EC 2.6.1.12)
Peltigera aphthosa					
Cephalodia	1.50	44.50	20.20	40.10	15.82
Digitonin-treated cephalodia (cyanobiont)	—	—	12.50	0.00	2.60
Thallus without cephalodia	2.90	56.00	47.00	—	—
Symbiotic *Coccomyxa*[b]	3.20	14.50	—	—	—
Mycobiont (medulla)	2.20	—	37.00	—	—
Whole thallus	—	—	58.40	—	—
Cultured *Nostoc* APH[c]	—	—	25.00	4.80	—
Free-living *Coccomyxa*[c]	—	—	520.00	—	—
P. canina					
Whole thallus	—	—	41.00	18.12	126.00
Free-living *Nostoc* CAN[c]	—	—	40.40	—	—
Digitonin-treated thallus (cyanobiont)	—	—	13.80—16.80	0.86	58.60

[a]   Data compiled from References 3, 27, 28, 59, 72, and from my own unpublished results.
[b]   Fresh isolates.
[c]   Cultured isolates.

were higher than GPT. In *Pseudevernia furfuracea*, however, GPT activity was higher than that of GOT (the respective reported activities of GPT and GOT are 10 and 8 units $\times$ 10^{-3} g^{-1} air-dried lichen material).[25] GOT and GPT have a K$_m$ value of 3.3 $\times$ 10^{-3} *M* (aspartic acid) and 22 $\times$ 10^{-3} *M* (alanine), respectively.

In *Peltigera canina* GOT, GPT and aspartate pyruvate transaminase (AsPT) activities have been reported (Table 8).[28,72] The GOT activities in the cultured *Nostoc* (*Nostoc* CAN)

Table 9
GLUTAMINE SYNTHETASE (EC 6.3.1.2)
ACTIVITY IN *PELTIGERA APHTHOSA* AND
***PELTIGERA CANINA*[a]**

Sample	Activity (nmol product formed min^{-1} mg^{-1} protein)
Peltigera aphthosa	
Whole thallus	25.70
Excised cephalodia	3.50
Thallus without cephalodia	17.50
Digitonin-treated cephalodia (cyanobiont)	4.00
Symbiotic *Coccomyxa*[b]	40.00
Free-living *Nostoc* APH[c]	70.00
Free-living *Coccomyxa*[c]	37.00
Mycobiont (medulla)	0.00
Peltigera canina	
Whole thallus	0.80
Cyanobiont	4.10
Free-living *Nostoc* CAN[c]	72.40
Mycobiont	0.00

[a] Data compiled from References 11, 27, 59 and 72.
[b] Fresh isolates.
[c] Cultured isolates.

and the *P. canina* thallus are rather similar (40.4 and 41 nmol product formed min^{-1} mg^{-1} protein, respectively). However, GOT activity in symbiotic *Nostoc* cells is significantly lower (16.8 nmol product formed min^{-1} mg^{-1} protein).[72] Similar activities are found when GOT is assayed in digitonin-treated thalli.[28] GPT activity in *P. canina* is reported to be 18.12 nmol product formed min^{-1} mg^{-1} protein. As in the case of GOT, the GPT activity in the symbiotic *Nostoc* (digitonin-treated thalli) is considerably lower (0.86 nmol product formed min^{-1} mg^{-1} protein). AsPT activity also is reported to be much lower in the digitonin-treated thalli than in untreated ones (58.6 and 126 nmol product formed min^{-1} mg^{-1} protein, respectively). Thus, it seems that all three aminotransferases mentioned above have much higher activities in the mycobiont than in the cyanobiont.

In *P. aphthosa* GOT, GPT, AsPT, glutamate glyoxylate transaminase and serine trans-hydroxymethylase activities have been reported.[3,27,59,72] Serine transhydroxymethylase activity is highest in the symbiotic *Coccomyxa* and lowest in the cephalodia indicating a very low activity or lack of activity in the cyanobiont. Glutamate glyoxylate transaminase activity is rather similar in cephalodia and the main thallus, but is much lower in the symbiotic *Coccomyxa* (Table 8). AsPT activity in cephalodia is mostly localized in the mycobiont.[27] GPT activity is about ten times higher in cephalodia as compared to that in cultured *Nostoc* APH and is located mostly in the mycobiont.[27] GOT activity in cephalodia is reported to be 20.2 nmol product formed min^{-1} mg^{-1} protein, which goes down to 12.5 on digitonin treatment of cephalodia indicating that though GOT was present in both the cyanobiont and the mycobiont, the latter has a higher specific activity.[72] The cultured *Coccomyxa* shows extremely high GOT activity (520 nmol product formed min^{-1} mg^{-1} protein).

GS activity and its localization in the lichen *P. canina* and *P. aphthosa* has been studied in detail.[27,28,59,72,75] GS activity of the cyanobionts has been reported to be reduced by over 90%, in both species, as compared to that in their cultured counterparts (Table 9). This reduction in activity was initially thought to be due to inactivation of the enzyme,[16] but later

Table 10
**GLUTAMINE SYNTHETASE (EC 6.3.1.2) ACTIVITY AND ANTIGEN LEVELS
IN SYMBIOTIC CYANOBACTERIA FROM *PELTIGERA APHTHOSA* AND *P.
CANINA*[5,11]**

Association	Glutamine synthetase activity of symbiotic cyanobacterium as % of activity of free-living isolate	Glutamine synthetase antigen level of symbiotic cyanobacterium as % of antigen level of free-living isolate
Peltigera canina and *Nostoc* CAN[a]	6	5
P. aphthosa and *Nostoc* APH[b]	5.5	6

[a] Isolate from *P. canina*.
[b] Isolate from *P. aphthosa*.

Table 11
**HETEROCYST
FREQUENCY IN
CYANOBIONTS OF SOME
LICHENS[10,11,72,82,83]**

Lichen	Heterocyst frequency (%)
Lobaria spp.	21—36
Peltigera aphthosa	21
P. canina	2—5
P. polydactyla	6
Sticta spp.	3—5

it was established that in symbiosis the enzyme is not synthesized in the cyanobiont (Table 10).[5,11] Whether this blockage of GS synthesis is at the level of transcription or translation is not yet clear. It has also not been determined yet what the effective mechanism/compound is which causes the reduction in GS synthesis in symbiosis. Recently, sarcosine has been proposed as one of the likely compounds.[80] The low GS levels present in the cyanobionts are not distributed evenly throughout the thallus. Instead, it has been suggested that much of the activity is localized in the cyanobiont in the growing regions of the thallus. Further implications of this with regard to the nitrogen transfer from cyanobiont to the mycobiont are discussed later. GS activity is undetectable in the mycobionts of both *P. aphthosa* and *P. canina*. The eukaryotic phycobiont *Coccomyxa* of *P. aphthosa*, shows normal GS activity and plays a significant role in regulation of nitrogenase which is located in the cyanobiont (see below).

NITROGEN FIXATION AND REGULATION OF NITROGENASE

Nitrogen fixation is restricted to cyanophilic lichens containing heterocystous cyanobionts (see Chapter II.B).

Under aerobic conditions heterocysts are the site of nitrogen fixation in heterocystous filamentous cyanobacteria. The heterocysts frequency is much higher in cyanobionts contained in cephalodia of tripartite lichens (22 to 36%) than in free-living cyanobacteria or in cyanobionts of bipartite lichens (Table 11). The possible explanations for this difference in heterocyst frequency include:

1.   Cyanobionts in cephalodia of tripartite lichens can develop a high percentage of their cells as heterocysts, which are nonphotosynthetic, because the main thallus has a different primary phycobiont which supplies the necessary fixed carbon to the mycobiont.[72]

2.   High heterocyst frequency in cephalodia of tripartite lichens may be caused by the primary phycobiont providing photosynthates, e.g., ribitol, to the cyanobiont and thus stimulating heterocyst formation.[82]

3.   In symbiosis, an effector may be produced by the mycobiont which causes an increase in heterocyst frequency of the cyanobiont by affecting heterocyst expression directly or via its effect on GS. However, this does not seem likely because in bipartite cyanophilic lichens the heterocyst frequency in cyanobionts is not different than in the free-living forms.

Heterocyst frequency in the cyanobionts also varies within the same lichen thallus. In *P. aphthosa*, heterocyst frequency in the cyanobiont of the apex region is lower (14%) than that in the central or basal region (21%) of the thallus.[10]

In the early 1920s, nitrogen fixation in lichens was thought to be due to the presence of *Azotobacter*.[84,85] However, since then nitrogen fixation in numerous lichens has been shown to be due to the presence of symbiotic cyanobacteria. Nitrogen fixation by lichens which contain nonheterocystous cyanobacteria as cyanobionts needs further critical investigation especially keeping in view that nonheterocystous forms, except *Gloeothece*,[86] fix nitrogen under anaerobic/microaerobic conditions only.[87] Acetylene reduction tests carried out on *Pyrenopsis* sp., a lichen reported to contain *Gloeothece*, have been negative.[88]

The first definitive reports of nitrogen fixation by lichens were those in *Collema auriculatum* and *Leptogium lichenoides* using ^{15}N-labeled molecular nitrogen.[89] This was followed by reports of nitrogen fixation in *Peltigera praetextata*,[90] *P. virescens*,[91] *Collema coccophorus*,[92] *C. pulposum*, *Stereocaulon* sp.,[93] *P. aphthosa*,[9] *Lichina confinis*, and *L. pygmaea*.[94] The development of acetylene reduction technique[95] enabled easy, rapid, and convenient testing for nitrogen fixation. Since then numerous lichen species have been tested and verified to be nitrogen-fixing: *Collema* sp.,[7,82,96] *Ephebe* sp.,[82] *Leptogium* sp.,[82] *Lichina* spp.,[7] *Lobaria* spp.,[82,96-99] *Massalongia* sp.,[82] *Nephroma* spp.,[88,100] *Pannaria* spp.,[82] *Parmeliella* spp.,[82] *Peltigera* spp.,[7,10,59,71,72,94,101,102] *Plecopsis* sp.,[7] *Placynthium* spp.,[82] *Polychidium* sp.,[82] *Pseudocyphallaria* sp.,[82] *Solorina* spp.,[82,100] *Stereocaulon* sp.,[103] and *Sticta* spp.[82]

Nitrogenase activity measured along the *Peltigera aphthosa*[10] and *P. canina*[111] thalli show that the highest nitrogenase activity is in the central part of the thallus. This has been correlated with heterocyst frequency in *P. aphthosa* which is lowest at the apex and at the base, and highest in the center.[10] Lower nitrogenase activity in the apex could be explained as due to lower heterocyst frequency, but lower nitrogenase activity at the basal region is due to nonactive heterocysts.[10] There has been some indication that in *P. canina* nitrogen fixation may occur even in vegetative cells of the cyanobiont.[12] This possibility was expressed because of high rates of nitrogen fixation in *P. canina* when calculated in terms of cyanobiont protein. However, there is reason to believe that cyanobiont content in *P. canina* was very much underestimated.[17] Recently, it has been shown that the cyanobiont content in *P. canina* is about 36%[16] rather than 2.7% reported earlier.[12] Keeping this in mind there is no reason to believe that vegetative cells of the cyanobiont in *P. canina* may be fixing nitrogen. Furthermore, it has been shown that nitrogenase activity in symbiotic *Nostoc* of *P. canina* and its cultured counterpart (*Nostoc* CAN) are very similar (7.4 and 9.4 nmol ethylene produced μg^{-1} chlorophyll *a* hr^{-1} for the symbiotic and cultured *Nostoc*, respectively). In *P. aphthosa* the rates of nitrogen fixation by the symbiotic and cultured *Nostoc* were 21 and 9 nmol ethylene produced μg^{-1} chlorophyll *a* hr^{-1}, respectively.[72] The high rates of nitrogen fixation in symbiotic *Nostoc* are due to high heterocyst frequency.[72]

In order to relate rates of acetylene reduction to actual nitrogen fixation it is necessary to determine the ratio of acetylene reduced: ^{15}N-labeled molecular nitrogen fixed. Thus, the actual ratios vary widely from the theoretical value of 3,[104,105] for which various reasons, including hydrogen evolution, have been proposed.[106-109] For *P. aphthosa*,[11] *P. canina*,[28] *Lobaria pulmonaria*,[99] and *L. oregana*[99] a ratio of about 4 has been reported, however, ratios up to 20 have been reported by Millbank.[110] The major difference between the studies of Millbank and others is that Millbank[110] correlated short-term (in minutes) acetylene reduction rates, over a period of 14 days, to the continuous incorporation of ^{15}N-labeled molecular nitrogen, while others[11,28,99] have done short-term experiments (in hours).

In *L. pulmonaria* and *L. oregana* nitrogen fixation (acetylene reduction) rates are reported to be enhanced at low molybdenum concentrations (1 ppm) while at high concentrations (10 ppm) molybdenum inhibited nitrogenase.[99] Regulation of nitrogenase in *P. canina* and *P. aphthosa* by combined nitrogen has been investigated.[11,59,72,76,102] Ammonium chloride (1 to 2 m*M*) fully inhibits acetylene reduction in cultured *Nostoc* CAN, but the inhibition of acetylene reduction in the *P. canina* thallus is only 40 to 45%.[72] Similarly, 1 to 2 m*M* KNO$_3$ inhibits 50 to 60% of the acetylene reduction rate in cultured *Nostoc* CAN, but virtually no inhibition of acetylene reduction occurs in the *P. canina* thallus. It has been proposed that symbiotic cyanobacteria have a mechanism to alleviate the inhibitory effect of combined nitrogen on nitrogenase and that low GS activity in the cyanobiont may be important in the sustenance of a partially active nitrogenase despite the presence of combined nitrogen.

Regulation of nitrogenase in *P. aphthosa* has also been studied in some detail.[59,71,102] Provision of 5 m*M* exogenous ammonia has no effect on the nitrogenase activity of excised cephalodia, whereas in cultured *Nostoc* APH nitrogenase is totally inhibited within 12 hr, the inhibition being detectable within 2 to 3 hr.[59] Such an inhibition is also observed when 5 m*M* ammonium is supplied to intact cephalodia (i.e., *P. aphthosa* disks containing main thallus as well as cephalodia). Total inhibition of nitrogenase is observed within 24 hr, the inhibition being detectable after 6 hr (Figure 3). This inhibition can be overcome by inhibiting main thallus GS by L-methionine-DL-sulphoximine (MSX), an irreversible inhibitor of GS. Addition of exogenous ammonium to thallus disks results in accumulation of large amounts of glutamine in cephalodia when main thallus GS is active, but not when it is inactivated by MSX or when ammonium is added to the excised cephalodia with or without MSX (Table 12). The lack of ammonia inhibition of nitrogenase in excised cephalodia is not due to nonutilization of ammonia because the exogenous ammonia was found to be taken up (Table 13). In view of these observations it has been suggested that a product of ammonia assimilation by GS in the *Coccomyxa* phycobiont (other components have none or negligible GS) affects nitrogenase activity in cephalodia when excess ammonium is available. The inhibitory compound is likely to be glutamine, which when added to excised cephalodia is a potent inhibitor of nitrogenase activity.[59] Furthermore, there is evidence to suggest that *Coccomyxa* phycobiont liberates glutamine when provided with exogenous ammonium (Table 14). When *Coccomyxa* cells were isolated from the thallus and exposed to 5 m*M* ammonium chloride in the presence of ^{14}C-labeled sodium bicarbonate, ^{14}C-labeled glutamine was liberated into the medium. The possibility of lack of an ammonium transport system in symbiotic cyanobacteria has been suggested.[63] If so, the lack of inhibition of nitrogenase by ammonium in symbiotic cyanobacteria may be due to the inability of ammonium to enter the cells. Although, it is not known at present whether cyanobionts of *P. canina* and *P. aphthosa* have an ammonium transport system, such a system has been found in the cyanobiont of *Azolla* and in the cultured *Nostoc* CAN.[64,111] Furthermore, both the mycobiont and the cyanobiont in cephalodia of *P. aphthosa* have been shown to assimilate exogenously supplied ammonium, although the cyanobiont does so at a slower rate due to low activity of the ammonia-assimilating enzyme GS.[59,102] Studies on *P. aphthosa* have indicated that ammonium and a product of ammonia assimilation via GS may exert independent inhibitory effects on nitro-

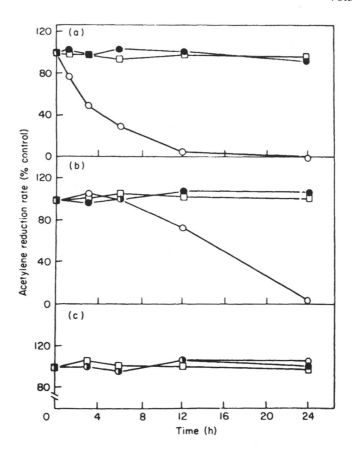

FIGURE 3.   The effect of combined-nitrogen on the rate of acetylene reduction by (a) cultured *Nostoc* APH, (b) disks, and (c) excised cephalodia of *P. aphthosa*. The differential treatments begun at time zero were: ○ = 5 m*M* ammonium chloride; ● = 5 m*M* ammonium chloride plus 0.25 m*M* MSX; and □ = 5 m*M* potassium chloride. The 100% acetylene reduction activity of cultured *Nostoc* APH was 9.3 nmol ethylene produced hr^{-1} μg^{-1} chlorophyll *a* and, of the disks and cephalodia, 0.40 and 6.35 nmol ethylene produced hr^{-1} mg^{-1} fresh weight, respectively. (From Rai, A. N. et al., *New Phytol.*, 85, 545, 1980. With permission.)

genase activity in the cyanobiont.[102] Such a conclusion has also been drawn from studies on the glutamine-auxotroph of the cultured cyanobacterium *Anabaena cycadeae*.[112] However, ammonium *per se* inhibits nitrogenase in the cyanobiont only when ammonia is allowed to accumulate in high concentrations, e.g., in the absence of carbon dioxide in darkness.[102] Normally, inhibition of nitrogenase by ammonia does not occur because: (1) ammonia does not accumulate since much of it is assimilated by the high level of GDH in the mycobiont hyphae associated with the cyanobiont; and (2) due to low GS activity in the cyanobiont very little assimilation of ammonia occurs via GS in the cyanobiont. Thus, both the ammonia level and the level of the product of its assimilation via GS remain below the inhibitory level. There is evidence to suggest that in free-living cyanobacteria, ammonia causes very little or no inhibition of nitrogenase when GS activity is low.[61,113,114]

Cyanobacteria fix nitrogen in light and in darkness, but the extent of dark nitrogen fixation is much lower compared to that in the light and is dependent on the supply of fixed carbon compounds previously accumulated in the light or supplied exogenously.[87] However, in *P.*

Table 12

EFFECTS OF EXOGENOUS AMMONIUM ON THE LEVELS OF FREE AMINO ACIDS PRESENT IN EXCISED AND ATTACHED CEPHALODIA OF *PELTIGERA APHTHOSA*[a]

| Amino acids | Untreated excised cephalodia | Ammonium treatment | | Ammonium treatment after preincubation with MSX | |
		Excised cephalodia	Cephalodia attached to disks during treatment	Excised cephalodia	Cephalodia attached to disks during treatment
Glutamate	14.90	12.60	22.20	10.70	14.50
Glutamine	4.60	8.50	27.90	2.90	3.90
Alanine	9.60	43.90	12.10	45.60	27.60
Others[b]	21.40	15.50	30.60	16.10	29.20
Total	50.50	80.50	92.80	75.30	75.20

[a]   The thalli were preincubated in the presence or absence of 0.25 mM MSX for 24 hr. Excised cephalodia and cephalodia-containing disks from such thalli were then incubated, as shown, in the presence or absence of MSX and 5 mM ammonium chloride for 24 hr, after which the free amino acid pools were extracted and determined. Amino acid concentrations are expressed as nanomole amino acid per milligram of fresh weight of cephalodia.

[b]   Mainly aspartate, threonine, serine, asparagine, isoleucine, glycine, and phosphoserine.

From Rai, A. N. et al., *New Phytol.*, 85, 545, 1980. With permission.

Table 13

UPTAKE OF AMMONIUM BY DISKS AND BY EXCISED CEPHALODIA OF *PELTIGERA APHTHOSA* IN THE PRESENCE AND ABSENCE OF MSX[a]

| Material | Material incubated in the absence of MSX | | Material preincubated in the presence of MSX | | Percent reduction in ammonium uptake on adding MSX |
	Rate of ammonium uptake	Glutamine synthetase activity	Rate of ammonium uptake	Glutamine synthetase activity	
P. aphthosa disks	2.22	25.70	1.15	not determined	48.19
Excised cephalodia	2.48	2.00	2.44	not determined	1.61

[a]   Samples were preincubated in the presence or absence of 0.25 mM MSX for 24 hr, then 5 mM ammonium chloride was added, and disappearance of ammonia determined over a further 24 hr period. The rate of ammonium uptake is expressed as nanomole ammonium assimilated per hour per milligram of fresh weight. Glutamine synthetase activity was determined at the end of the preincubation period with MSX and activities are expressed as nanomole product formed per minute per milligram of protein.

From Rai, A. N. et al., *New Phytol.*, 85, 545, 1980. With permission.

aphthosa the cyanobiont shows nitrogenase activity (acetylene reduction) for substantially longer periods in the darkness than it does when in the free-living state (Figure 4).[71] Dark nitrogenase activity both of the lichen disks and of the excised cephalodia continues at a similar rate to that in the light for 6 hr after which a slow decline is observed with 75% activity remaining after 12 hr and about 30% after 24 hr of darkness. In cultured *Nostoc* APH, nitrogenase activity ceases within 6 hr of darkness; the activity being less than 20%

Table 14
LIBERATION OF GLUTAMINE BY SYMBIOTIC
COCCOMYXA IN THE PRESENCE OF EXCESS
AMMONIA[a]

Incubation time (hr)	[14]C counts in glutamine liberated into the medium (Bq g^{-1} dry wt)	Amount of glutamine liberated into the medium[b] (nmol g^{-1} dry wt)
1	17×10^4	459.46
3	29×10^4	783.78
6	39×10^4	1054.05

[a]   Symbiotic *Coccomyxa* cells were isolated from *Peltigera aphthosa* thalli[11] and were immediately suspended into BG-11$_o$ medium[81] containing 5 m*M* ammonium chloride and ^{14}C-labeled sodiumbicarbonate (specific activity 37 kBq per nmol). At time intervals samples of the medium were analyzed using an LKB 4400 Amino Acid Analyzer and a lithium citrate buffer system. The amino acid peak fractions were collected off the column and the radioactivity counted in a Packard Tri-Carb 2660 Scintillation Spectrometer.

[b]   Calculated from the ^{14}C counts in glutamine liberated into the medium.

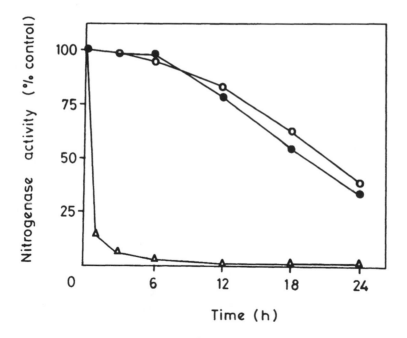

FIGURE 4.   Nitrogenase activity of disks (●), excised cephalodia (○), and cultured *Nostoc* APH (△) from *Peltigera aphthosa* placed in the dark at time zero and incubated under air.[11] Nitrogenase activities are expressed as percentages of control rates (air plus light) which were: disks, 0.38 nmol ethylene produced mg^{-1} fresh weight hr^{-1}; cephalodia, 6 nmol ethylene produced mg^{-1} fresh weight hr^{-1}; cultured *Nostoc* APH 9.5 nmol ethylene produced μg^{-1} chlorophyll *a* hr^{-1}.

within an hour in the darkness. Dark nitrogenase activity of the symbiotic *Nostoc* is supported by the catabolism of polyglucose accumulated in the light which in darkness serves to supply reductant and ATP. This is evidenced by the fact that the glucose content of cephalodia decreases as the dark incubation period increases and the glycogen granules which are present in the cephalodial *Nostoc* disappear after dark incubation.[71] The accumulation of enough polyglucose in the cyanobiont to support prolonged dark nitrogen fixation seems surprising at first considering the fact that in symbiosis a larger proportion of cyanobacterial cells becomes nonphotosynthetic heterocysts. Moreover, *Coccomyxa*, the only other photosynthetic symbiont in the thallus, seems unimportant as a source of fixed carbon for the *Nostoc* because the light and dark nitrogenase activities of cephalodia are unaffected by their attachment to the *Coccomyxa*-containing main thallus (see Figure 4). This, together with the fact that the cyanobiont is photosynthetically active and provides some fixed carbon to the mycobiont, suggests that the sustained nitrogenase activity of the symbiotic *Nostoc* is not dependent on the supply of fixed carbon from the *Coccomyxa*. Dark nitrogenase activity in digitonin-treated cephalodia is as high and prolonged as in intact cephalodia,[102] indicating that the mycobiont is not essential as a source of reductant or energy for the dark nitrogenase activity. This also rules out the involvement of a malate-aspartate type shuttle between the cyanobiont and the mycobiont for provision of reductant from the mycobiont, such as the one reported in *Alnus*.[117]

On the other hand, there are several factors which may facilitate accumulation of large amounts of polyglucose in the symbiotic *Nostoc* in the light, which will be sufficient for supporting dark nitrogenase activity over a prolonged period. First, the vegetative cells scarcely grow so that there is not much demand for fixed carbon for vegetative cell metabolism. Second, there is less demand, compared to the free-living *Nostoc*, for fixed-carbon skeletons for assimilation of ammonia produced during nitrogen fixation because ammonia assimilation occurs mainly in the mycobiont and very little in the cyanobiont. Thirdly, there is less of a drain by the mycobiont on fixed carbon produced by the cyanobiont because much of the fixed carbon required by the mycobiont is supplied by the *Coccomyxa*. Fourth, transfer of fixed carbon from cyanobiont to the mycobiont in darkness is very much reduced compared with that in the light.[48]

However, this availability of fixed carbon alone to supply adenosine triphosphate (ATP) and reductant for nitrogenase is not sufficient for prolonged nitrogenase activity in darkness. In the absence of carbon dioxide, dark nitrogenase activity declines much faster (Figure 5) although the rates of polyglucose degradation are similar in the presence and absence of carbon dioxide.[71] It has been shown that dark carbon dioxide fixation via phosphoenolpyruvate carboxylase (PEP-case) provides carbon skeletons for ammonia assimilation in the mycobiont by replenishing the tricarboxylic acid (TCA) cycle and is essential for prolonged dark nitrogenase activity. When PEP-case is specifically inhibited by maleic acid or malonic acid, dark nitrogenase activity declines much faster than the control (Figure 6), although these inhibitors do not show any effect on nitrogenase activity in the light, indicating that the inhibitors do not affect nitrogenase activity directly. The importance of dark carbon dioxide fixation in the mycobiont for the generation of carbon skeleton for ammonia assimilation is crucial since the mycobiont receives very little fixed carbon, if any, from the photosynthetic symbionts during darkness. Furthermore, unlike the mycobiont hyphae in the main thallus, the mycobiont hyphae in cephalodia synthesize and store very little mannitol.[116] Nitrogenase activity in the dark was inhibited in the absence of carbon dioxide, because in the absence of carbon dioxide the ammonia level in cephalodia goes up several-fold (due to lack of carbon skeletons to assimilate it) and this causes the inhibition of nitrogenase.[102] Such an inhibition is also caused by addition of exogenous ammonia when carbon skeletons are in short supply, e.g., in darkness and in the absence of carbon dioxide (Figure 7).

In the presence of digitonin, which disrupts mycobiont membranes and allows the release

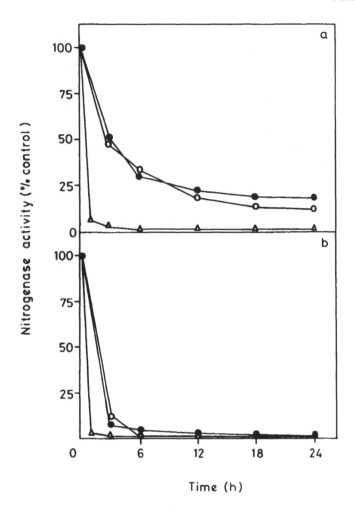

FIGURE 5.   Nitrogenase activity of disks (●), excised cephalodia (○), and cultured *Nostoc* APH (△) of *Peltigera aphthosa* placed in the dark at time zero and incubated under (a) N_2/O_2 (79/21, v/v) or (b) N_2/CO_2 (99.96/0.04, v/v).[11] Nitrogenase activities are expressed as percentages of control rates (light plus air) which were similar to those in Figure 4.

of nitrogen fixation-derived ammonia into the medium (i.e., when ammonia accumulation in the cephalodia was prevented) absence of carbon dioxide does not affect the dark nitrogen fixation (Figure 8).[102]

The question whether the inhibitory compound in the dark is ammonia or a product of ammonia assimilation has been addressed.[102] When GS activity of the cyanobiont (although low, the cyanobiont does contain some GS activity) is inhibited by MSX, the inhibition of dark nitrogenase activity caused by the absence of carbon dioxide persists, although MSX as such does not affect nitrogenase activity (Figure 9). Also, when amino acid pools of cephalodia incubated under various treatments were investigated, it was found that there is a consistent inverse correlation between nitrogenase activity and ammonia level, the pool of which increased with increasing inhibition of dark nitrogenase activity (see Figure 9 and Table 15). The glutamine pool also showed an increase as the dark nitrogenase activity declined, except in the presence of MSX, when it declined and yet nitrogenase remained inhibited. The above observations indicate that it is ammonia, and not a product of its assimilation via GS, that inhibits dark nitrogenase activity in the absence of carbon dioxide

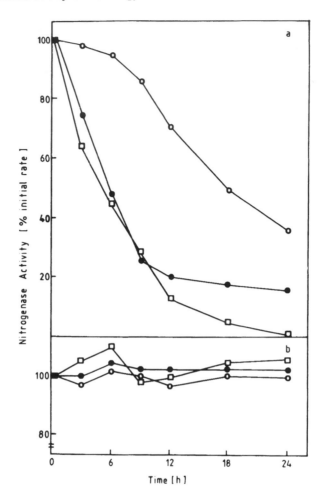

FIGURE 6.   The effect of PEP carboxylase inhibitors, maleic acid, and malonic acid on nitrogenase activity of excised cephalodia in (a) darkness and in (b) light. The nitrogenase activity at time zero was 20 to 25 nmol ethylene produced hr^{-1} mg^{-1} dry weight. Symbols: $\bigcirc$ = control; $\bullet$ = 5 mM maleic acid: $\square$ = 10 mM malonic acid. (From Rai, A. N. et al., *Physiol. Plant.*, 57, 285, 1983. With permission.)

because of its accumulation and increased level. Considering earlier findings that a product of ammonia assimilation if also involved in the inhibition of nitrogenase in cyanobacteria, it could be concluded that ammonia and a product of its assimilation via GS exert independent inhibitory effects on nitrogenase in cyanobacteria.[102] Dark nitrogenase activity in symbiotic cyanobacteria like those in *P. aphthosa* may be important in meeting the fixed nitrogen requirements of the other symbionts during the brief growth period, when favorable growth conditions prevail.

LIBERATION OF FIXED NITROGEN BY CYANOBIONTS AND ITS UTILIZATION BY OTHER SYMBIONTS

Substantial release of fixed nitrogen was shown to occur from *Nostoc* isolates of *Collema tenax* by Henriksson,[118] and it was speculated that the same may also occur in symbiosis, which was later confirmed.[90] Similar conclusions have been drawn in the case of *P. aphthosa*

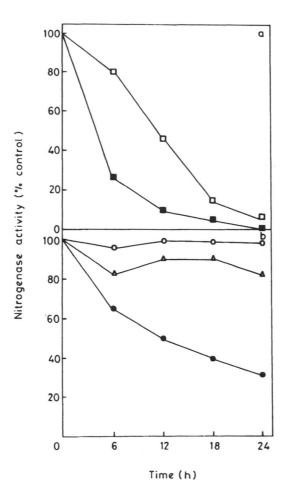

FIGURE 7.   Nitrogenase activity of excised cephalodia from *Peltigera aphthosa*.[11] Nitrogenase activity in (a) darkness under air plus 5 m*M* ammonium chloride (□) and N$_2$/O$_2$ (79/21, v/v) plus 5 m*M* ammonium chloride (■). Nitrogenase activity in the (b) light under air plus 5 m*M* ammonium chloride (○), under N$_2$/O$_2$ (79/21, v/v) plus 5 m*M* ammonium chloride (●), and under N$_2$/O$_2$ (79/21, v/v) (△). Nitrogenase activities are expressed as percentages of the control rates (light plus air) which were similar to those in Figure 4.

where ^{15}N-labeled molecular nitrogen fixed by the cyanobiont was found to be present in other parts of the thallus.[9] Later, using digitonin to disrupt the mycobiont membranes so that the actual release of fixed nitrogen by the cyanobiont could be followed, release of newly produced ammonia, i.e., ammonia produced during active nitrogen fixation, was shown to occur in *P. canina*.[72] A similar approach has been followed with *P. aphthosa* (Figure 10). Newly fixed ammonia was found to be liberated by the cyanobiont into the medium when the mycobiont membranes were disrupted by digitonin;[74] the ammonia thus liberated amounted to nearly 95% of the total fixed nitrogen.

The form(s) in which fixed nitrogen is released from the cyanobiont was first investigated by Millbank using *P. aphthosa*, *P. canina*, *P. polydactyla*, *Lobaria amplissima* and *Placopsis gelida*.[119] A mixture of nitrogenous compounds, largely peptides, was found to be released. A significant point to remember here is that in these experiments intact disks or cephalodia

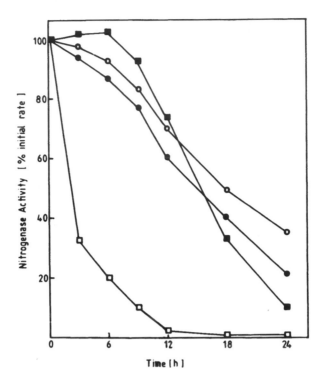

FIGURE 8.   The effect of digitonin on dark nitrogenase activity
of excised cephalodia in the presence and absence of CO_2. Nitrogenase activity at zero time was 20 to 25 nmol ethylene produced
mg^{-1} dry weight hr^{-1}. Symbols: $\bigcirc$ = air; $\bullet$ = air plus 0.01%
digitonin (w/v); $\square$ = N_2/O_2 (79/21, v/v); $\blacksquare$ = N_2/O_2 (79/21, v/
v) plus 0.01% digitonin (w/v). (From Rai, A. N. et al., *Physiol.
Plant.*, 57, 285, 1983. With permission.)

were used and, therefore, the released compounds may not truly reflect the nature of fixed
nitrogen released by the cyanobionts. In fact, it is most likely that the fungal metabolism
may modify the original nature of the fixed nitrogen compound released by the cyanobiont
before its release into the incubation medium. Detailed investigations in *Peltigera canina*,
using ^{15}N-labeled molecular nitrogen, show that a large proportion of the released nitrogen
is in the form of newly fixed ammonia.[72] This was shown by using digitonin-treated disks
of *P. canina* where fungal metabolism was eliminated. Therefore, whatever came out into
the incubation medium may represent the actual form of fixed nitrogen liberated by the
cyanobiont. Nearly 50% of the fixed nitrogen remained in the disks and the remaining 50%
was released. Of the latter, half was organic nitrogen and the other half ammonia. Rai et
al.[120] using ^{15}N-labeled molecular nitrogen also followed, in a more carefully designed experiment, the form of fixed nitrogen released by the cyanobiont of *P. canina* (Table 16).
Nearly 44% of the fixed nitrogen remained in the thallus while 56% was released into the
medium almost exclusively as ammonia. In these experiments digitonin treatment preceded
the incubation with ^{15}N-labeled molecular nitrogen and the disks were thoroughly washed
before they were exposed to the ^{15}N-labeled molecular nitrogen. In the experiments of Stewart
and Rowell[72] a larger proportion of organic nitrogen was found to be liberated by the
cyanobiont. This was probably due to fungal enzyme activities converting ammonia into
amino acids, because the thalli were not treated with digitonin beforehand. Instead, the
digitonin treatment and incubation with ^{15}N-labeled molecular nitrogen were simultaneous.
   Similar studies on the form of fixed nitrogen released by *P. aphthosa* cyanobiont indicate

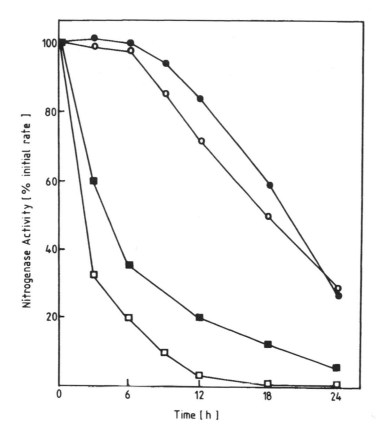

FIGURE 9. The effect of CO_2 on dark nitrogenase activity of excised cephalodia in the presence and absence of MSX. Nitrogenase activity at time zero was 20 to 25 nmol ethylene produced hr^{-1} mg^{-1} dry weight. Symbols: ○ = air; ● = air plus MSX (0.25 mM); □ = N_2/O_2 (79/21, v/v); ■ = N_2/O_2 (79/21, v/v) plus MSX (0.25 mM). (From Rai, A. N. et al., *Physiol. Plant.*, 57, 285, 1983. With permission.)

that nearly 95% of the total fixed nitrogen is released almost exclusively as ammonia; no organic nitrogen or amino acids could be detected.[59] In another experiment the digitonin-treated cephalodia were exposed to ^{14}C sodium bicarbonate and the medium was analyzed at time intervals for the liberation of any ^{14}C-labeled nitrogenous compound, but none were found. Such data indicate that at least in *P. canina* and *P. aphthosa* the fixed nitrogen released by the cyanobiont is in the form of ammonia. Other lichen cyanobionts also have to be tested critically, using digitonin, ^{15}N-labeled molecular nitrogen, and ^{14}C-labeled bicarbonate, for the release of any organic nitrogen. One must remember that to know the actual form of fixed nitrogen transferred by the cyanobiont, mycobiont metabolism has to be totally stopped, e.g., by digitonin treatment. Also, the digitonin-treated disks and cephalodia should be properly washed before suspending in the experimental medium for monitoring the release of fixed nitrogen. Loss of GDH activity may be used as a marker for disruption of the mycobiont metabolism. When using this method the time and concentration of digitonin incubation required for a particular lichen material should be established beforehand.

Now we came to the question of why fixed nitrogen is released by the cyanobiont. In cyanobacteria, assimilation of newly fixed ammonia occurs via the GS-GOGAT pathway.[62] GS is located both in heterocysts and vegetative cells and GOGAT only in vegetative

Table 15

**CHANGES IN THE POOLS OF AMMONIA
AND OF MAJOR AMINO ACIDS OF
EXCISED CEPHALODIA IN DARKNESS IN
THE PRESENCE AND ABSENCE OF CO$_2$
AND MSX[a]**

	nmol mg^{-1} Cephalodial dry weight			
Compound	Control (air)	Dark (air)	Dark (air-CO$_2$)	Dark (air-CO$_2$) plus MSX
Aspartate	8.40	8.70	12.90	12.20
Threonine	1.30	1.30	2.00	2.20
Serine	4.00	2.40	5.00	4.70
Glutamate	23.40	34.50	33.40	39.70
Glutamine	13.00	25.90	42.30	15.50
Alanine	20.30	12.00	25.00	57.00
Ammonia	21.10	23.20	38.90	41.50

[a]   Cephalodia were excised from the main thallus and were incubated for 3 hr in a tenfold dilution of the medium BG-11$_o$ under air or air-CO$_2$ (N$_2$/O$_2$, 79:21 v/v) as shown. MSX, where required, was added to a final concentration of 0.25 mM. Amino acid pool control values were obtained immediately before transfer to darkness.

From Rai, A. N. et al., *Physiol. Plant.*, 57, 285, 1983. With permission.

cells.[121,122] In lichens the GS and GOGAT activity of the cyanobiont is decreased by over 95%, with less than 5% of the normal activity remaining.[5,59,72] Based on such observations, it has been suggested that the liberation of fixed nitrogen by the cyanobiont which is almost exclusively in the form of ammonia, is due to the inability of the cyanobiont to assimilate all the newly fixed ammonia being generated by nitrogen fixation because of insufficient GS.[59,76,77,120] The fact that fixed nitrogen is liberated in the form of ammonia supports this view as well as do ^{15}N studies on *P. canina*.[120] In *P. canina* the flux of ^{15}N through the amide nitrogen of glutamine is considerably less than the rate of ^{15}N displacement from ammonia, and the flux of ^{15}N through glutamate is far greater than ^{15}N displacement from the amide nitrogen of glutamine. It is, therefore, evident that not all the ammonia generated by nitrogen fixation is assimilated in the cyanobiont of *P. canina*. A similar conclusion was reached in the case of *P. aphthosa* cephalodia in which routes of nitrogen metabolism were studied using ^{15}N-labeled molecular nitrogen.[27]

Both in *P. canina* and *P. aphthosa* the specific activities of GS and nitrogenase are comparable and thus may be expected to assimilate all the ammonia produced by nitrogenase.[59,76,77,120] However, as mentioned above, we know that not all the ammonia produced by nitrogenase is assimilated via GS and that over 50% of the newly fixed ammonia in *P. canina*[120] and over 90% in *P. aphthosa* cyanobionts are liberated. The inability of the GS-GOGAT pathway to assimilate all the ammonia produced by nitrogenase, despite their comparable levels, can be explained as follows: first, the heterocysts which are the site of nitrogen fixation are only a small percentage of the total cell population. Considering that nitrogenase is located only in heterocysts while GS is located both in heterocysts and vegetative cells, the total GS activity in heterocysts would be only a small percentage of that observed in whole filaments of the cyanobiont and would be far less than the total

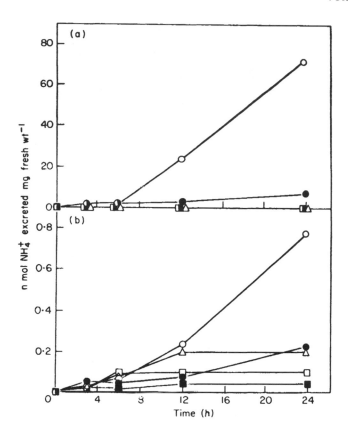

FIGURE 10. The effects of digitonin (0.01%, w/v) on the liberation of ammonia by (a) excised cephalodia and (b) disks of *Peltigera aphthosa*. Digitonin was added at time zero and ammonia production determined over 24 hr. Symbols: ○ = light, air, plus digitonin; ● = dark, air, plus digitonin; □ = light and air; ■ = light and argon/O_2/CO_2 (77.96/22/0.04, by volume); △ = light, argon/O_2/CO_2 plus digitonin. The rates of acetylene reduction by excised cephalodia and disks were 6.6 and 0.32 nmol ethylene produced hr^{-1} mg^{-1} fresh weight, respectively. (From Rai, A. N. et al., *New Phytol.*, 85, 545, 1980. With permission.)

nitrogenase activity. Second, GS would be required not only to assimilate newly fixed ammonia but also the ammonia produced during protein turnover. Third, as in *Azolla*,[123] there is no strict linear relationship between GS and nitrogenase throughout the lichen thallus. Maximum nitrogenase activity has generally been found in the center of the thallus, whereas GS activity is greatest towards the apex.[10,111] Since in most experiments whole thalli or randomly sampled large number of disks are used, any difference in GS or nitrogenase activity related to the age of the material is masked. Taken together, the above observations indicate insufficiency of GS to assimilate all the newly fixed ammonia in the cyanobiont. This leads to ammonia being liberated. Such a situation when created in free-living cyanobacteria by inhibiting GS with MSX also leads to liberation of newly fixed ammonia.[61]

Ammonium transport in cyanobacteria occurs in response to the electrical potential across the membrane ($\Delta\psi$).[64] It has been suggested that a decrease in $\Delta\psi$ by compounds like colicin K,[124] should lead to a considerable loss of ammonia from the cyanobiont cells in symbioses. However, one must remember that $\Delta\psi$ is not required only for ammonium transport but may be involved in many other processes necessary for cellular metabolism, including nitrogenase action,[125-128] and a general lowering of the electrical potential may not be desirable for the cyanobiont.

Table 16
INCORPORATION OF ^{15}N INTO DISKS,
EXTRACELLULAR AMMONIA, AND
EXTRACELLULAR ORGANIC-
NITROGEN OF DIGITONIN-TREATED
PELTIGERA CANINA[a]

Material	Percent of total fixed nitrogen in each sample as monitored by ^{15}N label
Disks	44.50
Extracellular amino acids (organic-nitrogen)	1.80
Extracellular ammonia	53.70

[a]   *Peltigera canina* thalli were incubated in a tenfold dilution of BG-11$_0$ medium[81] containing 0.01% (w/v) digitonin for 18 hr. One-cm-diameter disks were then cut, washed, resuspended in fresh medium, and then exposed to ^{15}N-labeled molecular nitrogen for 6 hr followed by determination of the ^{15}N content of the disks and of the ammonium and amino acids liberated into the medium. Data compiled from Reference 28.

Nitrogen fixed by the cyanobiont is transferred to other symbionts where it is utilized to fulfill the nitrogen requirements of the nonnitrogen-fixing symbionts. This aspect has been investigated in detail in *P. canina* and *P. aphthosa* using ^{15}N-labeled molecular nitrogen as tracer.[9,27,28,59,129] Based on preliminary experiments Millbank and Kershaw[9] suggested that virtually all of the nitrogen fixed by the cyanobiont is transferred to the rest of the thallus. This study was further extended by investigating the movement of ^{15}N over a period of 55 days. Labeled nitrogen which was fixed and liberated by the cyanobiont was found to be present in the thallus mycobiont and the primary phycobiont. However, when calculations were made considering the proportion of algal and fungal cells in the thallus, *Coccomyxa* cells received only 3% of the expected share of fixed nitrogen. This led to the conclusion that the cyanobiont was of minimal value as a source of fixed nitrogen to the primary phycobiont *Coccomyxa*.[129] Later, studies by Rai et al.[27,59] showed that nearly 95% of the newly fixed nitrogen was released by the cyanobiont, of which about 20% remained in cephalodia and the remaining 80% was transferred to the main thallus, probably mainly as alanine (see below). Of the total nitrogen fixed by the cyanobiont 5.2% was received by the primary phycobiont *Coccomyxa* (Figure 11). Since *Coccomyxa* constitutes about 9% of the thallus biomass,[4] it received about 60% of the fixed nitrogen expected on the assumption that the ^{15}N label would be distributed between the *Coccomyxa* and fungus in proportion to their contribution to the composition of the thallus. This is far higher than the value reported by Kershaw and Millbank,[129] who concluded that *Coccomyxa* received only 3% of the expected amount. The low labeling reported by Kershaw and Millbank[129] was most probably due to the loss of nitrogenous compounds from the leaky symbiotic *Coccomyxa* cells[130] during the prolonged procedure used to separate the algal cells from the remainder of the thallus. Even in the experiments of Rai et al.[27] where *Coccomyxa* showed about 60% of the expected level, some loss of nitrogenous material may have occurred during its separation from the main thallus although the separation procedure was much quicker than that of Kershaw and Millbank.[129] Thus, it can be said that the cyanobiont provides fixed nitrogen not only to the mycobiont, but that it provides nearly all the fixed nitrogen requirements of the primary phycobiont as well.

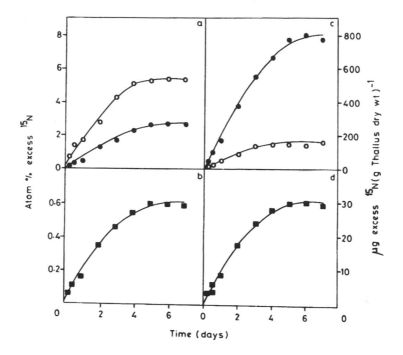

FIGURE 11. Distribution of fixed nitrogen between various components of *Peltigera aphthosa* after exposure of cephalodia-containing disks to ^{15}N-labeled molecular nitrogen for up to 6 days.[11] (a) and (b) = ^{15}N-labeling of each component (atom % excess ^{15}N); (c) and (d) = total ^{15}N label present in each component (μg ^{15}N g^{-1} cephalodial dry weight). Symbols: ● = disks without cephalodia (cephalodia detached at end of incubation period); ○ = cephalodia (attached to thallus disks during exposure and excised at end of exposure period to ^{15}N); ■ = *Coccomyxa* separated from disks at end of exposure to ^{15}N.

In *P. canina*, Stewart and Rowell[72] showed that 50% of the fixed nitrogen was liberated by the cyanobiont. This was confirmed by Rai et al.[28] who found that 55% of the newly fixed nitrogen was released by the cyanobiont and was metabolized by the mycobiont. By comparing the data obtained with *P. aphthosa* and *P. canina* it was found that the amount of fixed nitrogen liberated by the cyanobiont of *P. aphthosa* was higher both on a percentage basis (over 90% for *P. aphthosa* and 55% for *P. canina*) as well as per unit cyanobcaterial protein basis (0.53 and 0.12 μ nitrogen hr^{-1} mg^{-1} cyanobacterial protein for *P. aphthosa* and *P. canina*, respectively). However, since the cyanobiont of *P. canina* constitutes 36% of the thallus protein while that of *P. aphthosa* only about 3%, the fixed nitrogen liberated by the cyanobiont per unit of thallus protein is actually lower in *P. aphthosa* (17 ng nitrogen hr^{-1} mg^{-1} thallus protein) than in *P. canina* (45 ng nitrogen hr^{-1} mg^{-1} protein).[28] The cyanobiont in the cephalodia of *P. aphthosa* transfers fixed nitrogen during the daytime as well as at night, since large amounts of fixed carbon are available for dark nitrogen fixation.[71] This may not be possible in *P. canina* because the cyanobiont provides fixed nitrogen as well as fixed carbon to the mycobiont and, therefore, not enough fixed carbon is available for dark nitrogen fixation. Thus, the contribution of cyanobionts of *P. aphthosa* and *P. canina* regarding fixed nitrogen supply to the symbionts may be similar.

ROUTES OF NITROGEN METABOLISM

Pathways of nitrogen metabolism in *Peltigera canina* and *P. aphthosa* have been studied using ^{15}N-labeled molecular nitrogen as tracer.[11,27,28] The rate of ^{15}N incorporation in *P.*

canina was found to be 0.85 nmol hr^{-1} mg^{-1} dry weight. After a 3 hr exposure to ^{15}N molecular nitrogen, 75% of the total ^{15}N incorporated was found in the ethanol soluble fraction. Of this, 83% was present in the five major nitrogenous compounds; ammonia, glutamine, glutamate, alanine, and aspartate.

When kinetics of ^{15}N-labeling, from ^{15}N-labeled molecular nitrogen, into ammonia and major amino acids is followed over a 30-min period, it was found that the greatest initial labeling was into ammonia followed by glutamate and amide nitrogen of glutamine. Labeling of amino nitrogen of glutamine, aspartate, and alanine increased slowly. When a 30-min pulse of ^{15}N was followed by a ^{14}N chase over a 30-min period and ^{15}N-labeling of ammonia and amino acids was followed, it was found that the ^{15}N label of ammonia decreased most rapidly followed by the rates of decrease in label of glutamate and amide nitrogen of glutamine. On the other hand, labeling of alanine and aspartate increased during the chase period although label in aspartate declined after the increase during the first 5 min of the chase period. Such data indicate that glutamate and glutamine are synthesized by primary amination while aspartate and alanine were formed secondarily. Further analysis of data revealed that the displacement of ^{15}N from ammonia was far greater than the flux of ^{15}N through glutamine amide nitrogen. This indicated that GS was not responsible for assimilation of all the ammonia produced during nitrogen fixation. Also, the flux of ^{15}N through glutamate was far greater than the displacement from amide nitrogen of glutamine, indicating that only a small portion of glutamate was synthesized via GS. Overall, the above observations suggest that both GS and GDH may be involved in primary ammonia assimilation in *P. canina*. Since GS is localized in the cyanobiont and GDH in the mycobiont it could be assumed that the ammonia retained in the cyanobiont is assimilated via GS while the ammonia liberated by the cyanobiont is assimilated via GDH in the mycobiont. That ammonia assimilation in the mycobiont occurs via GDH was indicated by the fact: (1) that the flux of nitrogen through glutamate is higher than the rate of displacement from amide nitrogen of glutamine; (2) that glutamate is an early product of ammonia assimilation; (3) that when GS and GOGAT activity is inhibited using MSX and azaserine (Figure 12), the labeling into glutamate still continues although glutamine does not get labeled; and (4) that the mycobiont does not contain detectable activity of GS or GOGAT, but instead has a very high NADPH-dependent GDH activity which works in the biosynthetic direction.[28,73]

In view of the finding that the metabolic pools of amino acids, which may be small but turning over rapidly, and the existence of separate storage pools as well as the fact that activities of GS and nitrogenase are not uniformly distributed throughout the thallus,[111] one may speculate that much of the glutamine is produced in growing apical region of the thallus and subsequently transferred to the rest of the thallus and incorporated into storage pools. Extremely low GS activity in the central part of the thallus coupled with high mycobiont GDH activity suggest that much of the ammonia liberated in the center of the thallus is transferred to the mycobiont and assimilated there via GDH.

P. aphthosa readily incorporates ^{15}N-labeled molecular nitrogen via nitrogen fixation by the cyanobiont in cephalodia and transfers fixed nitrogen from *Nostoc*-containing cephalodia to the main thallus.[9-11,27,59,129] Excised cephalodia continue to fix and transfer nitrogen for up to 3 days, indicating that there is no obligate dependence on the main thallus for nitrogenase to function.[27]

^{15}N kinetic studies carried out over a 30-min period using excised cephalodia show that the highest initial label is into ammonia. After 12 min, however, little further increase in ammonia label occurs while that in the amide group of glutamine and glutamate continues to increase. The ^{15}N-labeling of the amino group of glutamine and aspartate increases more slowly, followed by an increase in the labeling of alanine. The lag in the labeing of aspartate and alanine suggest that they may be formed secondarily after ammonia, glutamate, and glutamine.

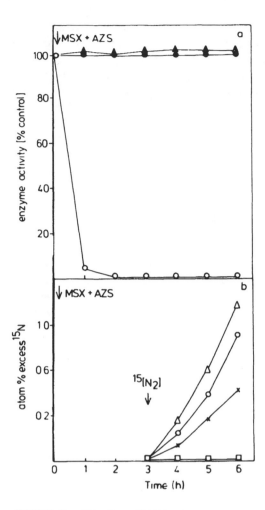

FIGURE 12. The effect of MSX and azaserine on activities of (a) nitrogenase (▲), glutamine synthetase (○), and glutamate dehydrogenase (●); and on (b) the ^{15}N-labeling pattern of the major amino acids in *Peltigera canina*. The thalli were incubated in BG-11₀ medium containing 0.25 mM MSX and 0.20 mM azaserine, and the activities of nitrogenase, glutamine synthetase, and glutamate dehydrogenase were monitored over a 6 hr period. In a second series of experiments (b), the thalli were incubated as above and after 3 hr, when glutamine synthetase activity had declined to zero, ^{15}N-labeled molecular nitrogen was introduced and the subsequent ^{15}N-labeling of glutamate (△), glutamine (□), aspartate (X), and alanine (○) was determined. The initial enzymic activities were nitrogenase, 0.75 nmol ethylene produced min^{-1} mg^{-1} protein; glutamine synthetase, 0.8 nmol product formed min^{-1} mg^{-1} protein; glutamate dehydrogenase, 195.8 nmol product formed min^{-1} mg^{-1} protein. (From Stewart, W. D. P. et al., *Ann Microbiol. (Inst. Pasteur)*, 134B, 205, 1983. With permission.)

When a 30-min pulse of ^{15}N-labeled molecular nitrogen is followed by a 30-min chase there is rapid depletion of ^{15}N label from ammonia. This was followed by decreases in the labeling of glutamate and amide nitrogen of glutamine. Such data indicate that glutamine and glutamate are the primary products of ammonia assimilation. Aspartate shows a slight increase in its ^{15}N label during the first 5 min of the chase. Its labeling, thereafter, decreased. [15]N label in alanine increased throughout the 30-min chase period, suggesting that alanine is derived secondarily from other amino acids and that ADH is not important in primary ammonia assimilation in cephalodia. Such a conclusion has been made also in the case of the free-living cyanobacterium *Anabaena cylindrica*.[131] That aspartate and alanine are produced via aminotransferases is further supported by the fact that addition of aminooxyacetate, an inhibitor of aminotransferases,[132] inhibits alanine and aspartate synthesis.[27] High levels of aminotransferase activities have been found in the mycobiont.[27] Virtually all the GPT activity is located in the mycobiont and much of the alanine is likely to be synthesized in the mycobiont.

The ^{15}N enrichment of and dilution from ammonia is biphasic (a rapid initial dilution followed by a slower secondary dilution) and the level of enrichment remains low.[134] Both of these observations are consistent with the occurrence of a small metabolic pool of ammonia, which turns over fast and in which newly fixed nitrogen is incorporated.

The fact that both glutamate and glutamine are primary amination products suggests operation of both GS-GOGAT and GDH pathways. Thus, the ammonia retained by the cyanobiont (less than 10%) is assimilated via the GS-GOGAT pathway in the cyanobiont and the ammonia released by the cyanobiont is assimilated by the mycobiont via GDH (more than 90%). Evidence for the latter are (1) total ^{15}N incorporation into glutamate is much higher than the displacement from glutamine amide nitrogen; (2) during the chase period, labeling of glutamate decreased more rapidly than that from glutamine; and (3) in the presence of MSX and azaserine, ^{15}N-labeling into glutamate continues, while that into glutamine stops. One may argue that the assimilation of ammonia via GDH in the latter experiment may occur only because the ammonia pool builds up on inhibition of GS by MSX. However, there is evidence to indicate that this does not happen.[134] Also, it should be borne in mind that the mycobiont does not have GS or GOGAT while it has extremely high GDH activity especially in the mycobiont hyphae close to the cyanobiont cells.

It has been suggested that ammonia released by the cyanobiont in cephalodia is assimilated by the cephalodial mycobiont into organic nitrogen which then moves to the main thallus. Alanine has been suggested as the most likely amino acid transported from cephalodia to the main thallus, since alanine is a secondary product of ammonia assimilation, by being a major pool constituent and by accumulating in relation to other amino acids when cephalodia are excised and deprived of a sink. However, it may be that alanine is not the normal compound transported and that lack of transfer of other nitrogenous compounds results in accumulation of newly fixed nitrogen in the alanine pool of cephalodia.

Based on the ^{15}N studies, routes of nitrogen metabolism in *P. aphthosa* cephalodia have been proposed (Figure 13).[12]

It is important that ammonia assimilation in the cephalodial mycobiont occurs via GDH, that glutamine production in the cyanobiont is low, that it is alanine which moves to the main thallus from cephalodia, that no primary ammonia assimilation occurs via GS of *Coccomyxa*, that *Coccomyxa* is not in a direct contact with the cyanobiont, and that all the transfer of nitrogen occurs via the mycobiont. This arrangement keeps the nitrogenase functioning despite the presence of ammonia.[59] If ammonia released by the cyanobiont were to be assimilated by the *Coccomyxa* GS, production of glutamine in excess may have caused inhibition of nitrogenase. Also, if the mycobiont had GS for ammonia assimilation, a similar inhibition would have occurred.

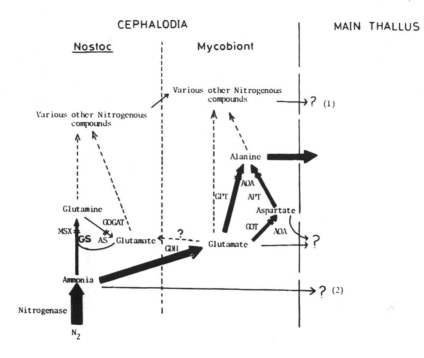

FIGURE 13.   Major and minor routes (bold and thin lines, respectively) of initial metabolism of ammonia in the cephalodia of *Peltigera aphthosa* (based on ^{15}N kinetics and enzymic studies).[11] Broken arrows indicate possible minor pathways operating but not studied in detail. Question marks indicate possible transferable compounds suggested by others.[10,119]

FUTURE PROSPECTS

Although during the past decade considerable work has been done regarding nitrogen metabolism in lichens, answers to many questions raised by these investigations remain to be found. We now know that in cyanophilic lichens a number of the nitrogen metabolizing enzymes are modified in the cyanobiont. However, we do not yet know how and in what order these changes are brought about. Another important aspect is the investigation regarding inhibition of GS synthesis. It is yet to be determined whether this inhibition is at the level of transcription or translation. Monitoring of the GS mRNA levels in the cyanobionts and their free-living cultivated isolates should answer this question. A detailed analysis of lichen compounds to pinpoint the compound/factor involved in the regulation of GS synthesis would be the next logical step. Such studies would provide information regarding manipulation of GS in free-living cyanobacteria for photobiological production of ammonia from nitrogen fixation.[135,136]

High heterocyst frequency is a property which, if incorporated into free-living forms, would be highly desirable for increased nitrogen-fixation rates. Investigations into the mechanisms by which cyanobionts increase their heterocyst frequency would, therefore, be of great value.

In studies on nitrate reductase (NR) in cyanobacteria, including *Anabaena cycadeae*, the isolate from *Cycas* root nodules has yielded interesting results. This includes mechanisms of regulation of NR by combined nitrogen[137,138] evidence of a common genetic regulation of NR and GS,[139] and absence of NR in heterocysts.[140] Cyanophilic lichens, with more than 90% of the GS being absent, should be good models through which to study NR regulation and its interaction with GS. There are some recent reports on the regulation of NR in *Evernia prunastrii*, suggesting involvement of phytochrome,[141] and on the presence of NR and GS in various lichens.[142] This is a welcome sign and should merit further research.

Nitrogen fixation is a vital process in the utilization of atmospheric molecular nitrogen, and most organisms lack this property except some prokaryotes (however, see Yamada and Sakaguchi[143] and Weathers et al.[144]). Recently, certain archaebacteria have been shown to fix nitrogen.[145,146] Nitrogen-fixing cyanobacteria form symbioses with algae to angiosperms but none with crop plants.[5,147] Attempts to establish artificial nitrogen-fixing symbioses between nitrogen-fixing cyanobacteria and crop plants would be highly desirable. Although early attempts have failed, with newer information further attempts would be worthwhile.[148-150]

Recently, akinetes of a cyanobacterium *Anabaena doliolum* have been shown to lack GS and NR.[151] These could be used to study some of the problems regarding GS regulation/ expression, NR regulation/expresion, and heterocyst-nitrogenase expression/regulation.[152] We hope that by studying the above aspects in cyanobacterial akinetes some light will be thrown on heterocyst differentiation and GS and NR regulation in symbioses.

ACKNOWLEDGMENT

I am thankful to Dr. R. N. Sharan for his help in the preparation of this manuscript. My own work reported here has been supported by financial assistance from Ministry of Education (India Government), INSA, EEC, University of Hyderabad, and North-Eastern Hill University. Permission from various publishers for reproduction of material published earlier is also thankfully acknowledged.

REFERENCES

1. **Millbank, J. W. and Kershaw, K. A.,** Nitrogen metabolism, in *The Lichens*, Ahmadjian, V. and Hale, M. E., Eds., Academic Press, New York, 1973, 289.
2. **Millbank, J. W.,** Associations with blue-green algae, in *The Biology of Nitrogen Fixation*, Quispel, A., Ed., North-Holland, Amsterdam, 1974, 238.
3. **Stewart, W. D. P., Rowell, P., and Rai, A. N.,** Symbiotic nitrogen-fixing cyanobacteria, in *Nitrogen Fixation*, Stewart, W. D. P. and Gallon, J. R., Eds., Academic Press, New York, 1980, 239.
4. **Stewart, W. D. P., Rai, A. N., Reed, R. H., Creach, E., Codd, G. A., and Rowell, P.,** Studies on the N₂-fixing lichen *Peltigera aphthosa*, in *Current Perspectives in Nitrogen Fixation*, Gibson, A. H. and Newton, W. E., Eds., Australian Academy of Sciences, Canberra, 1981, 237.
5. **Stewart, W. D. P., Rowell, P., and Rai, A. N.,** Cyanobacteria-eucaryotic plant symbioses, *Ann. Microbiol. (Inst. Pasteur)*, 134B, 205, 1983.
6. **Stewart, W. D. P., Preston, T., Rai, A. N., and Rowell, P.,** Nitrogen cycling, in *Nitrogen as an Ecological Factor*, Lee, J. A., McNeill, S., and Rorison, I. H., Eds., Blackwell Scientific, Oxford, 1983, 1.
7. **Hitch, C. J. B. and Stewart, W. D. P.,** Nitrogen fixation by lichens in Scotland, *New Phytol.*, 72, 509, 1973.
8. **Goas, G. and Bernard, T.,** Contribution a l'étude du metabolisme azote des lichens; les différentes formes d'azote de quelques espéces de la famille des Stictacées, *C. R. Acad. Sci. Ser. D*, 265, 1187, 1967.
9. **Millbank, J. W. and Kershaw, K. A.,** Nitrogen metabolism in lichens. I. Nitrogen fixation in the cephalodia of *Peltigera aphthosa*, *New Phytol.*, 68, 721, 1969.
10. **Englund, B.,** The physiology of the lichen *Peltigera aphthosa*, with special reference to the blue-green phycobiont (*Nostoc* sp.), *Physiol. Plant.*, 41, 298, 1977.
11. **Rai, A. N.,** *Studies On the Nitrogen-Fixing Lichen Peltigera aphthosa Willd.*, Ph.D. thesis, University of Dundee, Dundee, Scotland, 1980.
12. **Millbank, J. W.,** Nitrogen metabolism in lichens. IV. The nitrogenase activity of the *Nostoc* phycobiont in *Peltigera canina*, *New Phytol.*, 71, 1, 1972.
13. **Smith, D. C.,** Studies in the physiology of lichens. I. The effects of starvation and of ammonia absorption upon the nitrogen content of *Peltigera polydactyla*, *Ann. Bot. (London)*, 24, 52, 1960.
14. **Scott, G. D.,** Further investigations of some lichens for nitrogen fixation, *New Phytol.*, 55, 111, 1956.

15. **Massé, L. C.,** Etude comparée des teneur en azote des lichens et de leurs substrate. Les espéces ornith-ocoprophiles, *C. R. Acad. Sci. Ser. D,* 262, 1721, 1966.

16. **Sampaio, M. J. A. M., Rai, A. N., Rowell, P., and Stewart, W. D. P.,** Occurrence, synthesis and activity of glutamine synthetase in N_2-fixing lichens, *FEMS Microbiol. Lett.,* 6, 107, 1979.

17. **Millbank, J. W.,** Aspects of nitrogen metabolism in lichens, in *Lichenology: Progess and Problems,* Brown, D. H., Hawksworth, D. L., and Bailey, R. H., Eds., Academic Press, New York, 1976, 441.

18. **Ramakrishnan, S. and Subramanian, S. S.,** Amino acids of *Rocella montegnei* and *Parmelia tinctorium, Indian J. Chem.,* 2, 467, 1964.

19. **Ramakrishnan, S. and Subramanian, S. S.,** Amino acids of *Peltigera canina, Curr. Sci.,* 33, 522, 1964.

20. **Ramakrishnan, S. and Subramanian, S. S.,** Amino acids of *Cladonia rangiferina, Cladonia gracilis* and *Lobaria isidiosa, Curr. Sci.,* 34, 345, 1965.

21. **Ramakrishnan, S. and Subramanian, S. S.,** Amino acids of *Lobaria subisidiosa, Umbilicaria pustulata, Parmelia nepalensis* and *Ramalina sinensis, Curr. Sci.,* 35, 124, 1966.

22. **Ramakrishnan, S. and Subramanian, S. S.,** Amino acids of *Dermatocarpon moulinsii, Curr. Sci.,* 35, 284, 1966.

23. **Solberg, Y. J.,** Studies on the chemistry of lichens. VIII. An examination of the free sugars and ninhydrin positive compounds of several Norwegian lichen species, *Lichenologist,* 4, 271, 1970.

24. **Solberg, Y. J.,** Studies on the chemistry of lichens. XIV. Chemical investigations of the lichen species *Anaptychia fusca, Peltigera canina* and *Omphalodiscus spodochrous, Z. Naturforsch.,* 30c, 445, 1975.

25. **Jäger, H. J. and Weigel, H. J.,** Amino acid metabolism in lichens, *Bryologist,* 81, 107, 1978.

26. **Bernard, R. A. and Oaks, A.,** Metabolism of proline in maize root tips, *Can. J. Bot.,* 48, 1155, 1970.

27. **Rai, A. N., Rowell, P., and Stewart, W. D. P.,** $^{15}N_2$-incorporation and metabolism in the lichen *Peltigera aphthosa* Willd., *Planta,* 152, 544, 1981.

28. **Rai, A. N., Rowell, P., and Stewart, W. D. P.,** Interactions between cyanobacterium and fungus during $^{15}N_2$-incorporation and metabolism in the lichen *Peltigera canina, Arch. Microbiol.,* 134, 136, 1983.

29. **Fabian-Galan, G., Atanasiu, L., and Salageanu, N.,** Organic substances produced by photosynthesis in lichens, *Prog. Photosyn. Res.,* 3, 1553, 1969.

30. **Fabian-Galan, G. and Anatasiu, L.,** Products of $^{14}CO_2$ incorporation in the lichens *Peltigera horizontalis* Baumg. and *Usnea florida* Wigg., *Rev. Roum. Biol. Ser. Biol. Veg.,* 23, 23, 1978.

31. **Bednar, T. W.,** Physiological Studies On The Isolated Components Of The Lichen *Peltigera Aphthosa,* Ph.D. thesis, University of Wisconsin, Madison, 1962.

32. **Bernard, T. and Goas, G.,** Contribution à l'étude du metabolisme azote des lichens. Caracterisation et dosages des methylamines de quelques espèces de la famille des Stictacées, *C. R. Acad. Sci.,* 267, 622, 1968.

33. **Bernard, T. and Larher, F.,** Contribution a l'étude du metabolisme azote des lichens. Role de la glycine $^{14}C_2$ dans la formation des methylamines chez *Lobaria laetevirens, C. R. Acad. Sci.,* 272, 568, 1971.

34. **Smith, T. A.,** Putrescine, spermidine and spermine in higher plants, *Phytochemistry,* 9, 1479, 1970.

35. **Cohen, S. S.,** *Introduction to the Polyamines,* Prentice-Hall, Englewood Cliffs, N. J., 1971.

36. **Smith, T. A.,** Recent advances in the biochemistry of plant amines, *Phytochemistry,* 14, 865, 1975.

37. **Lange, O. L.,** Versuche zur Hitze- und Trockenresistenz der Flechten, *Flora,* 140, 39, 1953.

38. **Lange, O. L.,** Die funktionellen Anpassungen der Flechten an die Ökologischen Bendingungen arider Gebiete, *Ber. Dtsch. Bot. Ges.,* 82, 3, 1969.

39. **Bernard, T. and Goas, G.,** Biosynthèse de la sticticine chez lichen *Lobaria laetevirens, Physiol. Plant.,* 53, 71, 1981.

40. **Cohen-Bazire, G. and Bryant, D. A.,** Phycobilisomes; composition and structure, in *The Biology of Cyanobacteria,* Carr, N. G. and Whitton, B. A., Eds., Blackwell Scientific, Oxford, 1982, 143.

41. **Smith, D. C.,** The physiology of *Peltigera polydactyla* (Neck.) Hoffm., *Lichenologist,* 1, 209, 1961.

42. **Vicente, C., Legaz, M. E., Arruda, E. C., and Xavier Filho, L.,** The utilization of urea by the lichen *Cladonia sandstedei, J. Plant Physiol.,* 115, 397, 1984.

43. **Blanco, M. J.,** Incorporacion metabolica de urea por talo de Evernia prunastri, Ph.D. thesis, Complutense University, Madrid, 1983.

44. **Vicente, C., Azpiroz, A., Estevez, M. P., and Gonzalez, M. L.,** Quaternary structure changes and kinetics of urease inactivation by L-usnic acid in relation to nutrient transfer between lichen symbionts, *Plant Cell Environ.,* 1, 29, 1978.

45. **Xavier Filho, L. and Vicente, C.,** Exo- and endourease from *Parmelia roystonea* and their regulation by lichen acids, *Bol. Soc. Broteriana,* 52, 55, 1978.

46. **Vicente, C. and Xavier Filho, L.,** Urease regulation in *Cladonia verticillaris, Phyton (Buenos Aires),* 37, 137, 1979.

47. **Ciluentes, R., Estevez, M. P., and Vicente, C.,** *In vivo* protection of urease of *Evernia prunastri* by dithiothreitol, *Physiol. Plant.,* 53, 245, 1981.

48. **Smith, D. C., Muscatine, L., and Lewis, D. H.,** Carbohydrate movement from autotrophs to heterotrophs in parasitic and mutualistic symbioses, *Biol. Rev.,* 44, 17, 1969.

49. **Syrette, P. J.**, Nitrogen assimilation, in *Physiology and Biochemistry of Algae*, Levin, R. A., Ed., Academic Press, New York, 1962, 171.

50. **Schofield, E. A.**, A Cultural Comparison of Free-living and Lichenized Fungi, M. A. thesis, Clark University, Worcester, Mass., 1964.

51. **Ahmadjian, V.**, Lichens, in *Symbiosis*, Vol. 1, Henry, S. M., Ed., Academic Press, New York, 1966, 35.

52. **Ahmadjian, V.**, Algal/fungal symbioses, *Prog. Phycol. Res.*, 1, 179, 1982.

53. **Garcia, I., Cifuentes, B., and Vicente, C.**, L-usnate-urease interactions: binding site for the ligand, *Z. Naturforsch.*, 35c, 1098, 1980.

54. **Legaz, M. E. and Brown, D. H.**, Factors affecting urease activity in the lichen *Evernia prunastri*, *Ann. Bot. (London)*, 52, 261, 1983.

55. **Legaz, M. E., Cifuentes, B., and Vicente, C.**, Mecanismos de sintesis e inactivacion de urease en *Evernia prunastri*, in *Estudios sobre Biologia*, Vincente, C. and Municio, A. M., Eds., Editorial Universidad Complutense, Madrid, 1982, 101.

56. **Magaña-Plaza, I. and Ruiz-Herrera, J.**, Mechanisms of regulation of urease biosynthesis in *Proteus rettgeri*, *J. Bacteriol.*, 93, 1294, 1967.

57. **Salomon, H.**, Über das Vorkommen und die Aufnahme einiger wichtiger Nährsalze bei den Flechten, *Jahrb. Wiss. Bot.*, 54, 309, 1914.

58. **Macmillan, A.**, The entry of ammonia in fungal cells, *J. Exp. Bot.*, 7, 113, 1956.

59. **Rai, A. N., Rowell, P., and Stewart, W. D. P.**, NH_4^+ assimilation and nitrogenase regulation in the lichen *Peltigera aphthosa* Willd., *New Phytol.*, 85, 545, 1980.

60. **Dharmawardene, M. W. N., Haystead, A., and Stewart, W. D. P.**, Glutamine synthetase of the nitrogen-fixing alga *Anabaena cylindrica*, *Arch. Mikrobiol.*, 90, 281, 1973.

61. **Stewart, W. D. P. and Rowell, P.**, Effects of L-methionine-DL-sulphoximine on the assimilation of newly fixed NH_3, acetylene reduction and heterocyst production in *Anabaena cylindrica*, *Biochem. Biophys. Res. Commun.*, 65, 846, 1975.

62. **Wolk, C. P., Thomas, J., Shaffer, P. W., Austin, S. M., and Galonsky, A.**, Pathways of nitrogen metabolism after fixation of ^{13}N-labelled nitrogen gas by the cyanobacterium, *Anabaena cylindrica*, *J. Biol. Chem.*, 251, 5027, 1979.

63. **Kleiner, D., Phillips, S., and Fitzke, E.**, Pathways and regulatory aspects of N_2 and NH_4^+ assimilation in N_4-fixing bacteria, in *Biology of Inorganic Nitrogen and Sulfur*, Bothe, H. and Trebst, A., Eds., Springer-Verlag, Berlin, 1981, 131.

64. **Rai, A. N., Rowell, P., and Stewart, W. D. P.**, Evidence for an ammonium transport system in free-living and symbiotic cyanobacteria, *Arch. Microbiol.*, 137, 241, 1984.

65. **Smith, D. C.**, Studies in the physiology of lichens. II. Absorption and utilisation of some simple organic nitrogen compounds by *Peltigera polydactyla*, *Ann. Bot. (London)*, 24, 172, 1960.

66. **Smith, D. C.**, Studies in the physiology of lichens. III. Experiments with dissected discs of *Peltigera polydatyla*, *Ann. Bot. (London)*, 24, 188, 1960.

67. **Galinou, M. A.**, Sur la mise en évidence de quelques biocatalyseurs chez les lichens, Proc. Int. Bot. Congr. 8th, Sect. XVIII, 1954, 2.

68. **Moissejeva, E. M.**, Biochemical properties of lichens and their practical importance, *Izd. Akad. Nauk. USSR Moscow*, 82, 82, 1961.

69. **Massé, L. C.**, Quelques aspects de l'uricolyse enzymatique chez les lichens, C. R. Acad. Sci. Ser. D, 268, 2896, 1969.

70. **Shapiro, I. A.**, Urease activity in lichens, *Fiziol. Rast.*, 24, 1135, 1977.

71. **Rai, A. N., Rowell, P., and Stewart, W. D. P.**, Nitrogenase activity and dark CO_2 fixation in the lichen *Peltigera aphthosa* Willd., *Planta*, 151, 256, 1981.

72. **Stewart, W. D. P. and Rowell, P.**, Modifications of nitrogen-fixing algae in lichen symbioses, *Nature (London)*, 265, 371, 1977.

73. **Bernard, T. and Goas, G.**, Glutamate deshydrogenases du lichen *Lobaria laetevirens* (Lightf.) Zahlbr. Caractéristiques de l'enzyme du Champignan, *Physiol. Veg.*, 17, 535, 1979.

74. **Chambers, S., Morris, M., and Smith, D. C.**, Lichen physiology. XV. The effect of digitonin and other treatments on biotrophic transport of glucose from alga to fungus in *Peltigera polydactyla*, *New Phytol.*, 76, 485, 1976.

75. **Rai, A. N., Rowell, P., and Stewart, W. D. P.**, The activities of glutamine synthetase and glutamate synthase in the N_2-fixing lichens *Peltigera aphthosa* and *Peltigera canina*, *SGM Q.* 8, 261, 1981.

76. **Rai, A. N., Rowell, P., and Stewart, W. D. P.**, Glutamate synthase activity in symbiotic cyano bacteria, *J. Gen. Microbiol.*, 126, 515, 1981.

77. **Rai, A. N., Rowell, P., and Stewart, W. D. P.**, Interrelations of carbon and nitrogen metabolism in the N_2-fixing lichens *Peltigera canina* and *Peltigera aphthosa*, Paper B-39, in Proc. IV Int. Symp. Photosynthetic Procaryotes, Bombannes (Bordeaux), France, 1982.

78. **Bernard, T.,** Contribution à l'étude du metabolisme azote des lichens. Activité de la glutamate-decarboxylase de cinq espèces de la famille des Stichtacées, *C. R. Acad. Sci.*, 269, 823, 1969.
79. **Bernard, T. and Goas, G.,** Contribution à l'étude du metabolisme azote des lichens. Mise en evidence de quelques transaminases; activite de la glutamate oxaloacetate transaminase dans cinq espèces de la famille des Stictacées, *C. R. Acad. Sci.*, 269, 1657, 1969.
80. **Hällbom, L.,** Sarcosine: a possible regulatory compound in *Peltigera praetextata-Nostoc* symbiosis, *FEMS Microbiol. Lett.*, 22, 119, 1984.
81. **Rippka, R., Deruelles, J., Waterbury, J. B., Herdman, M., and Stanier, R. Y.,** Generic assignments, strain historics and properties of pure cultures of cyanobacteria, *J. Gen. Microbiol.*, 111, 1, 1979.
82. **Hitch, C. J. B. and Millbank, J. W.,** Nitrogen metabolism in lichens. VII. Nitrogenase activity and heterocyst frequency in lichens with blue-green phycobionts, *New Phytol.*, 75, 239, 1975.
83. **Griffiths, H. B., Greenwood, A. D., and Millbank, J. W.,** The frequency of heterocysts in the *Nostoc* phycobiont of the lichen *Peltigera canina*, *New Phytol.*, 71, 11, 1972.
84. **Cengia-Sambo, M.,** *Atti Soc. Ital. Sci. Nat. Mus. Civ. Stor. Nat. Milano*, 62, 226, 1923.
85. **Cengia-Sambo, M.,** *Atti Soc. Ital. Sci. Nat. Mus. Civ. Stor. Nat. Milano*, 64, 191, 1926.
86. **Gallon, J. R.,** Nitrogen fixation by photoautotrophs, in *Nitrogen Fixation*, Stewart, W. D. P. and Gallon, J. R., Eds., Academic Press, New York, 1980, 197.
87. **Stewart, W. D. P.,** Some aspects of structure and function in N_2-fixing cyanobacteria, *Annu. Rev. Microbiol.*, 34, 497, 1980.
88. **Snyder, J. M. and Wullstein, L. H.,** The role of desert cryptogames in nitrogen fixation, *Am. Midl. Nat.*, 90, 257, 1973.
89. **Bond, G. and Scott, G. D.,** An examination of some symbiotic systems for fixation of nitrogen, *Ann. Bot. (London)*, 19, 67, 1955.
90. **Scott, G. D.,** Further investigations of some lichens for fixation of nitrogen, *New Phytol.*, 55, 111, 1956.
91. **Watanabe, A. and Kiyohara, T.,** Symbiotic blue-green algae of lichens, liverwarts and cycads, in *Studies in Micro-Algae and Photosynthetic Bacteria*, University of Tokyo Press, Tokyo, 1963, 189.
92. **Rogers, R. W., Lange, R. T., and Nicholas, D. J. D.,** Nitrogen fixation by lichens of arid soil crusts, *Nature (London)*, 209, 96, 1966.
93. **Fogg, G. E. and Stewart, W. D. P.,** *In situ* determination of biological nitrogen fixation in Antaractica, *Bull. Br. Antarct. Surv.*, 15, 39, 1968.
94. **Stewart, W. D. P.,** Algal fixation of atmospheric nitrogen, *Plant Soil*, 32, 555, 1970.
95. **Stewart, W. D. P., Fitzgerald, G. P., and Burris, R. H.,** *In situ* studies on N_2-fixation using acetylene reduction technique, *Proc. Natl. Acad. Sci. U.S.A.*, 58, 2071, 1967.
96. **Henriksson, E. and Simu, B.,** Nitrogen fixation by lichens, *Oikos*, 22, 119, 1971.
97. **Millbank, J. W. and Kershaw, K. A.,** Nitrogen metabolism in lichens. III. Nitrogen fixation by internal cephalodia of *Lobaria pulmonaria*, *New Phytol.*, 69, 595, 1970.
98. **Denison, W. C.,** *Lobaria oregana*, a nitrogen-fixing lichen in old-growth Douglas fur forests, in *Symbiotic N_2-fixation in the Management of Temperate Forests*, Gordon, J. C., Wheeler, C. T., and Perry, D. A., Eds., Oregon State University, Corvallis, 1979, 266.
99. **Horstmann, J. L., Denison, W. C., and Silvester, W. B.,** $^{15}N_2$-fixation and molybdenum enhancement of acetylene reduction by *Lobaria* spp., *New Phytol.*, 92, 235, 1982.
100. **Kallio, P., Suhonen, S., and Kallio, H.,** The ecology of nitrogen fixation in *Nephroma arcticum* and *Solorina crocea*, *Rep. Kevo Subarct. Res. Stn.*, 9, 7, 1972.
101. **Hitch, C. J. B.,** A Study of Some Environmental Factors Affecting Nitrogenase Activity in Lichens, M.Sc. thesis, University of Dundee, Dundee, Scotland, 1971.
102. **Rai, A. N., Rowell, P., and Stewart, W. D. P.,** Mycobiont-cyanobiont interactions during dark nitrogen fixation by the lichen *Peltigera aphthosa*, *Physiol. Plant.*, 57, 285, 1983.
103. **Kallio, S.,** The ecology of nitrogen fixation in *Stereocaulon paschale*, *Rep. Kevo Subarct. Res. Stn.*, 10, 34, 1973.
104. **Hardy, R. W. F., Burns, R. C., and Holsten, R. D.,** Application of the acetylene-ethylene assay for measurement of nitrogen fixation, *Soil Biol. Biochem.*, 5, 47, 1973.
105. **Rennie, R. J., Rennie, D. A., and Fried, M.,** Concepts of ^{15}N usage in dinitrogen fixation studies, in *Isotopes in Biological Dinitrogen Fixation*, International Atomic Energy Agency, Vienna, 1978, 107.
106. **Bergersen, F. J.,** The quantitative relationship between nitrogen fixation and the acetylene reduction assay, *Aust. J. Biol. Sci.*, 23, 1015, 1970.
107. **Schubert, K. R. and Evans, H. J.,** Hydrogen evolution: a major factor affecting nitrogen fixation in nodulated symbionts, *Proc. Natl. Acad. Sci. U.S.A.*, 73, 1207, 1976.
108. **Graham, B. M., Hamilton, R. D., and Campbell, N. E. R.,** Comparison of the nitrogen-15 uptake and acetylene reduction methods for estimating the rates of nitrogen fixation by fresh water blue-green algae, *Can. J. Fish. Aquat. Sci.*, 37, 488, 1980.

109. **Bothe, H., Neuer, G., Kalbe, I., and Eisbrenner, G.,** Electron donation and hydrogenase in nitrogen-fixing microorganisms, in *Nitrogen Fixation,* Stewart, W. D. P. and Gallon, J. R., Eds., Academic Press, New York, 1980, 83.

110. **Millbank, J. W.,** The assessment of nitrogen fixation and throughput by lichens. I. The use of a controlled environment chamber to relate acetylene reduction estimates to nitrogen fixation, *New Phytol.,* 89, 647, 1981.

111. **Rowell, P., Rai, A. N., and Stewart, W. D. P.,** Studies on the nitrogen metabolism of the lichens *Peltigera aphthosa* and *Peltigera canina,* in *Lichen Physiology and Cell Biology,* Brown, D. H., Ed., Plenum Press, New York, 1985, 145.

112. **Singh, H. N., Rai, U. N., Rao, V. V., and Bagchi, S. N.,** Evidence for ammonia as an inhibitor of heterocyst and nitrogenase formation in the cyanobacterium *Anabaena cyacdeae, Biochem. Biophys. Res. Commun.,* 111, 180, 1983.

113. **Ownby, J. D.,** Effects of amino acids on methionine-sulphoximine-induced heterocyst formation in *Anabaena, Planta,* 136, 277, 1977.

114. **Ladha, J. K., Rowell, P., and Stewart, W. D. P.,** Effects of 5-hydroxylysine on acetylene reduction and NH_4 assimilation in the cyanobacterium *Anabaena cylindrica, Biochem. Biophys. Res. Commun.,* 83, 688, 1978.

115. **Richardson, D. H. S., Hill, D. J., and Smith, D. C.,** The role of the alga in determining the pattern of carbohydrate movement between lichen symbionts, *New Phytol.,* 67, 469, 1968.

116. **Feige, G. B.,** Investigations on the physiology of cephalodia from the lichen *Peltigera aphthosa* (L.) Willd. II. The photosynthetic labeling pattern and the movement of carbohydrate from phycobiont to mycobiont, *Z. Pflanzenphysiol.,* 80, 386, 1976.

117. **Akkermans, A. D. L., Huss-Danell, K., and Roelofsen, W.,** Enzymes of the tricarboxylic acid cycle and the malate aspartate shuttle in the N_2-fixing endophyte of *Alnus glutinosa, Physiol. Plant.,* 53, 289, 1981.

118. **Henriksson, E.,** Nitrogen fixation by a bacteria-free symbiotic *Nostoc* strain from *Collema, Physiol. Plant.,* 4, 542, 1951.

119. **Millbank, J. W.,** Nitrogen metabolism in lichens. V. The forms of nitrogen released by the blue-green phycobiont in *Peltigera* spp., *New Phytol.,* 73, 1171, 1974.

120. **Rai, A. N. Rowell, P., and Stewart, W. D. P.,** Interactions between cyanobiont and fungus during $^{15}N_2$-incorporation and metabolism in the lichen *Peltigera canina, Arch. Microbiol.,* 134, 136, 1983.

121. **Rai, A. N., Rowell, P., and Stewart, W. D. P.,** Glutamate synthase activity of heterocysts and vegetative cells of the cyanobacterium *Anabaena variabilis* Kütz., *J. Gen. Microbiol.,* 128, 2203, 1982.

122. **Thomas, J., Meeks, J. C., Wolk, C. P., Shaffer, P. W., Austin, S. M., and Chien, W. S.,** Formation of glutamine from (^{13}N) ammonia, (^{13}N) dinitrogen and (^{14}C) glutamate by heterocysts isolated from *Anabaena cylindrica J. Bacteriol.,* 129, 1545, 1977.

123. **Peters, G. A., Mayne, B. C., Ray, T. B., and Toia, R. E., Jr.,** Physiology and biochemistry of the *Azolla-Anabaena* symbiosis, in *Nitrogen and Rice,* International Rice Research Institute Las Banos, Philippines, 1979, 325.

124. **Weiss, M. G. and Luria, S. E.,** Reduction of membrane potential, an immediate effect of colicin K., *Proc. Natl. Acad. Sci. U.S.A.,* 75, 2483, 1978.

125. **Reed, R. H., Rowell, P., and Stewart, W. D. P.,** Characterization of the transport of potassium ions in the cyanobacterium *Anabaena variabilis* Kütz., *Eur. J. Biochem.,* 116, 323, 1981.

126. **Reed, R. H., Rowell, P., and Stewart, W. D. P.,** Uptake of potassium and rubidium ions by the cyanobacterium *Anabaena variabilis, FEMS Microbiol. Lett.,* 11, 223, 1981.

127. **Hawkesford, M. J., Reed, R. H., Rowell, P., and Stewart, W. D. P.,** Nitrogenase activity and membrane electrogenesis in the cyanobacterium *Anabaena variabilis* Kütz., *Eur. J. Biochem.,* 115, 519, 1981.

128. **Hawkesford, M. J., Reed, R. H., Rowell, P., and Stewart, W. D. P.,** Nitrogenase activity and membrane electrogenesis in the cyanobacterium *Plectonema boryanum, Eur. J. Biochem.,* 127, 63, 1982.

129. **Kershaw, K. A. and Millbank, J. W.,** Nitrogen metabolism in lichens. II. The partitioning of cephalodial-fixed nitrogen between the mycobiont and the phycobionts of *Peltigera aphthosa, New Phytol.,* 69, 75, 1970.

130. **Green, T. G. A. and Smith, D. C.,** Lichen physiology. XIV. Differences between lichen algae in symbiosis and in isolation, *New Phytol.,* 73, 753, 1974.

131. **Rowell, P. and Stewart, W. D. P.,** Alanine dehydrogenase of the N_2-fixing blue green alga *Anabaena cylindrica, Arch. Microbiol.,* 107, 115, 1976.

132. **Hopper, S. and Segal, H. L.,** Kinetic studies of rat liver glutamic-alanine transaminase, *J. Biol. Chem.,* 237, 3189, 1962.

133. **Kennedy, I. R.,** Primary products of symbiotic nitrogen fixation. I. Short term exposure of *Serradella* nodules to $^{15}N_2$, *Biochim. Biophys. Acta.,* 130, 285, 1966.

134. **Rhodes, D., Sims, A. P., and Folkes, B. F.,** Pathways of ammonia assimilation in illuminated *Lemna minor, Phytochemistry,* 19, 357, 1980.

135. **Musgrave, S. C., Kerby, N. W., Codd, G. A., and Stewart, W. D. P.,** Sustained ammonia production by immobilized filaments of the nitrogen-fixing cyanobacterium *Anabaena* 27893, *Biotechnol. Lett.,* 4, 647, 1982.

136. **Stewart, W. D. P., Codd, G. A., and Rai, A. N.,** H_2 production from sunlight, air and water by N_2-fixing systems involving cyanobacteria, in *Photochemical, Photoelectrochemical and Photobiological Processes,* Vol. 2, Hall, D. O., Palz, W., and Pirrwitz, D., Eds., D. Reidel Publ., Dortrecht, Holland, 1983, 214.

137. **Bagchi, S. N., Rai, A. N., and Singh, H. N.,** Regulation of nitrate reductase in cyanobacteria. Repression-derepression control of nitrate reductase apoprotein in the cyanobacterium *Nostoc muscorum, Biochim. Biophys. Acta,* 838, 370, 1985.

138. **Bagchi, S. N., Rai, U. N., Rai, A. N., and Singh, H. N.,** Nitrate metabolism in the cyanobacterium *Anabaena cycadeae:* regulation of nitrate uptake and reductase by ammonia, *Physiol. Plant.,* 63, 322, 1985.

139. **Singh, H. N., Rai, A. N., and Bagchi, S. N.,** Evidence for a common genetic regulation of glutamine synthetase and nitrate uptake and reductase in the cyanobacterium *Anabaena cycadeae, Mol. Gen. Genet.,* 198, 367, 1985.

140. **Kumar, A. P., Rai, A. N., and Singh, H. N.,** Nitrate reductase activity in isolated heterocysts of the cyanobacterium *Nostoc muscorum. FEBS Lett.,* 179, 125, 1985.

141. **Avalos, A. and Vicente, C.,** Phytochrome enhances nitrate reductase activity in the lichen *Evernia prunastri, Can. J. Bot.,* 63, 1350, 1985.

142. **Shapiro, I. A.,** Activities of nitrate reductase and glutamine synthetase in lichens, *Fiziol. Rast.,* 30, 699, 1983.

143. **Yamada, T. and Sakaguchi, K.,** Nitrogen fixation associated with a hotspring green alga, *Arch. Microbiol.,* 124, 161, 1980.

144. **Weathers, P. J., Danielli, J. F., Bradley, P. M., Hebb, P. M., Miller, J. E., and Pesano, R. L.,** Possible evidence for novel metabolism of nitrogen gas by a green alga, *Physiol. Plant.,* 61, 441, 1984.

145. **Murray, P. A. and Zinder, S. H.,** Nitrogen fixation by a methanogenic archaebacterium, *Nature (London),* 312, 284, 1984.

146. **Belay, N., Sparling, R., and Daniels, L.,** Dinitrogen fixation by a thermophilic methanogenic bacterium, *Nature (London),* 312, 286, 1984.

147. **Rai, A. N. and Kumar, H. D.,** Nitrogen-fixing cyanobacteria in the phyllosphere, *SGM Q.,* 7, 106, 1980.

148. **Davey, M. R. and Power, J. B.,** Polyethylene glycol-induced uptake of microorganisms into higher plant protoplasts: an ultrastructural study, *Plant Sci. Lett.,* 5, 269, 1975.

149. **Burgoon, A. C. and Bottino, P. J.,** Uptake of nitrogen-fixing blue green alga *Gloeocapsa* into protoplasts of tobacco and maize, *J. Hered.,* 67, 223, 1976.

150. **Meeks, J. C., Malmberg, R. L., and Wolk, C. P.,** Uptake of auxotrophic cells of a heterocyst-forming cyanobacterium by tobacco protoplasts, and the fate of their associations, *Planta,* 139, 55, 1978.

151. **Rao, V. V., Rai, A. N., and Singh, H. N.,** Metabolic activities of akinetes of the cyanobacterium *Anabaena doliolum:* oxygen exchange, photosynthetic pigments and enzymes of nitrogen metabolism, *J. Gen. Microbiol.,* 130, 1299, 1984.

152. **Rai, A. N., Rao, V. V., and Singh, H. N.,** Akinete germination in the cyanobacterium *Anabaena doliolum:* role of light and metabolic changes, *J. Gen. Microbiol.,* in press.

Chapter VI.C

LICHEN ENZYMOLOGY

Carlos Vicente and María Estrella Legaz

I. INTRODUCTION

All physiological events in the living world are mediated by chemical reactions catalyzed by enzymes. Enzymology, which is studied as a matter of course in most plant groups, has been neglected in lichens. With the exception of a few specialized areas, such as dinitrogen fixation (Chapter VI.B), little is known about the general enzymology of many types of metabolic pathways in lichens. This aspect of lichen biology is especially intriguing since two distinct organisms may contribute to the production of metabolites, especially to secondary compounds such as lichen acids. (See also Chapter IX.A.)

In this chapter, we wish to summarize available data on lichen enzymes and relate this to various metabolic processes in lichens.

II. ENZYMES OF PHOTOSYNTHESIS AND CARBOHYDRATE METABOLISM

A. Enzymatic Use of Photosynthetic Reducing Power

1. Ferredoxins

Ferredoxins constitute a family of low molecular weight, strongly electronegative electron carriers containing both nonheme iron and labile sulfur.[1] Based on their source and absorption spectra, they can be divided into two main classes, bacterial and plant ferredoxins.

Only two ferredoxins have been purified from lichens.[2,3] The absorption spectrum of *Evernia prunastri* ferredoxin shows peaks at 280, 320, 326, and 386 nm in its oxidized form.[2] Its molecular weight has been estimated at 5.5×10^3, and it contains four atoms of iron and four acid-labile sulfides. The oxidation-reduction potential has been calculated at about -480 mV.

Oxidized ferredoxin of *Lobaria pulmonaria* is a brown protein with absorption maxima at 280, 328, 370, and 400 nm and a broad peak at 420 nm (Figure 1). Its reduced form, which is yellow, shows a single peak at 268 nm.[3] Although this ferredoxin shows four maxima in the visible zone of the spectrum, it differs from ferredoxin of spinach,[4] alfalfa,[5] and fern[6] in showing a maximum at 370 nm, which is absent from ferredoxins of higher plants. Such a distinct absorption maximum was also demonstrated in ferredoxin from *Evernia*.[2] In addition, the absorption spectrum does not show the displacement in the ultraviolet (UV) zone which is characteristic of ferredoxin from certain algae.[7] The estimated oxidation-reduction potential of -440 mV is very similar to that reported for spinach ferredoxin.[4,8] *Lobaria* ferredoxin contains equivalent amounts of iron (5 atoms) and acid labile sulfide, and a calculated molecular mass of 6×10^3.

2. Ferredoxin-NADP+ oxidoreductase

Ferredoxin-NADP+ oxidoreductase (EC 1.18.1.2) is a component of the photosynthetic electron transport chain which catalyzes the reduction of NADP+ by photosystem I-reduced ferredoxin. This reduction is carried out by an one-electron, FAD-mediated reaction, as deduced from electron paramagnetic resonance (EPR) studies on the reductase of *Anabaena*.[9]

A 220-fold purified ferredoxin-NADP+ oxidoreductase of *Evernia prunastri*[2] showed absorption maxima at 280, 335, and 410 nm in its oxidized form and optimal activity at pH 8.0. The purified enzyme was able to reduce NADP+ using several electron donors, such

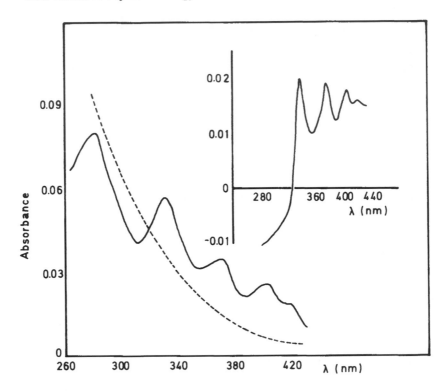

FIGURE 1. Absorption spectrum of oxidized ferredoxin (solid line) and H_2-reduced protein (broken line). The insert shows the difference between the spectrum of oxidized minus reduced forms. (From Vicente, C. and Requena, V., *Photosynthetica*, 18, 57, 1984. With permission.)

as methyl viologen, PMS, or dithionite, although maximal reduction was achieved when the enzyme was coupled to a reduced ferredoxin. The K_m value for $NADP^+$ reduction has been estimated at 0.33 mM, whereas the K_m value for NAD^+ was 5.0 mM. Oxidoreductase was able to reduce NAD^+ by using NAPH as electron donor, with a K_m value for NADPH of 2 mM; the enzyme did not reduce $NADP^+$ at NADH concentrations ranging from 0.16 to 10 mM.

Diaphorase activity has also been examined using methylene blue as an oxidant. In this case, K_m values for both NADPH and NADH have been estimated at 0.2 and 10 mM, respectively.

Ferredoxin-$NADP^+$ oxidoreductase was subjected to control by nucleotides.[10] Feedback inhibition of reductase activity was achieved when NADH concentration exceeded 0.5 mM. In addition, both adenosine 5'-monophosphate (AMP) and adenosine 5'-diphosphate (ADP) acted as competitive inhibitors of the enzyme. Inversely, adenosine 5'-triphosphate (ATP) behaved as an enzyme activator at concentrations ranging from 0.1 to 1.8 mM. There may be a possible relationship between NADPH inhibition and ATP activation of this enzyme. As shown in Figure 2, 1.2 mM ATP reversed feedback inhibition of the reductase produced by NADPH. Thus, ATP, a product of photophosphorylation, accelerated the formation of the second product of the noncyclic electron transport, NADPH, by activating ferredoxin-$NADP^+$ oxidoreducatase.[10]

3. Ribulose Bisphosphate Carboxylase-Oxygenase

Available evidence suggests that photosynthetic carboxylations in lichens, as in all photosynthetic eukaryotes, are catalyzed by ribulose-1,5-P-carboxylase-oxygenase using ribulose-1,5-bisphosphate as the carbon dioxide acceptor (e.g., References 11 to 16). To date,

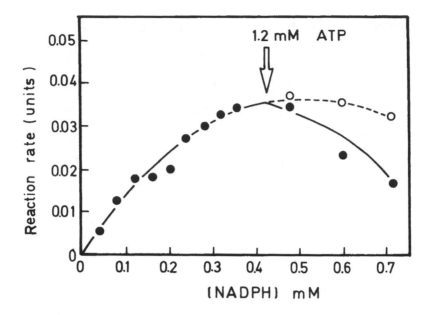

FIGURE 2.   Kinetics of ferredoxin-NADP⁺ reductase using NADPH as substrate and DCPIP as electron acceptor for NADPH concentrations varying from 0 to 0.72 mM (●). In (○) 1.2 mM ATP was added for NADPH concentrations varying from 0.48 to 0.72 mM. (From Ramirez, R. et al., *Phyton (Buenos Aires)*, 37, 81, 1979. With permission.)

this enzyme from lichens has not been thoroughly characterized, although there is no reason to suspect that it differs significantly from the characterized enzymes from other sources.

B. Photoassimilate Interconversions
1. Polyol Metabolism
a. Occurrence of Polyols

Polyols are commonly encountered in most groups of plants. Polyols and their glycosides are the predominant soluble carbohydrates found in lichens; sugars are generally found only in small amounts. In lichens with eukaryotic phycobionts, polyols are the usual mobile carbohydrate, the most commonly encountered are ribitol and erythritol.[17-20] Irrespective of the nature of the transferred carbohydrate, most are immediately converted into fungal polyols such as mannitol[12,21-23] or arabitol.[21,24-27] Less commonly found fungal polyols include volemitol[25,28,29] and siphulitol.[30] In some lichens, appreciable amounts of various hexose polyol glucosides have been found. These include an arabitol-galactoside,[31] mannitol-galactoside,[21] and a mannose-mannitol-galactoside[32] (see also Chapter IX.B).

Table 1 summarizes the names, formulas, and distribution of polyols in algae, fungi, and lichens. It should be noted that sorbitol, which has been detected in algae and fungi, has not been found in lichens.

2. The Intermediary Metabolism and Enzymology

Ingram and Wood[34] considered fungal polyols to be derived from intermediates of the pentose phosphate pathway by the action of nonspecific phosphatases and polyol dehydrogenases. Alternatively, sugars could be produced by the action of an acid phosphatase of broad specificity on the sugar phosphates of the hexone monophosphate pathway.[32] In green plants, polyols appear to be direct products of photosynthesis.

Enzymes catalyzing interconversions of polyols and sugars have been isolated from a range of fungi. In the case of hexitols, both polyol-sugar and polyol-phosphate-sugar-phos-

Table 1
DISTRIBUTION OF POLYOLS IN ALGAE, FUNGI, AND LICHENS

Tetritols and Pentitols

	Erythritol	D-Threitol	D-Arabitol	L-Arabitol	Ribitol	Xylitol
Formulas	CH_2OH / HCOH / HCOH / CH_2OH	CH_2OH / HOCH / HCOH / CH_2OH	CH_2OH / HOCH / HOCH / HCOH / CH_2OH	CH_2OH / HCOH / HOCH / HOCH / CH_2OH	CH_2OH / HCOH / HCOH / HCOH / CH_2OH	CH_2OH / HCOH / HOCH / HCOH / CH_2OH
Algae	+	—	—	—	+	—
Fungi	+	+	+	+	—	+
Lichens	+	—	+	—	+	—

Hexitols, Heptitols, and Octitols

	Sorbitol	Mannitol	Galactitol	Volemitol	Siphulitol
Formulas	CH_2OH / HCOH / HOCH / HCOH / HCOH / CH_2OH	CH_2OH / HOCH / HOCH / HCOH / HCOH / CH_2OH	CH_2OH / HCOH / HOCH / HCH / HCOH / CH_2OH	CH_2OH / HOCH / HOCH / HCOH / HCOH / HCOH / CH_2OH	CH_2OH / HOCH / HOCH / HOCH / HCOH / HCOH / CH_2OH
Algae	+	+	+	+	—
Fungi	+	+	+	+	—
Lichens	—	+	—	+	+

Modified from Lewis, D. H. and Smith, D. C., *New Phytol.*, 66, 143, 1967.

phate oxidoreduction systems are known; for pentitols, only the direct polyol-sugar inter-conversion systems have been found.

Apart from dehydrogenases, two other types of enzymes are involved in polyol metabolism: polyol kinases and polyol phosphate phosphatases. The former has not been isolated from fungi (but from *Escherichia coli*[35]); examples of the latter (e.g., a mannitol phosphate phosphatase) from *Pyricularia oryzeae* have been reported by Yamada et al.[36]

Polyhydroxy alcohols are formed when the aldehyde group of an aldose or the keto group of a ketose is reduced. This reduction is presumably brought about by NAD(P)H-requiring reductase.

To date, none of the enzymatic pathways of polyol interconversion in lichens has been elucidated. Figure 3 summarizes the probable pathways by which polyols and sugars are metabolized. This pathway, as well as the enzymes involved, should be the focus of future studies in lichen intermediate metabolism. Information on this pathway and its enzymology in other organisms may be found in References 37 to 47.

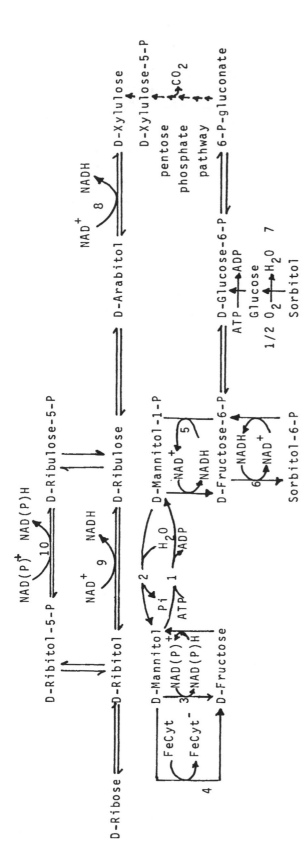

FIGURE 3. Interconversions between sugars and polyols. (1) Mannitol kinase (EC 2.7.1.57.); (2) mannitol-1-phosphatase (EC 3.1.3.22.); (3) D-mannitol dehydrogenase (NAD⁺) (EC 1.1.1.67.) and (NADP⁺) (EC 1.1.1.138.); (4) D-mannitol dehydrogenase (cytochrome) (EC 1.1.2.2.); (5) D-mannitol-1-P dehydrogenase (EC 1.1.1.17.); (6) D-sorbitol-6-P dehydrogenase (EC 1.1.1.140); (7) sorbitol oxidase; (8) D-arabitol dehydrogenase (EC 1.1.1.11.); (9) D-ribitol dehydrogenase (EC 1.1.1.56.); (10) D-ribitol-5-P dehydrogenase (EC 1.1.1.137.).

C. Biosynthesis of Polysaccharides

1. Enzymatic Production of Chitin Precursors

The innermost layer of the cell wall of filamentous fungi is composed of both cellulose and chitin. Cellulose molecules are unbranched chains of D-glucopyranose residues linked by β-1 → 4 glycosidic bounds; chitin molecules are long, unbranched chains of N-acetyl-D-glucosamine residues linked by β-1 → 4 glycosidic linkages. The biosynthesis of chitin polymers, which involves several enzymes, is initiated by both glucosaminephosphate isomerase (EC 5.3.1.19.) and glucosaminephosphate acetyl transferase (EC 2.3.1.4.).

Evernia prunastri possesses glucosaminephosphate isomerase activity which develops when thalli are floated on media containing glucose or both glucose and ammonium sulfate.[48] The inclusion of 8-azaguanine (inhibitor of transcription) or cycloheximide (inhibitor of translation) increases enzyme activity. This unusual effect is similar to the effect of actinomycin D on glucokinase,[49] acid phosphatase,[50] or on glucosamine 6-phosphate isomerase of *Proteus mirabilis*.[51] The enzyme is usually found (71%) in the mycobiont cells; the remaining 30% in algal cells can be explained by the participation of this enzyme in the biosynthesis of hexosamines and glycoproteins.[52]

Cifuentes and Rapsch[53] reported the existence of a glucosaminephosphate acetyl transferase in *Evernia prunastri* thalli which synthesized N-acetylglucosamine 6-phosphate through a transacetylation using glucosamine 6-phosphate and acetyl CoA as substrates. The enzyme developed activity when the samples were incubated on 50 m*M* glucose in either darkness or white light for 16 hr. Inclusion of 8-azaquanine or cycloheximide in the medium greatly depressed enzyme activity suggesting *de novo* enzyme synthesis. Specific activity was found primarily in the mycobiont (88%).

2. Other Polysaccharides

The typical polysaccharides of lichens are discussed in Chapter IX.B. Unfortunately, little is known about the enzymes involved in the synthesis of such polysaccharides. It is assumed that they do not differ radically from other known organisms.

D. Polysaccharide-Degrading Enzymes

Polysaccharide-degrading hydrolases have been found in lichens which epiphytically colonize other plants. Epiphytic lichens appear to penetrate phorophyte tissues by partial dissolution of intracellular material.[54] Lichen hyphae penetrate the wood, not only through wounds or lenticels, but also by passing actively through intact tissues, reaching as far as the xylem.[55] This process seems analogous to that found in fungal pathogens which can enzymatically decompose primary cell walls.[56,57]

Yagüe et al.[58] found that carboxymethylcellulose induced a β-1,4-glucanase in epiphytic *Evernia prunastri* thalli. Activity was maintained in the light, but decayed in the dark. The enzyme was secreted to the media, and the secreted activity was five times higher than that retained in the thallus. β-1,4-Glucanase activity was not affected by cycloheximide, but strongly inhibited by 8-azaguanine; the inhibition by the latter was apparently due to the interference with enzyme secretion (Table 2). This lichen species also produced an endo-polygalacturonase which was induced by sodium polygalacturonate. The activity is completely suppressed with either cycloheximide or 8-azaguanine.

III. FATTY ACID AND LIPID METABOLISM

A. Carboxylations Other Than RUBP-Carboxylase

Some enzymes in the carboxy-lyase subgroup have secondary reactions associated with decarboxylations followed by phosphorylation of the decarboxylated residue. The phosphorylation of oxaloacetate to phosphoenolpyruvate (and CO_2) can take place at the expense of

Table 2
EFFECT OF 8-AZAGUANINE ON THE LEVEL OF β-1,4-
GLUCANASE ACTIVITY OF *EVERNIA PRUNASTRI* THALLUS
INCUBATED ON CARBOXYMETHYLCELLULOSE (CMC)

Incubation conditions		Control Specific activity[a]	50 μM 8-Azaguanine		100 μM 8-Azaguanine	
			Specific activity[a]	Inhibition (%)	Specific activity[a]	Inhibition (%)
White light	Thallus	0.33	0.34	0	0.12	63.64
	Media	1.84	1.34	27.18	0.43	76.64
	Total	2.17	1.68	22.59	0.55	74.66
Darkness	Thallus	0.45	0.40	11.12	0.27	40.00
	Media	1.88	1.34	28.73	0.60	68.09
	Total	2.33	1.74	25.33	0.87	62.67

[a] The specific activity is expressed as milligrams of CMC consumed mg^{-1} protein hr^{-1}.

From Yagüe, E., Orús, M. I., and Estévez, M. P., *Planta,* 160, 212, 1984. With permission.

orthophosphate (EC 4.1.1.31.), pyrophosphate (EC 4.1.1.38.), ATP (EC 4.1.1.49.), or guanosine 5′-triphosphate (GTP) (EC 4.1.1.32.) decarboxylases. With diphosphomevalonate decarboxylase (EC 4.1.1.33.), a dehydration of the substrate occurs as well as a decarboxylation; this reaction is coupled with the hydrolysis of ATP to ADP and orthophosphate and can be regarded as the converse of a ligase reaction.

In *Peltigera aphthosa,* some enzymes of the carboxy-lyase subgroup have been described.[14] Phosphoenopyruvate carboxylase (EC 4.1.1.31.) was active both in excised cephalodia and disks, with activity being approximately 20 and 10%, respectively, of that of RuBP carboxylase. This activity was sufficiently high to account completely for the observed rates of dark $^{14}CO_2$ incorporation into disks and cephalodia after incubation of the samples in the presence of $NaH^{14}CO_3$ at 20°C in the dark (Table 3). Low activity of carbamoylphosphate synthetase (ammonia) (EC 6.3.4.16.) or glutamine-hydrolyzing (EC 6.3.5.5.), and negligible activity of both phosphoenolpyruvate carboxykinase (EC 4.1.1.49.) and phosphoenolpyruvate carboxyphosphotransferase (EC 4.1.1.38.) were detected in extracts of disks or cephalodia of *P. aphthosa.*

B. Fatty Acid Biosynthesis

The route of fatty acid biosynthesis may be divided into three distinct phases. First, acetyl-CoA is carboxylated to yield malonyl CoA; this reaction is catalyzed by acetyl-CoA carboxylase (EC 6.4.1.2.). Second, C_2 units derived from malonyl-CoA condense to form a fatty acid of intermediate chain length (usually palmitic acid). This is a multistep process and is catalyzed by a multienzyme complex called fatty acid synthase. Third, a range of long-chain saturated and unsaturated fatty acids is derived from palmitic acid by the concerted action of fatty acid elongation and desaturation systems.

Very little information is available about the above scheme in lichens. Protolichesterinic acid, obtained from *Cetraria islandica* and *Cladonia papillaria,* is of interest for its antibiotic properties as well as its structural relationship to the fatty acid cycle and the tricarboxylic acid cycle, although it represents a very minor metabolic pathway in these species. Bloomer and Hoffman[59] indicated that C-1, C-2, and C-5 of protolichesterinic acid were derived from a C_3 unit and that the key steps in the biosynthesis were catalyzed by the phycobiont.[60] When glucose and sodium [1-^{14}C]acetate were hydroponically administered to whole *Cetraria islandica* thalli for 1 week, there was an equal initial incorporation of radioactivity into C-

Table 3
ACTIVITIES OF VARIOUS CARBOXYLATING
ENZYMES IN EXCISED CEPHALODIA AND
THALLI OF *PELTIGERA APHTHOSA*

	Activity (nmol CO_2 fixed min^{-1} mg^{-1} protein)[a]	
	Excised cephalodia	**Thallus without cephalodia**
RuBP carboxylase[b]	16.4 (17.3)	39.9 (6.6)
PEP carboxylase	3.2	2.3
PEP carboxykinase	N.D.[c]	N.D.
PEP carboxyphosphotransferase	N.D.	N.D.
CP synthase[d]	1.7 (0.25)	0.1 (0.05)

[a] Each value is the mean of triplicate determinations.
[b] The values are for the 30 min 35,000 × *g* supernatants; those in parentheses for the pellet fractions.
[c] N.D. = Not detectable.
[d] CP synthase was assayed using NH_4Cl or glutamine (values in parentheses) as substrates and 30 min 35,000 × *g* supernatant fractions.

From Rai, A. N. et al, *Planta*, 151, 256, 1981. With permission.

2 and C-5 suggesting a symmetrical intermediate such as succinate. However, the relatively high degree of incorporation into C-2 and C-5 would not be expected from [1-^{14}C]acetate into [2,3-^{14}C]succinate. The authors suggested that protolichesterinic acid arose instead from pyruvate, or that one of its C_3 precursors arose from the glycolytic pathway.

C. Lipoxygenases

Lipoxygenase (EC 1.13.11.12.), linoleate:oxygen oxidoreductase, catalyzes the oxidation of polyunsaturated fatty acids containing cis,cis-1,4-pentadinene systems (linoleic acid, linolenic acid, and arachidonic acid of natural origin) to form fatty acids such as cis,trans-1,3-butadiene hydroperoxide.

In *Evernia prunastri* thalli, two lipoxygenase isozymes, LOX I and LOX II, have been reported by Cifuentes and Gómez.[61] Both isozymes exhibited a temperature optimum at 35°C and a bimaximal pH optimum at 7.0 and 8.5. Enzyme activity as a function of linoleic acid concentration showed a classical hyperbolic dependence; the K_m for the fatty acid was estimated at 1.25 m*M* for LOX I and 1.58 m*M* for LOX II. The molecular weight of the two isozymes has been estimated by gel filtration as 54 × 10^3 and 40 × 10^3 for LOX I and LOX II, respectively. Neither of the isozymes exhibited a closed specificity for linoleic acid, since linolenic and arachidonic acid were oxidized at similar reaction rates.

IV. ENZYMES OF PHENOL METABOLISM

A. Orsellinate Synthase

Monocyclic units forming lichen phenolics arise from a polyacetate chain through two different cyclization processes; one, which has been described at enzymatic level, produces orsellinic acid as the simplest product, whereas the other produces phloroacetophenone through a hypothetical phloroglucinolic cyclization.

Orsellinic acid synthase, similar to many aromatic acid synthetases, catalyzes the cyclization of a C_8 polyacetate chain to orsellinic acid without further modification. The enzyme has been characterized from *Penicillium madriti*[62] as a particulate fraction and does not show any requirement for acetyl-CoA and malonyl-CoA in phenol synthesis. In fact, both orsellinic acid and fatty acids were labeled from 2-^{14}C-malonyl-CoA when NADPH was included in the reaction mixture; fatty acids were not synthesized when NADPH was omitted. As shown in Figure 4, the synthase complex contained two transacetylase activities, an acyl carrier protein, a condensing enzyme, a cyclization enzyme, and hydrolase activity. Lichens are able to form orsellinic acid in agreement with this pathway, although the multienzyme system has not yet been described.

B. Enzymes of Depside Biosynthesis
1. Outlines of Depside Biosynthesis

Twenty years ago, Mosbach[63] began studies on lichen phenolics biosynthesis using radioactive precursors supplied to thalli maintained under laboratory conditions. When malonate-2[14C] was supplied to *Umbilicaria pustulata* thalli, gyrophoric acid (a tridepside) incorporated radioactivity into the 2 and 4 positions of the phenolic rings as well as into the carboxyl group at the 1 position. However, when the free carboxyl group of malonyl-CoA was labeled, $^{14}CO_2$ was evolved from lichen thalli and gyrophoric acid was not labeled. Chemical degradation of gyrophoric acid, after a 20 hr thallus exposure to $^{14}CO_2$, revealed that all the carbon atoms were uniformly labeled. Thus, it was obvious that randomization of radioactivity has taken place.[64] This pattern of labeling was in agreement with the mechanism proposed by Gaucher and Shepherd[62] for an orsellinic acid synthase in which three molecules of malonyl-CoA were decarboxylated and condensed with an acetyl-CoA bound to an acyl carrier protein. This mechanism did not require any dehydrogenating reaction as opposed to that found for methyl-6-salicylate-forming aromatic acid synthase, characteristic of other species of free-living fungi.[65] Orsellinic acid synthase then catalyzed a final dehydration to cyclize the polyketide chain to give orsellinic acid.

Methyl-β-orsellinic acid, used to produce β-orcinol depsides, could be synthesized by lichens using ^{14}C-acetate or ^{14}C-formate. In the latter case, atranorin, a depside from β-orcinol series, incorporated radioactivity into the −CHO and −CH₃ groups.[66] The pathway involved in the transfer of the C_1 unit did not utilize the cyclized polyketide chain since tritium-labeled methyl-β-orsellinate was incorporated into atranorin, whereas tritiated orsellinic acid was not.[67]

2. Orsellinate Depside Hydrolase

Depsides are synthesized from monocyclic precursors by an esterification reaction, and it is possible that several esterases exist. Mosbach and Ehrensvärd[68] found that *Lasallia pustulata (Umbilicaria pustulate)* contained an esterase which hydrolyzed gyrophoric, umbilicaric, and evernic acids (Figure 5). This orsellinate depside hydrolase (EC 3.1.1.40.) has been purified and characterized.[69] The molecular weight of a 135-fold purified enzyme was about 42×10^3 as determined by sodium dodecylsulfate-polyacrylamide gel electrophoresis (SDS-PAGE). The hydrolase was an exceptionally stable protein, withstanding incubation at 57°C for 10 min without any loss of activity. In addition, it could be stored frozen for 6 months or at 25°C for 4 days with practically no loss of activity. The K_m value for lecanoric acid was estimated as about 56 μ*M*. The enzyme hydrolyzed gyrophoric, lecanoric, and evernic acids, but no hydrolysis was achieved when iso-evernic, digallic acids, or phenylbenzoate were used as substrates. This indicated that a free hydroxy ortho to the depside linkage was required for activity.[69]

A similar esterase has been purified from *Evernia prunastri*.[70] Optimal temperature varied from 30 to 50°C, optimum pH over 8 to 9 and its molecular weight was about 12×10^4 by SDS-PAGE. The K_m value for evernic acid was estimated at 21.3 m*M*. Orcinol, as well as both L- and D-usnic acids, behave as competitive inhibitors of the enzyme with K_m values

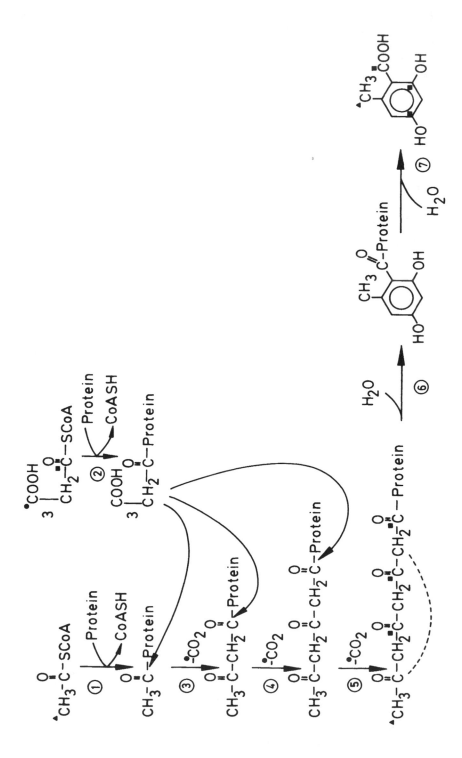

FIGURE 4. Orsellinic acid biosynthesis catalyzed by orsellinic acid synthase. ① and ② = two transacetylases and an acyl carrier protein; ③ to ⑤ = condensing enzyme; ⑥ = a cyclasation enzyme; ⑦ = a hydrolase. (From Gaucher, G. M. and Shepherd, M. G., *Biochem. Biophys. Res. Commun.*, 32, 664, 1968. With permission.)

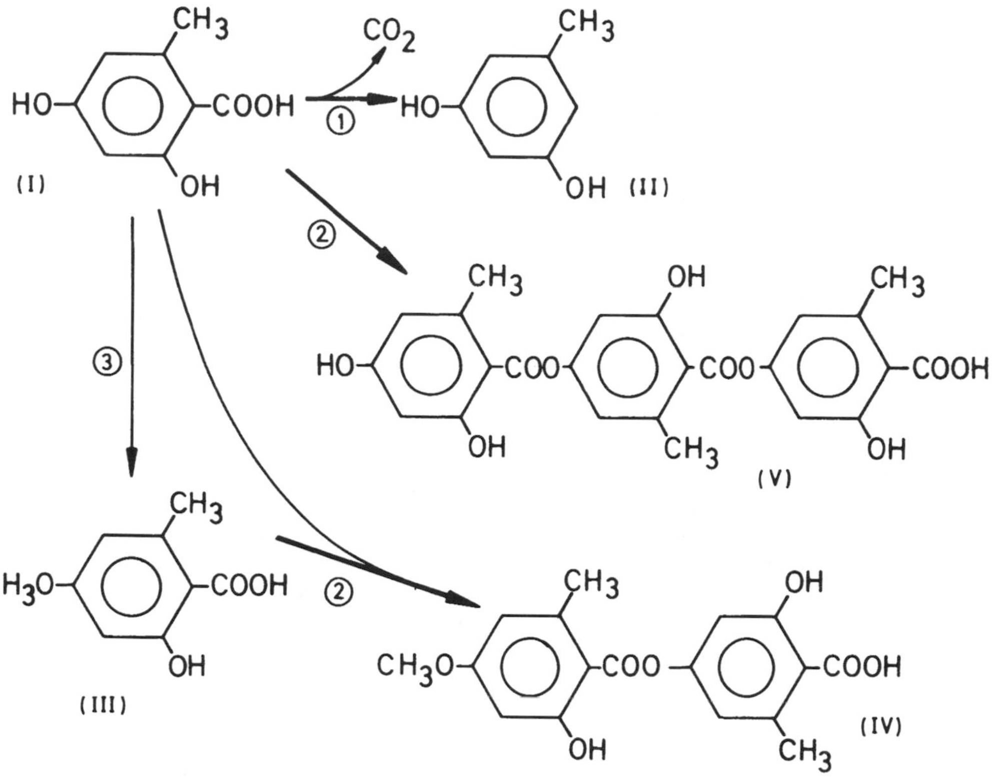

FIGURE 5. Biosynthesis of the depside, evernic acid. (I) Orsellinic acid; (II) orcinol; (III) everninic acid; (IV) evernic acid; (V) gyrophoric acid. The enzymes involved are ① = orsellinate decarboxylase; ② = orsellinate depside hydrolase; ③ = methyltransferase. (From Mosbach, K. and Ehrensvärd, U., *Biochem. Biophys. Res. Commun.*, 22, 145, 1966. With permission.)

of 0.72, 2.87, and 6.0 mM, respectively. Salicylic acid was a noncompetitive inhibitor with a K$_i$ value of 29.6 mM.[70]

The *Evernia* hydrolase behaved as a constitutive protein since increases in enzyme activity following incubations of the thalli on buffer, 2% bicarbonate or 35 μM evernic acid solutions, were not inhibited by 100 μM cycloheximide.[71] On the other hand, a complete loss of enzyme activity was achieved when the thalli were floated for 5 hr on 2% bicarbonate and then dried in air flow for 3 hr at room temperature. Thus, it seems that water content was the major factor affecting hydrolase activity in this lichen.

3. Location and Biological Role of Depside Biosynthesis Enzymes

Mosbach and Ehrensvärd[68] found that hydrolase activity of cultured *Trebouxia*, the phycobiont of *Lasallia pustulata*, was similar to that found in the holobiont. However, the lichen contained an orsellinic acid decarboxylase (EC 4.1.1.58.) which was not found in the isolated alga. The 440-fold purified enzyme had a molecular weight of about 72 × 10^3. The K$_m$ for orsellinic acid was 0.21 mM; the enzyme could also decarboxylate both 3- and 5-chloro-orsellinic acids. However, isoevernic, salicylic, p-amino- and 6-methyl salicylic, as well as α-, β- and γ-resorcylic acids were not substrates for this enzyme. In fact, β-resorcylic acid inhibits decarboxylase with a K$_i$ value of 0.8 mM.[72]

Both enzymes, hydrolase and decarboxylase, were viewed as catabolic enzymes.[72] It has been suggested that these proteins will carry out their functions only when lichen thalli are damaged. This hypothesis was based on the rationale that thalli *in situ* show low enzymatic

Table 4
DISTRIBUTION OF RADIOACTIVITY IN THE FOUR PHENOLS OF *EVERNIA PRUNASTRI* THALLUS INCUBATED ON ^{14}C-UREA IN THE LIGHT

Compound[a]	Radioactivity (%)	Specific activity[b] KBq × mmol^{-1}, 10^2
Evernic acid	72.5	n.a.
Usnic acid	1.2	n.a.
Atranorin-chloroatranorin	26.3	n.a.
Evernic acid monomethyl ester	n.d.	39.96
Everninic acid	n.d.	3.01
Orsellinic acid monomethyl ester	n.d.	36.95

[a] The four phenols of *E. prunastri* were analyzed after isolation by thin-layer chromatography from thalli incubated on ^{14}C-urea. Evernic acid monomethyl ester as well as the monocyclic phenols have not been detected (n.d.) in the thalli. These last compounds were chemically obtained from recovered ^{14}C-evernic acid. Thus, lichen phenolics have not been assayed (n.a.) for estimation of specific activity.

[b] Specific activity was estimated as the ratio of counts per minute vs. concentration of the phenol obtained after chemical treatment.

From Blanco, M. J. et al., *Planta*, 162, 305, 1984. With permission.

activities due to the slow rate of growth, whereas thalli extracts show high enzymatic activities. On the other hand, Scott[73] postulated that the hydrolysis of depsides cannot be taken as evidence that the synthesis of these compounds follows the reverse pathway; he hypothesized a biosynthetic exo-enzyme which could produce the depside from precursors secreted to the lichen cortex.

In our laboratory, we have been studying the assimilation of urea by lichens. This assimilation was initiated by the hydrolysis of urea to carbon dioxide and ammonia. In a similar way to phenols biosynthesis from $^{14}CO_2$ and $^{14}CO_3H^-$, we hypothesized that if a lichen thallus containing urease was supplied with ^{14}C-urea, a part of the $^{14}CO_2$ produced could be used in the synthesis of phenols. Approximately half of the ^{14}C obtained from ^{14}C urea was incorporated into lichen phenolics and about 70% of this radioactivity was recovered as evernic acid[74] (Table 4). When incubations were carried out on 40 mM urea in 2% bicarbonate, orsellinic acid depside hydrolase reached a value higher than when *Evernia prunastri* thalli were floated on bicarbonate alone. Since this enzyme behaved as a constitutive protein which was reactivated following thallus rehydration and since urea did not affect the enzyme activity as measured in vitro conditions,[70] it could be assumed that the increase in activity was due to a higher rehydration of the thallus by the osmotic action of urea, as well as to an increased supply of evernic acid precursors.

Although these data can be taken as an indication of the biosynthetic role of this enzyme, the most direct evidence comes from the close parallel between evernic acid content and hydrolase activity during a year.[75] As shown in Table 5, a relatively low amount of evernic acid and low orsellinate depside hydrolase activity (about 2.0 mU) was found in *E. prunastri* thalli collected during July, whereas the same lichen species collected during the winter showed a high amount of the depside as well as high esterase activity (about 160 mU). Precursors were present in the July samples although their concentration was not more than 4% of the total orsellinate derivatives. In lichens collected during the winter, everninic acid was not found, although the amount of orsellinic acid was similar to that found in the summer

Table 5
RELATIONSHIP BETWEEN
ORSELLINATE DEPSIDE HYDROLASE
ACTIVITY AND ACCUMULATION OF
EVERNIC ACID PRECURSORS IN
EVERNIA PRUNASTRI **THALLI**

	nmol $\times$ g^{-1} dry weight[b]	Percent
Lowest hydrolase[a] activity (2.0 mU)		
Orsellinic acid	0.52 ± 0.03	1.78
Everninic acid	0.46 ± 0.05	1.58
Evernic acid	28.04 ± 3.20	98.63
Highest hydrolase[a] activity (160 mU)		
Orsellinic acid	0.78 ± 0.06	1.89
Everninic acid	—	—
Evernic acid	40.49 ± 3.80	98.11

[a]   The lowest and the highest hydrolase activities found in lichen samples collected during the year. 1 mU of activity is equivalent to 1.0 nmol of evernic acid hydrolyzed per milligram protein per minute.

[b]   Values are the mean ± standard error of four replicates.

samples. This lack could be explained by a higher production of everninic acid, in which orsellinate was used in a "double" way, either to produce evernic acid or to be esterified to give evernic acid. In fact, radioactivity incorporation into orsellinic acid from $^{14}CO_2$ derived from urea was ten times higher than that found in everninic acid.[74] Thus, it is possible that the amount of everninic acid was a limiting factor in depside biosynthesis and this could be reversed when high values of esterase activity were reached. In addition, decarboxylation of orsellinic and everninic acids to orcinol and orcinol monomethylester[72] could achieve a feedback inhibition of orsellinic acid depside hydrolase.[70]

C. Synthesis of Depsidones

Although the origin of depsidones remains uncertain, the existence of coupled depside-depsidones in the same lichen species, such as olivetoric-physodic acid in *Cetraria ciliaris*,[76] indicates that these compounds could be biogenetically related. In fact, it has been proposed that depsidones are formed by dehydrogenative coupling of depsides. Such a synthesis strategy was used when a naturally occurring depsidone was synthesized in the laboratory (Figure 6). In any case, formation of monocyclic units seems to be a common pathway for both depsides and depsidones. This conclusion was derived from labeling experiments in which the distribution of radioactivity from $^{14}CO_2$ into the depsides evernic acid, atranorin and chloroatranorin, synthesized by *E. prunastri,* and into the depsidones, physodic and physodalic acids, from *Hypogymnia physodes,* were almost identical and coincided with that expected from a conventional aromatic synthetase activity.[64]

Several authors have recognized that the secondary structural differences between known depsides and depsidones need not necessarily have occurred after cyclization. In other words, the attachment of a fumaric residue on 3-substituted of methyl-3-orsellinate to produce the second monocyclic precursor of fumarprotocetraric acid (a depsidone) can be achieved on the monocyclic phenol, on atranorine (a possible precursor of the depsidone) or even on the depsidone itself (Figure 6). Vicente et al.[77] found that the time course of fumarprotocetraric

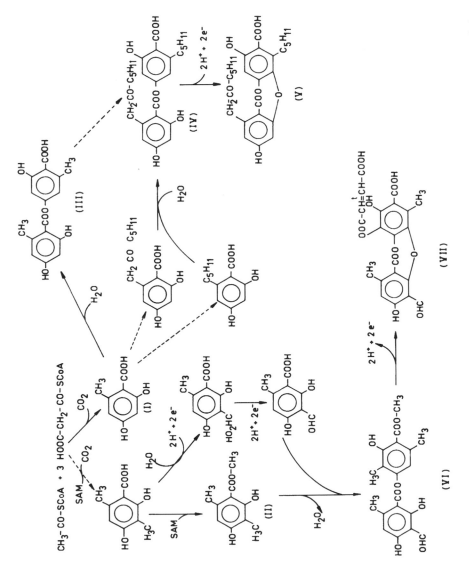

FIGURE 6.   Proposed synthesis of the depsidones, physodic, and fumarprotocetraric acid. (I) Orsellinic acid; (II) methyl-β-orcinol carboxylate; (III) lecanoric acid; (IV) olivetoric acid; (V) physodic acid; (VI) atranorin; (VII) fumarprotocetraric acid.

acid production in *Cladonia sandstedei* floating on urea was inversely related to that of atranorin accumulation. This may imply a biogenetical relation between both compounds, but it does not provide any evidence about the structural change described above. However, norstictic or salazinic acids are produced by *C. sandstedei* thalli floating on ammonia, whereas fumarprotocetraric acid and atranorin are completely remobilized. This may be considered as evidence supporting the suggestion that depsidones can be modified after the formation of the ether bond.

D. Enzymes Involved in the Metabolism of Usnic Acids
1. Usnic Acid Biosynthesis

The biosynthesis of usnic acid has been studied by Taguchi et al.[78] using several species of lichens and four labeled precursors: ^{14}C-1-acetate, ^{14}C-2-malonate, $^{14}CH_3$-CO-phloroacetophenone, and $^{14}CH_3$-CO-methylphloroacetophenone. When ^{14}C-1-acetate was used, the constitutive units of usnic acid incorporated radioactivity at the 1, 3, 5, and 7 positions whereas position 1 was not labeled when ^{14}C-2-malonate was used as a precursor. This implies that the CO-CH$_3$ group in each phenolic unit derives from the acetate moiety. Consistent results were found by Fox and Mosbach[64] for the synthesis of usnic acid by *Cladonia sylvatica* supplied with $^{14}CO_2$ (Figure 7).

Labeling experiments further demonstrated that usnic acid was formed from the condensation of two methylphloroacetophenone molecules. Usnic acid retained radioactivity in the acetyl groups when $^{14}CH_3$-CO-methylphloroacetophenone was supplied to thalli; this group was unlabeled when $^{14}CH_3$-CO-phloroacetophenone was used.[78] Even when ^{14}C-formate was supplied, radioactivity incorporated into usnic acid was negligible.[79] Although methylphloroacetophenone has not been isolated from lichens, its presence was suggested by isotopic dilution. Nonactive methylphloroacetophenone was added to a cell-free extract of *Cladonia mitis* which had been exposed to $^{14}CO_2$ for 10 days; recovered methylphloroacetophenone showed considerable radioactivity.[79] The condensation pathway would involve successive dehydrogenase- and dehydrase-catalyzing reactions,[79] although attempts to prepare soluble enzymes capable of such reactions were unsuccessful. The oxidative coupling of methylphloroacetophenone in lichens should be stereospecific, since usnic acid occurs mainly in an optically active form.[80]

2. Usnic Acid Dehydrogenase

It has been observed that *Evernia prunastri* thalli maintained under starvation conditions remobilized usnic acid, and the loss of this phenolic was enhanced in the light.[81] The enzyme responsible for usnic acid remobilization, DL-usnic acid dehydrogenase, has been isolated and characterized. The 150-fold purified enzyme had a molecular weight of about 45×10^4 and showed an activity optimum at pH 8.0 and a temperature optimum at 30°C.[82] The substrate saturation kinetics were sigmoidal for both L-usnic acid and NADH; the cofactor requirement was absolutely restricted to NADH. D-Usnic acid also served as a substrate, although activity on the D-isomer was about three times less than on L-usnic acid. It seems probable that this enzyme and the dehydrogenase predicted by Taguchi et al.[79] are two different proteins, since the oxidative coupling of two units of methylphloroacetophenone might be stereospecific, as mentioned above, whereas both D- and L-usnic acids appear to be reduced by this latter dehyrogenase.

In addition, a second dehydrogenase has recently been described.[83] The substrate saturation kinetics were sigmoidal for both D-usnic acid and NADH. The K_m value for the phenolic was about 0.1 mM and 0.13 mM for NADH. There were two classes of interaction sites for D-usnic acid on the enzyme. One class of binding sites was by the ligand. Preincubation of protein and substrate for 5 min before NADH addition resulted in a strong inhibition of enzyme activity (Figure 8). There was no activity when L-usnic acid was used as a substrate, but this isomeric form behaved as a competitive inhibitor of the enzyme. In this case, K_i

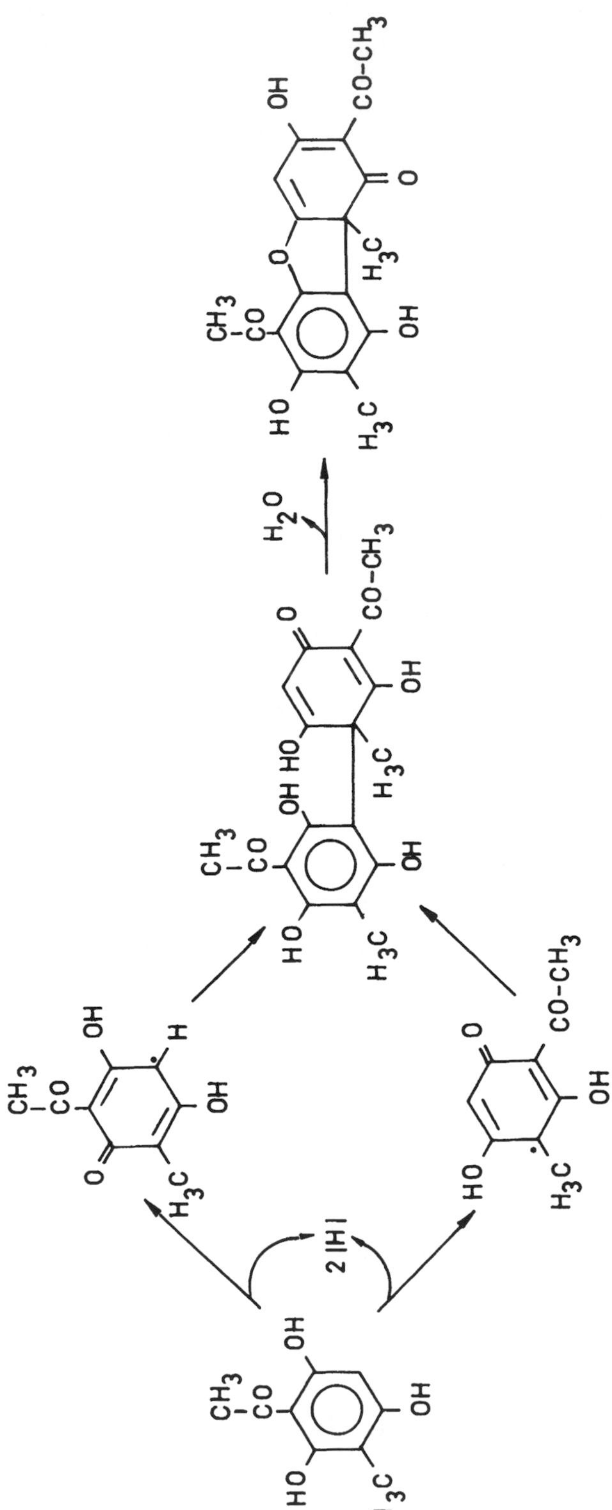

FIGURE 7.   Proposed synthesis of usnic acid. (From Taguchi, H. et al., *Chem. Pharm. Bull.*, 17, 2054, 1969. With permission.)

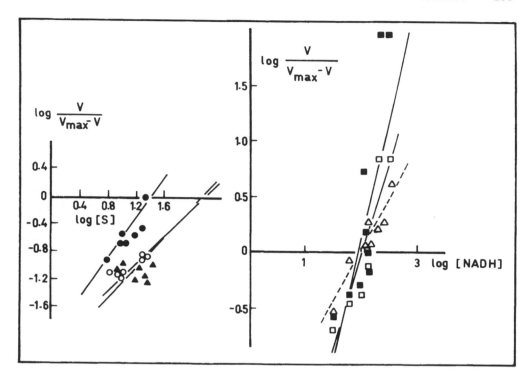

FIGURE 8. Hill plots of the effect of several ligands on D-usnic acid dehydrogenase. (●) D-Usnic acid added at time zero; (○) enzyme preincubated with D-usnic acid prior to NADH addition; (▲) preincubation with 18 μM L-usnic acid prior to the addition of substrates; (■) NADH added at time zero; (□) enzyme preincubated with NADH prior to D-usnic acid addition; (△) preincubation with NAD⁺ prior to the addition of substrates. (From Vicente, C. and González, A., *Biochim. Biophys. Acta*, in press. With permission.)

value was about 8.5 μM and the number of interaction sites for the substrate diminished to $n_H = 1$, similar to that found after preincubation of the enzyme with D-usnic acid.

D-Usnic acid dehydrogenase could use both NADH and NADPH as cofactors ($K_m = 4.0$ mM) although the affinity for NADH is about 30 times higher than that for NADPH. Both forms of reduced nucleotides, as well as NAD⁺, behaved as positive effectors of the enzyme. Cofactors interacted with the protein through two interaction sites which were not modified after preincubation of the enzyme with the ligands in absence of D-usnic acid.

The dehydrogenases seem to be produced in response to different stimuli. One of them, DL-usnic acid dehydrogenase, was found when lichen thalli were exposed to drying conditions and nutrient starvation under white light. The same response was obtained when thalli irradiated with low photon flux rates were submitted to high values of osmotic pressure by floating thalli on polyethylene glycol solutions. This increased enzyme activity was not the result of *de novo* synthesis since it was not inhibited by cycloheximide. The second dehydrogenase was clearly an inducible enzyme since D-usnic acid dehydrogenase activity appeared and increased only when lichen thalli were supplied with an excess of exogenous D-usnic acid. We have previously speculated about the remobilizing role of these enzymes,[81] which would promote the synthesis of acetyl-CoA from the accumulated phenols when exogenous nutrients are limited. The environmental conditions under which these enzyme activities are enhanced lends support to this hypothesis.

The structure of the products of both dehydrogenases has been elucidated by analysis of infrared (IR) spectra of the substrate and reduced products (Table 6). Reduced D-usnic acid shows an unchelated hydroxy group, as revealed by the bands of 3320 and 910 cm⁻¹. The displacement of the peak characteristic of the 4-carbonyl group from 1535 to 1550 cm⁻¹

Table 6

SPECTRAL CHARACTERISTICS OF USNIC ACID AND ITS ENZYMATICALLY REDUCED PRODUCTS

	Usnic acid		Reduced product from L-usnic acid		Reduced product from D-usnic acid
υ (cm⁻¹)	Function	υ (cm⁻¹)	Function	υ (cm⁻¹)	Function
—		3400	Unchelated -OH group	3350	Unchelated -OH group
—		—		3280	Unchelated -OH group
3080	Chelated 5-OH group	—		3100	Chelated 5-OH group
3000	Chelated 5-OH group	—		3000	Chelated 5-OH group
2920	Chelated 7-OH group	2920	Chelated 7-OH group	2920	Chelated 7-OH group
2850	Aromatic C-acetyl group	2850	Aromatic C-acetyl group	2840	Aromatic C-acetyl group
2750	Chelated -OH groups	—		2760	Chelated -OH groups
				2700	
				2600	
2350—2310	Very strongly hydrogen-bounded enolic -OH group	2310	Very strongly hydrogen-bounded enolic -OH group	2360	Very strongly hydrogen-bounded enolic -OH group
—		1735	Diketone	—	
1690	Displacement of the ether band in relation to the furanic heterocycle	—		—	
1610—1630	Enol ether double bond which coalesces with aromatic C-acetyl group	—		—	
1535	Conjugated chelated 4-carbonyl group	—		1550	Conjugated chelated 4-carbonyl group
1450	Methyl group	1450	Methyl group	1460	Methyl group
1420	-OH flexion	—		1400	-OH flexion
1370	Symmetrical formation of methyl group	1370	Symmetrical formation of methyl group	—	
1350	Aromatic C-acetyl group	1350	Aromatic C-acetyl group	1350	Aromatic C-acetyl group
1285	Ether bond	1265	Keto group	1290	Ether bond
—				—	
1065—1035	Oxygen in cyclopentane	1060-1030	Oxygen in cyclopenthane	1050—1035	Oxygen in cyclopenthane
990—950-930—835	Isolated hydrogen	895-840	Isolated hydrogen	970-910-890	Isolated hydrogen

seems to indicate that this function has been reduced to unchelated, free hydroxyl. However, a peak at 3000 to 3080 cm^{-1} indicates that 5-hydroxy remains as chelate. Chelated hydroxy groups are also revealed by several bands at 2760, 2700, and 2600 cm^{-1}, as well as that at 1400 cm^{-1}. Possibly, these peaks define 5-hyroxy as well as 7-hydroxy functions. There is a characteristic dissappearance of the peak at 1690 cm^{-1}, typical of the displacement of the ether band in relation to the furanic nature of the heterocycle. The stability of the furanic ring is not altered by action of D-usnic acid dehydrogenase, since the bands at 1215, 1060, and 1030 cm^{-1} remain unchanged. Joint occurrence of the stability of the peak at 3080 cm^{-1} (chelated 5-hydroxy group[84]) and the disappearance of the band at 1690 cm^{-1} is difficult to evaluate (Table 6). However, it is possible that the analyzed product was a mixture of two successive states of reduction of D-usnic acid, as is shown in Figure 9. This may be possible since many dehydrogenases, which use FAD as carrier, transfer the electrons one-by-one to the acceptor through a stable semiquinone state.[9]

The nature of the product of D-usnic acid dehydrogenase reaction, originally described by Estévez et al.,[82] must now be modified. A more detailed IR analysis revealed that the reduced product contained an unmodified furanic ring, as deduced from the peak at 1210 cm^{-1} (characteristic of an ether bond) as well as those at 1060 and 1030 cm^{-1}, which defined an oxygen atom in a cyclopentane. Both bands at 1690 and 1610 cm^{-1} disappeared suggesting that the 4-carbonyl group has been converted to an unchelated hydroxy group whereas the double bond between C_9 and C_{10} positions has been reduced. Both dehydrogenase reactions are summarized in Figure 9.

The catabolism of usnic acids in lichens is, thus, very different from that performed by soil microorganisms.[85,86] In the soil, microorganisms such as *Pseudomonas* or *Mortierella isabellina*[87,88] biodegrade D-usnic acid to 6-desacetylusnic and 6-acetyl-8,9B-α-dimethyl-1,9b,2,3-tetrahydro-1α,hydroxy-2-desacetylusnic acid, (+)-2-desacetylusnic acid, and 6-acetyl-8,9α-1-oxodibenzofuran, respectively, but the 4-carbonyl group always remains unaffected.

E. The Role of Symbiosis in the Synthesis of Lichen Phenolics

Depsides, depsidones, and dibenzofurans have always been considered as products of the metabolism of the fungal partner in lichens, although the symbiotic state was required for synthesis. Isolated mycobionts rarely produce acids characteristic of the symbiotic state, although several workers claim to have some success in this area. Castle and Kubsch[89] claimed that the isolated mycobiont of *Cladonia cristatella* produced usnic, didymic, and rhodocladonic acids, but no experimental evidence was shown. In addition, these results could not be repeated by other authors.[90] Squamatic acid has been synthesized by the cultured mycobiont of *C. crispata* following irradiation with UV light.[91] Reports on the synthesis of D-usnic and salazinic acid by the mycobiont of *Ramalina crassa* are contradictory.[92,93]

It is possible that the synthesis of these compounds was readily achieved by isolated mycobionts, but the products could not be detected due to rapid hydrolysis. This was suggested by Umezawa et al.,[94] who reported that lecanoric acid was produced by the free-living fungus *Pyricularia* sp., but was hydrolyzed to orsellinic acid and then decarboxylated to orcinol.

Some free-living fungi produce secondary metabolites characteristic of lichens. *Aspergillus terreus* produced the depside, 4-0-dimethylbarbatic acid;[95] *Auricularia delicata* produced atranorin, lecanoric, and salazinic acids in a symbiotic association with yeasts and bacteria.[96] To date, there are no references to the production of lichen phenolics by free-living algae nor isolated photobionts.

F. Environmental Modulation of the Metabolic Potential to Synthesize Lichen Phenolics

Light appears to be a dominant environmental factor affecting the synthesis of lichen phenolics. Usnic acid increased linearly in response to sunlight intensity increase on the

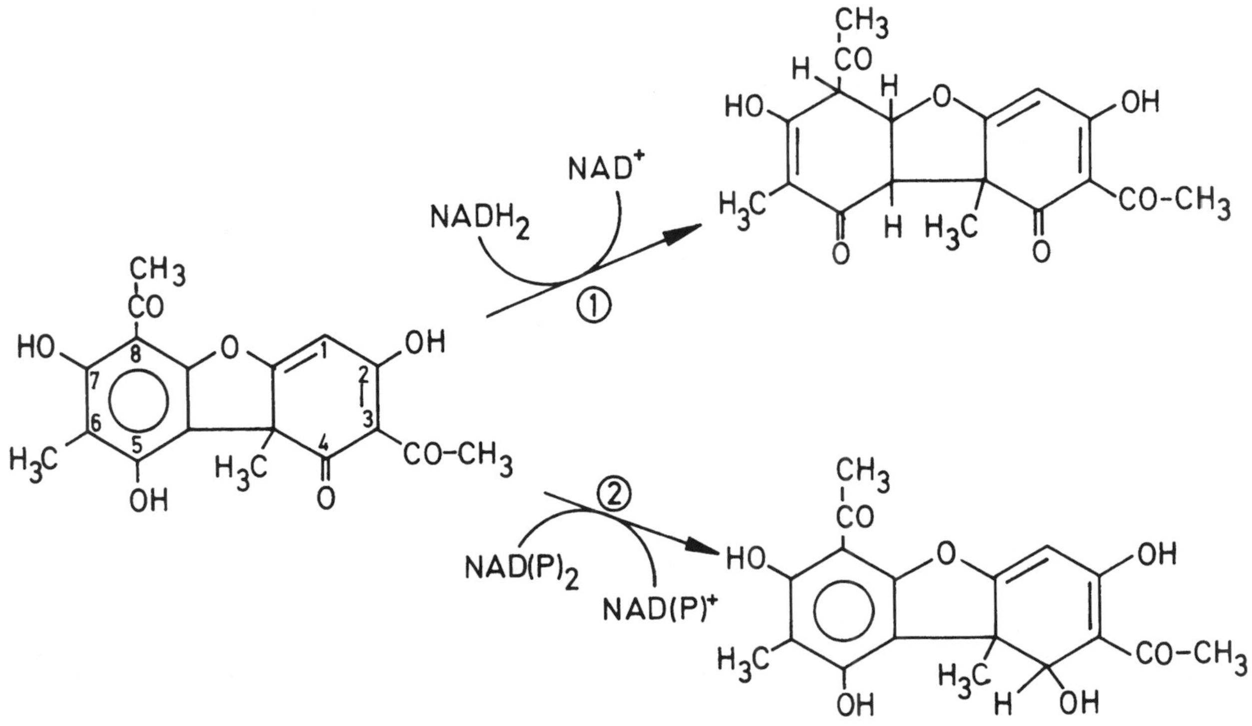

FIGURE 9.   Reduction of D-usnic acid by the action of ① = DL-usnic acid dehydrogenase and ② = D-usnic acid dehydrogenase.

surface of the podetial apices of *Cladonia subtenuis*.[97] Similar results have been observed for fumarprotocetraric acid in *C. rangiferina*[98] and *C. verticillaris*.[99]

Few studies have been made on the effect of temperature on the synthesis and accumulation of lichen phenolics. Hamada[100] studied the content of salazinic acids in *Ramalina siliquosa* and demonstrated that its content was higher in thalli growing on dark-colored rocks (higher mean temperature) than that found in thalli growing on light-colored rocks (lower mean temperature). It is probable that temperature near the dark surface increased both production and accumulation of the depsidone. Similar results were obtained for divaricatic and salazinic acids from *R. subbreviuscula*.[101] These data were confirmed in lichens cultured under controlled conditions in the laboratory.[102]

Culberson et al.,[103] using four cloned lines of *C. cristatella*, have found that barbatic acid and its derivatives reached the highest concentration at low temperature values whereas didymic acid and related compounds were not affected by temperature variations.

Age of the thallus is often considered as a determinant of phenolic concentration although the relationship between age and phenols need not be a direct one. Vulpinic acid concentration was lowest in old basal branches of *Letharia vulpina* thalli and increased towards the young branch tips,[104] whereas the reverse was true for atranorin. Culberson and Culberson[105] found that *Lasallia papulosa* contained gyrophoric acid and other lichen compounds that were quantitatively equally distributed among thalli of all ages. Later, Culberson et al.[103] showed that cloned *C. cristatella* mycobionts produced barbatic acid in all ages and at similar concentration, but that didymic acid content increased from young to mature squamules.

V. ENZYMES OF NITROGEN METABOLISM

A. Nitrogenase

Nitrogenase is an oxidoreductase which catalyzes the transfer of six electrons from reduced ferredoxin to dinitrogen to form ammonia. The enzymology and metabolic control of this and related enzymes have been summarized in Chapter VI.B. of this volume.

B. Reduction of Inorganic Nitrogen

Enzymes which reduce inorganic nitrogen belong to the oxidoreductases. The substrate which is oxidized is regarded as hydrogen donor and the systematic name is based on donor:acceptor oxidoreductase. Some enzymes, including nitrate and nitrite reductase, are exceptional in that reduced forms of coenzymes act as donors. Enzymes 1.6.1—4 are concerned with the reduction of nitrate to ammonia or aminogroups in plants and bacteria.

Nitrate reductase (EC 1.6.6.1.) is a well documented, phytochrome-regulated enzyme in higher plants;[106-108] there is only one paper which describes its existence in lichens. Avalos and Vicente[109] demonstrated that dry thalli of *Evernia prunastri* floated on 10 m*M* potassium nitrate in distilled water developed nitrate reductase activity which increased logarithmically as a function of time in the dark. This increase was totally blocked by the addition of 100 μ*M* cycloheximide to the incubation media. This observation suggested the possible induction of the enzyme by its substrate, as reported for green algae[110,111] or free-living fungi.[112] The induction of nitrate reductase activity was slightly enhanced when thalli were irradiated with red light for the first 10 min of incubation; the enhancement was more dramatic with repeated light treatments. Red-light illumination for 10 min at the beginning of each hour of incubation gave an increase in the enzyme activity 2.3 times higher than that obtained after only one initial illumination. This suggested a high irradiance reaction, since even the reversal of the red effect by far-red light was only significant after a repeated irradiation (Figure 10). All these results suggest the action of the photoconverted phytochrome (Pfr) as a positive effector of nitrate reductase in *E. prunastri*. Finally, the increase of the enzyme activity was a direct function of the percent of Pfr actually present in algal cells isolated from the thallus samples.

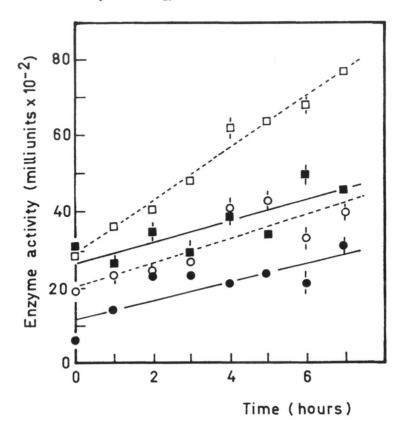

FIGURE 10. Time course of nitrate reductase activity in *Evernia prunastri* thalli floated on 10 m*M* KNO$_3$ when the incubations are started by (○) 10 min red light (y = 0.03 × + 0.206; r^2 = 0.83); (●) 10 min red light + 10 min far-red light (y = 0.025 × + 0.117; r^2 = 0.79); both treatments, (□) red (y = 0.068x + 0.29; r^2 = 0.98) and (■) red + far-red (y = 0.028 x + 0.26; r^2 = 0.75) are repeated at the beginning of each hour of incubation. (From Avalos, A. and Vicente, C., *Can. J. Bot.*, 63, 1350, 1985. With permission.)

C. The Biosynthesis of Glutamic Acid and Related Compounds

Much work has been done on the nitrogen metabolism of lichens with cyanobionts. The nature of the process and its enzymology has been discussed by Rai (Chapter VI.B) in this volume. Several other enzymes from this general pathway will be described here.

Bernard[113] demonstrated glutamate decarboxylase activity (EC 4.1.1.15) from five species of the Stictaceae; *Lobaria laetevirens, L. pulmonaria, Sticta sylvatica, S. limbata,* and *S. fuliginosa.* The enzyme had an optimal pH of 5.0 and a temperature optimum at 33°C; the addition of 1.2 m*M* cysteine to the reaction mixtures increased enzyme activity. The highest glutamate decarboxylase activity was found in *S. sylvatica* and the lowest in *L. laetevirens.* This appeared to correlate with the presence of the amine, 8-aminobuthyrate, which was abundant in *Sticta* species but not commonly encountered in *Lobaria.*

D. Metabolism of L-Arginine

In general, the chief nitrogen transport compounds in plants are asparagine, aspartate, glutamate, glutamine, and arginine, but variation among these nitrogen species is considerable. For both the storage and transport of nitrogen, an important criterion is the N:C atomic ratio of the compound involved-the higher the better. For asparagine it is 1:2 and for canavanine, 4:5. All stored and transported amino acids are closely related to either 2-

oxoglutarate or oxaloacetate which are easily diverted away from or into the tricarboxylic acid cycle. A final important requirement of transport and storage of amino acids is that their metabolism should be quickly and effectively controlled, and under the regulation of ammonia availability.

It is well known (see Chapter VI.B) that in lichens containing cyanobionts, glutamate and glutamine are primary products of ammonia assimilation. However, lichens containing a green alga accumulated a high concentration of L-arginine (e.g., *Evernia prunastri*), while in other lichens (e.g., *Rocella montagnei*, *Cladonia rangiferina*, *C. gracilis*, *Parmelia tinctorum*, *P. nepalensis*[114-118]) L-glutamic or L-aspartic acids were abundant in the free amino acid pool. Such amino acid pools appear to vary depending upon the season.[119]

1. Arginase and L-Ornithine Decarboxylase

Studies on polyamine content of several lichen species indicated that L-arginine was rapidly mobilized from the pool.[120] The hydrolysis of L-arginine by arginase produces L-ornithine and urea. The enzyme was induced by L-arginine in *Evernia prunastri* thalli incubated in the dark,[121] although some activity was detected during light incubation. The addition of urea to the culture media prevented, or lessened, the induction of arginase caused by the amino acid. A 158-fold purified enzyme (mol wt 18×10^4) had a K_m value of 0.20 mM for L-arginine. The optimum pH was bimaximal, the main peak being approximately 9.0 and showing a secondary maximum at 6.5. Urea was a competitive inhibitor with a K_i value of 2.58 mM and agmatine behaved as a noncompetitive inhibitor with a K_i value of 21.54 mM. L-Ornithine and putrescine behaved as activators of the enzyme (apparent K_m 0.13 and 0.14 mM, respectively). The strong inhibition of arginase by urea provides a feedback mechanism that assures the complete and effective regulation.

A second form of arginase, a constitutive one, was present in *E. prunastri* thalli (Figure 11). The preexistent but inactive protein was activated by the liberation of L-arginine from the vacuoles to the cytosol.[122] These results are in agreement with those reported for the arginase of *Neurospora crassa*.[123] The enzyme has been purified 920-fold from thalli incubated on cycloheximide.[124] The relation between the reaction rate vs. substrate concentration showed a typical Michaelis-Menten relationship with a K_m value for L-arginine of 2.5 mM. L-Ornithine, agmatine, and putrescine (from 1.0 to 3.0 mM) behaved as nonessential activators of the enzyme. Apparent K_a values, estimated by the Dixon plot, were 1.1 mM for L-ornithine, 5.88 mM for agmatine, and 2.7 mM for putrescine. The constitutive arginase, as well as the inducible one, were dependent on Mn^{2+} for activity.

L-Ornithine, produced as a result of arginine hydrolysis, can be further decarboxylated to produce putrescine and urea. Escribano and Legaz[125] reported the existence of an L-ornithine decarboxylase from *E. prunastri* thalli floated on 40 mM L-ornithine in the dark. The enzyme was also detectable when thalli were floated on Tris buffer alone, although the final level of activity was 2.5 times lower than that found in ornithine-floated thalli. Addition of 40 μM cycloheximide to an L-ornithine-containing medium did not eliminate ornithine decarboxylase activity; when 100 μM chloramphenicol was added, there was a loss of about 90% of activity. From these data, the authors speculated the existence of two enzymes synthesized by both mycobiont and phycobiont (Figure 12). Approximately 80% of the total decarboxylase activity in both symbionts depended upon protein synthesized in organelle ribosomes and the remaining 20% by the cytosol.

2. L-Arginine Decarboxylase, Agmatine Amidinohydrolase, and Agmatine Iminohydrolase

An alternative pathway to L-arginine hydrolysis is the decarboxylation of arginine to produce agmatine. Thalli of *Evernia prunastri* floated on 40 mM L-arginine in the dark developed an inducible arginine decarboxylase for the first 4 hr of incubation which decreased with time. Vicente and Legaz[126] purified the enzyme 117-fold. Its optimum pH, which was

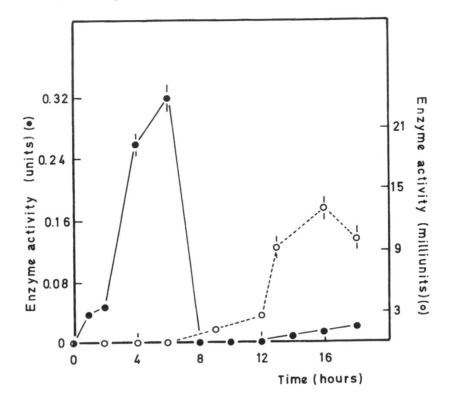

FIGURE 11. Time course of inducible arginase activity (●) in *Evernia prunastri* thalli incubated on 40 m*M* L-arginine in 0.1 m Tris-HCL buffer, pH 9.15, and of constitutive arginase activity (○) in thalli floated on 40 μ*M* cycloheximide in 0.1 *M* Tris-HCL buffer, pH 9.15. (From Martín-Falquina, A. and Legaz, M. E., *Plant Physiol.*, 76, 1065, 1984. With permission.)

quite sharp, showed a maximum at 7.1 and a temperature optimum at 26°C. The molecular weight of the protein, estimated by filtration through Sepharose 6B, was about 30×10^4. Double-reciprocal plots of the activity vs. L-arginine concentration showed a typical Michaelis-Menten relationship with an apparent K_m value of 12.5 m*M*. Thus, at least in *E. prunastri*, the synthesis of urea is more effective via arginase reactions than by decarboxylases.

Both putrescine and urea significantly inhibited decarboxylase, suggesting a possible feedback mechanism in the pathway of urea synthesis. Agmatine (40 m*M*) produced only a moderate inhibition of the decarboxylase, but was a potent noncompetitive inhibitor of arginase.[121]

Agmatine produced by decarboxylation of arginine can be hydrolyzed by an agmatine amidinohydrolase to give putrescine and urea. A 485-fold purified enzyme from *E. prunastri* showed an optimum pH at 6.9 and an optimal temperature from 35 to 40°C, although about 6% of the activity remained when the reaction was carried out at 0°C and about 20% at 70°C.[127] The molecular weight of this enzyme was approximately at 32×10^4. From Woolf plots, the K_m value for agmatine was estimated at 6.4 m*M*. The enzyme showed specificity for agmatine and did not hydrolize L-arginine, L-ornithine, or putrescine. These latter three analogs behaved as activators of the hydrolase for agmatine concentrations lower than 14 m*M*, but the same compounds seem to be inhibitors of the enzyme when agmatine concentration were higher than 14 m*M*. Urea, from 10 to 40 m*M*, completely inhibited agmatine amidinohydrolase.

As mentioned above, L-arginase induces L-arginine decarboxylase and agmatine amidi-

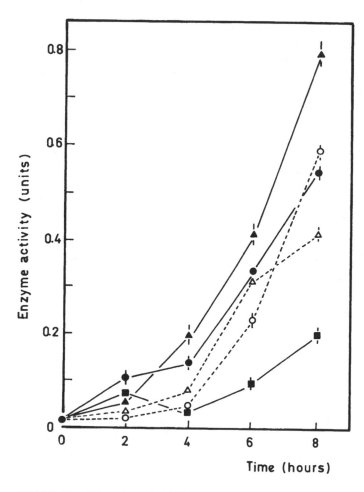

FIGURE 12.   Effect of protein inhibitors on L-ornithine decarboxylase activity of *Evernia prunastri*. Thalli floated on (○) buffer alone (y = 0.07x − 0.05; r^2 = 0.91); (●) buffer containing 40 μm cycloheximide (y = 0.08x − 0.06; r^2 = 0.98); (▲) 40 mM L-ornithine (y = 0.02x$^{1.64}$; r^2 = 1.00); (△) amino acid-containing media + 40 μM cycloheximide (y = 0.01x$^{2.05}$; r^2 = 0.94); or (■) 100 μM chloramphenicol (y = 0.02x + 0.001; r^2 = 0.90). (From Escribano, M. I. and Legaz, M. E., *Phyton (Buenos Aires)*, 44, 171, 1984. With permission.)

nohydrolase. Urea acts as a catabolite repressor for both enzymes. It is well known that the addition of urea to bacterial cultures inhibits the expression of the operons sensitive to catabolite repression.[128] This is in contrast to the repression caused by glucose, which is reversed by AMP even if the microorganism does not produce this nucleotide.[129] Inhibition of operons in prokaryotes by urea cannot be reversed by AMP. Vicente and Legaz[130] reported that the gene for light-arginase in *E. prunastri* was subject of catabolite repression, since the addition of 100 mM glucose to incubation media containing 40 mM L-arginine produced arginase activity three times lower than that reached in the absence of the sugar. Glucose did not have any effect on the activity of the purified enzyme. Addition of 0.5 mM AMP to the incubation medium completely reversed, and even stimulated the synthesis of this enzyme. Although agmatine amidinohydrolase behaved in a similar way, it also showed important differences. The maximal activity was reached by thallus samples floated for 8 hr on 40 mM L-arginine. Addition of 50 or 100 mM glucose to the incubation medium completely inhibited the appearance of enzyme activity. When these treatments were performed in the presence of 0.5 m*M* AMP, the values of hydrolase activity were almost identical to those obtained in the control assay (Figure 13). Reversal by AMP of the repression by glucose of light-arginase and agmatine amidinohydrolase activity might imply inactivation of an adenylcyclase. However, this does not explain the same reversal when urea acts as a repressor of both enzymes, as opposed to that found in prokaryotes[128] where apparently urea interacts with a specific sequence in the promoter.

Agmatine produced by the action of L-arginine decarboxylase can also be converted to *N*-carbamyl putrescine by an agmatine iminohydrolase. This pathway has been found in *Hordeum vulgare*,[131] *Zea mays*,[132] and *Glycine max*.[133] Agmatine iminohydrolase appeared in *E. prunastri* thalli floated on Tris-HCl buffer during the first 2 hr of incubation. However, addition of 40 m*M* L-arginine to the media supported a greater increase of activity up to the 4th hr of incubation. The appearance of hydrolase activity was dependent on the synthesis of protein and was considered as an induction of the enzyme by the amino acid. The hydrolase has been 35.4-fold purified from *E. prunastri* thalli.[134] The relation between the reaction rate vs. substrate concentration was typically hyperbolic, with a K_m value for agmatine of 0.8 m*M*. Urea behaved as an activator of the enzyme whereas L-arginine inhibited the hydrolase at concentrations of agmatine higher than 4 m*M*; at low substrate concentrations (to 2 m*M*), L-arginine appeared to activate the enzyme.

E. Metabolism of Amines and Polyamines

1. Monoamines

Certain species of the family Stictaceae possess soluble nitrogen compounds in a methylamine form instead of amino acids or amides. Bernard and Goas[135] determined these compounds from thalli of *Lobaria pulmonaria*, *L. laetevirens*, *Sticta sylvatica*, *S. fuliginosa*, and *S. limbata*. The predominant amine was always trimethylamine (70%); dimethylamine was less abundant (10%) and monomethylamine was the least important.

Some years later, Bernard and Larher[136] investigated the synthesis of these methylamines by using radioactive precursors. When ^{14}C-glycine was supplied to *L. laetevirens* thalli, they found that after 5 min, labeled methylamines appeared. The most abundant was monomethylamine which constituted 93% of the total radioactivity. Labeled trimethylamine appeared after 24 hr and accounted for 98.8% of the total radioactivity. Labeled dimethylamine increased with the time, but never represented more than 7% of the total. They concluded that monomethylamine was the first product formed, followed by dimethylamine and trimethylamine. The relatively high labeling of dimethylamine after 24 hr suggested the existence of another synthetic pathway, possibly from decarboxylation of sarcosine. Trimethylamine might also be formed by *N*-oxidation from choline or betaine. A representative scheme is shown in Figure 14.

2. Other Amines

Another soluble nitrogen form, sticticin, has been isolated from thalli of *L. laetevirens*. Its concentration was about 1 *M* when thallus water content reached 10 to 12%, suggesting it might play an efficient role in osmoregulation. The biosynthetic pathway first involved *N*-methylation of tyrosine followed by an orthohydroxylation of the ring, and finally an esterification of the acid function.[137] The pathway is shown in Figure 15.

The same lichen also contained DOPA betaine, tyrosine, betaine, and *N*-dimethyltyrosine which appeared to be intermediates in the biosynthesis of sticticin.[138] The enzymes implicated in these metabolic routes have not yet been described in lichens.

3. Polyamines

During the past 20 years, interest has been increasing in the naturally occurring polyamines, putrescine, spermidine and spermine, although their specific functions are still obscure.

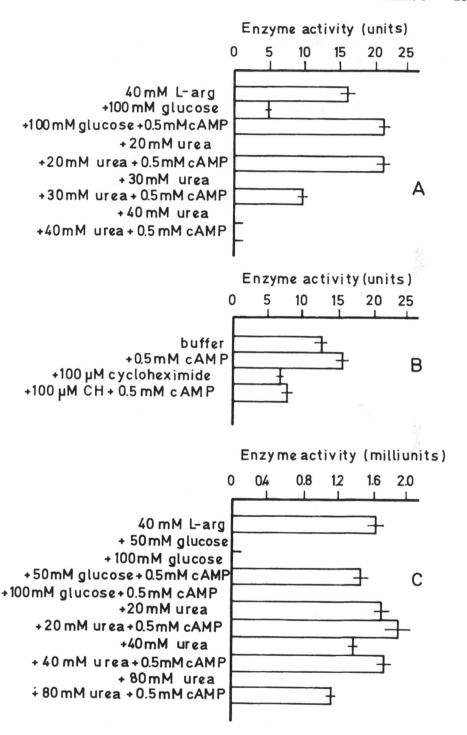

FIGURE 13. Catabolite repression of arginase and agmatine amidinohydrolase of *Evernia prunastri*. (A) Effect of cAMP on the differential rate of light-arginase synthesis in the presence or absence of glucose or urea. (B) Effect of cAMP on heavy-arginase activity revealed by cycloheximide treatment. (C) Effect of cAMP on the differential rate of agmatine amidinohydrolase activity in the presence or absence of glucose of urea. (Modified from Vicente, C. and Legaz, M. E., *Phytochemistry*, 24, 217, 1985.)

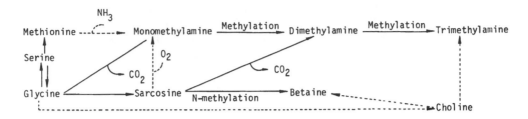

FIGURE 14.   Possible pathway of methylamine formation in *Stictaceae*. (From Bernard, T. and Larher, F., *C. R. Acad. Sci. Paris*, 272, 568, 1971. With permission.)

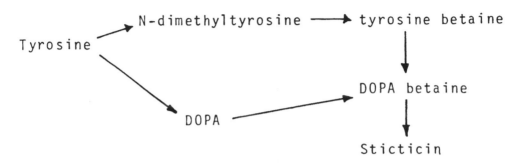

FIGURE 15.   Formation of sticticin in *Lobaria laetevirens*. (From Bernard, T. and Goas, G., *Physiol. Plant.*, 53, 71, 1981. With permission.)

Several workers indicate that polyamines may have an important regulatory role in plant growth and development.[139-142]

It is generally accepted that in plants, the diamine putrescine is synthesized by ornithine decarboxylase and arginine decarboxylase. In *Evernia prunastri*, both enzymes have been characterized.[125,126]

While the synthesis of putrescine, at least in *E. prunastri*, is well established, nothing is known about the biosynthesis of spermidine and spermine or about their catabolism in lichens.

F. Urea Catabolism

1. Regulation of Lichen Urease

Urease is an enzyme with very restricted specificity, which hydrolyzes urea, hydroxyurea, and hydroxamic acids according to the equations:[143]

$$H_2N-C-NH_2 + H_2O \rightarrow 2NH_3 + CO_2$$
$$\parallel$$
$$O$$

$$HOHN-C-NH_2 + H_2O \rightarrow NH_2OH + NH_3 + CO_2$$
$$\parallel$$
$$O$$

$$HOHN-C-R + H_2O \rightarrow NH_2OH + R-COOH$$
$$\parallel \qquad\qquad H^+$$
$$O$$

It has been found in several lichen species and may be related to the ability of these lichens to use urea as an organic nitrogen source. Lichens can be divided into two groups based on their ability to synthesize the enzyme:

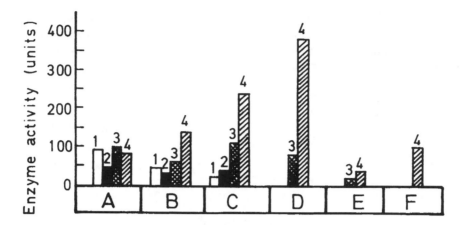

FIGURE 16. Control of urease activity in several lichen species. (A) *Peltigera aphthosa;* (B) *Cladonia alpestris;* (C) *Cladonia rangiferina;* (D) *Cladonia mitis;* (E) *Cetraria islandica;* (F) *Hypogymnia physodes.* 1 = Control, starting lichen; 2 = floating on water; 3 = on NH_4NO_3 solution; 4 = on urea solution. (From Shapiro, I. A., *Fiziol. Rast.,* 24, 1135, 1977. With permission.)

1. Lichens containing cyanobacteria produce urease, the activity of which remains at a constant level and does not increase when urea is supplied to the thalli (e.g., *Peltigera aphthosa*[144] and *P. canina*[145]). The level of urease activity remains unaltered even when other sources of nitrogen, such as ammonia or nitrate, are included in the culture media[144] (Figure 16).

2. Lichens containing a green alga synthesize urease as a response to a supply of exogenous urea. This nutritional induction has been found in 12 lichen species and it is reversed by inhibitors of protein synthesis. Differences in the biosynthetic ability for this enzyme coincide with that reported by Massé[146] for uricase activity. It is necessary to emphasize here that both enzymes have often been mistaken as a single protein.

Urease is synthesized by both free-living fungi and algae. It is probable that both symbionts are able to produce urease in many lichen species. For example, the mycobiont of *Cladonia verticillaris* retains about 25% of the total urease activity[99] and that of *Parmelia roystonea* about 54% of the enzyme.[147] However, urease of *Evernia prunastri* is exclusively located in the phycobiont cells.[148]

Many bacteria grow on lichen surfaces and it is, therefore, possible that urease induction may have been due to the presence of this enzyme in contaminating bacteria. Brown et al.[149] found that *E. prunastri* extracts used for urease activity assays were significantly contaminated with actively growing bacteria. However, changes in bacterial numbers produced by addition of suitable penicillin concentrations did not parallel alterations in urease activity and the bacteria which were tested were found to be urease negative.

The biological role of urease in lichens appears to be complex and multifaceted. Undoubtedly, urease is induced in green alga-containing lichens and is related to the amount of available urea from the substrate. In addition, many lichens can accumulate L-arginine[119,120,150,151] from which urea may be produced through hydrolysis or decarboxylation reactions.

When induction of urease is achieved in the laboratory by floating thalli on urea solutions, enzyme activity increases during the first 5 to 7 hr of incubation followed by a decrease in activity,[99,152] although, in some cases, activity may increase for up to 15 to 16 hr.[77,153]

Three hypotheses can be forwarded to explain this behavior:

1. Simultaneous to enzyme induction, a repressor of urease can be mobilized, which would result in a transient decrease in enzyme activity. Repressor concentration would be diluted during incubation allowing new enzyme to be synthesized as a response to the presence of the inducer in the incubation medium.[153]

2. Urease is a protein which contains nickel strongly bound in its active site and which is required for the synthesis of protein;[154,155] an apoenzyme can be produced without nickel,[156] but it does not possess enzymatic activity. It is possible that a continuous synthesis of urease cannot be achieved if endogenous nickel is depleted during incubation of the thalli.

3. The hydrolysis of urea produces carbon dioxide which is used to synthesize lichen phenolics.[74] These phenols can inhibit the enzyme by blocking the thiol groups of the protein.[157] Secretion of these phenols to the lichen cortex, as a function of the time of incubation, would reverse this effect.

2. The Existence of a Specific Repressor of Urease

The drop of urease activity over time can be due to the synthesis of a repressor of the enzyme which accumulated or diluted during thallus incubation. This is supported by the fact that inducible urease appeared after incubation of the thalli on L-arginine, since urea can be produced from the amino acid and used to induce its L-arginine hydrolase. However, the level of urease activity in thalli floated on urea was always higher than that obtained when the thalli were floated on L-arginine.

Pseudevernia furfuracea thalli develop high urease activity after incubation on urea but only very low activity levels when thalli are incubated on phosphate buffer. This suggests that a small but sufficient amount of urea exists in the thallus for enzyme induction. Cycloheximide (100 μM) added to both phosphate buffer and urea-containing medium eliminates urease activity. Enzyme activity is also very low when the thalli are floated on 40 mM L-arginine. This activity is enhanced by addition of 100 μM cycloheximide, but depleted when 300 μM cycloheximide is added to the media (Figure 17). This may be interpreted as evidence for the synthesis of a urease repressor effected by L-arginine. Cycloheximide, at the lowest concentration used, would totally eliminate the synthesis of this repressor and then urease could be synthesized. However, 300 μM, cycloheximide would inhibit the synthesis of both urease and its repressor.[158] Similar results have been found for *Evernia* urease.

3. Role of Nickel

The addition of 1.5 μM nickel chloride to urea-containing media enhanced urease activity of *Evernia prunastri* thalli in the dark; addition of 10 mM citrate to the media resulted in a drop in induction of urease activity. These results suggest that available nickel may play a role in the regulation of urease activity.

4. Regulation by Lichen Phenolics

The decrease of urease activity after a time of incubation of lichen thalli on media containing an inducer could be explained on the basis of enzyme inhibition by lichen phenolics. Such inhibition can be demonstrated in vitro by including small amounts of several lichen phenols in standard reaction mixtures. Thus, L-usnic acid inactivated urease by the formation of high molecular weight aggregates. The substrate, urea, produced temporary aggregation states which were either active or inactive, the latter being reversible.[159] However, these inactive aggregates (82 × 10⁴ mol wt) were irreversibly stabilized by L-usnic acid (Table 7). By including variable concentrations of L-cysteine in the reaction mixture, inhibition could be partially reversed. This reversal was due to the appearance of active high molecular weight polymers, which were inactive in the absence of the amino acid, rather than a depolymerizing action.[160] In addition, other lichen phenolics, such as

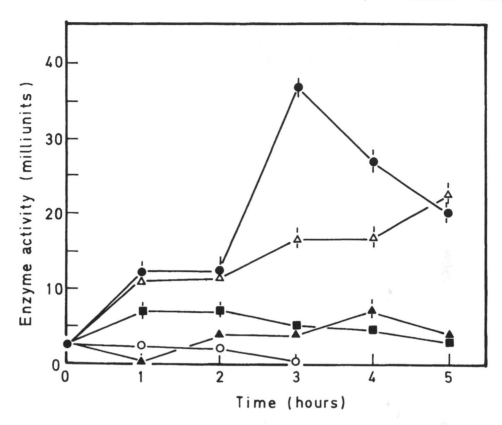

FIGURE 17.   Time course of urease activity in *Pseudevernia furfuracea* thalli floated on (●) 40 m*M* urea; (○) 40 m*M* urea + 100 μ*M* cycloheximide; (▲) 40 m*M* L-arginine; (△) 40 m*M* L-arginine + 100 μ*M* cycloheximide; (■) 40 m*M* L-arginine + 300 μM cycloheximide. (From Avalos, A. et al., *Endocyt. Cell Res.*, 2, 15, 1985. With permission.)

fumarprotocetraric acid from *Cladonia verticillaris*[99] or evernic acid from *E. prunastri*[161,162] inhibited the enzyme without appreciable protein polymerization. This was interpreted as a double process in which: (1) lichen phenols blocks -SH groups in the protein and caused inhibition by preventing the formation of the enzyme-substrate complex; and (2) lichen phenols polymerized the protein, the solubility of which diminished inversely to the molecular weight. This implies that at least two classes of binding sites of the phenolics exist on the urease molecule. Scatchard plots of the interaction between L-usnic acid and urease gave a biphasic curve which can be interpreted as indicating a limited number of high affinity binding sites for the ligand together with a larger number of lower affinity sites. In addition, the data provided evidence for positive cooperation during the interaction of L-usnic acid with urease.[157] Inclusion of L-cysteine in the incubation mixtures decelerated the cooperative binding of L-usnic acid to high affinity sites without any modification of the binding pattern on low affinity sites. Since the integrity of -SH groups is necessary for urease activity,[163] it can be postulated that L-cysteine protects these groups against L-usnic acid and that -SH groups are involved in the high affinity binding sites of the ligand.

Binding sites for polymerization were studied by including L-alanine and L-proline in reaction mixtures containing L-usnic acid. Under these conditions, the loss of activity caused by L-usnic acid was reduced by about 50%. Polymers produced by incubation with amino acids retained enzymatic activity as opposed to those polymerized by phenol.[164] Figure 18 shows a Scatchard plot where the existence of two kinds of sites for the ligand are indicated. Inclusion of both alanine and proline in the incubation mixtures decreased the relative values

Table 7

REVERSIBLE AND IRREVERSIBLE INACTIVATION OF UREASE BY UREA AND USNIC ACID

Treatment	Polymer ($\times 10^4$)	Specific activity (units)	Inactivation (%)
—	48	48.5	—
Incubation with urea[a]	82	<0.005	100.0
Incubation with urea[b] and subsequent dialysis[c]	82	39.6	18.4
Incubation with urea and L-usnic acid	82	<0.005	100.0
Incubation with both compounds and subsequent dialysis	82	<0.005	100.0
Incubation with L-usnic acid	70	<0.005	100.0
Incubation with L-usnic acid and subsequent dialysis	70	<0.005	100.0

[a]   0.2 mmol of urea in 10 mℓ of incubation mixture.

[b]   18.0 μmol of L-usnic acid in 10 mℓ of incubation mixture.

[c]   Dialysis was carried out for 20 hr at 4°C in 75 mM buffer phosphate, pH 6.9.

From Vicente, C. et al., *Plant Cell Environ.*, 1, 29, 1978. With permission.

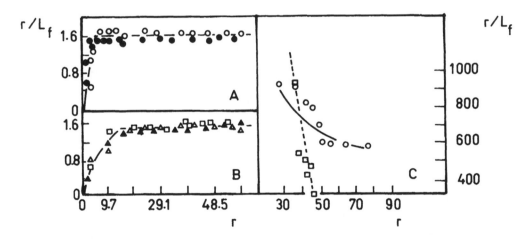

FIGURE 18.   Scatchard plots for the binding of L-usnic acid on urease molecule. (A) L-Usnic acid binds on urease as a function of two concentrations of the enzyme, (●) 0.166 and (○) 0.033 mg protein per milliliter in absence of another effector. (B) L-Usnic acid binds to urease molecule (0.166 mg protein per milliliter) in the presence of 10 mM (▲), 30 mM (△), and 50 mM (□) L-cysteine in the incubation mixtures. Linkage was estimated after 5 min incubation of the enzyme with the ligand (usnic acid) and (where added) L-cysteine. (C) L-Usnic acid binds on urease molecule after 20 min incubation in absence of any other effectors (○) or in the presence of 30 mM L-alanine and L-proline (□). Dissociation constants were 0.102 μM in the absence of amino acids and 0.033 μM in the presence of both L-alanine and L-proline. r = Micromole ligand per milligram of protein; L_f = micromole free ligand. (Plots A and B from García, I. et al., *Z. Naturforsch.*, 35c, 1098, 1980; plot C from Cifuentes, B. et al., *Z. Naturforsch.*, 38c, 273, 1983. With permission.)

of binding of L-usnate to the low affinity sites. Direct binding studies supported the conclusion that L-usnic acid binds less tightly to the enzyme incubated with both amino acids than to untreated enzyme since the value of the apparent dissociation constant for L-usnate increased about threefold when amino acids were absent from the incubation mixtures. This may explain the observation that reversal of the inactivation by both L-alanine and L-proline was lower than that achieved by L-cysteine[166] which protects the active site (high affinity site) of the enzyme; low affinity sites are, therefore, only involved in the polymerization process.

Much experimental data exist which suggest that inhibition of the enzyme by lichen phenolics can act as an in vivo regulation mechanism of urease activity. Data can be summarized as follows:

1. Active urease extracted from *E. prunastri* thalli floated on 40 mM urea shows molecular weight heterogeneity, varying from 72×10^4 to 1.6×10^6. This pattern of molecular mass diversity did not change when 1 mM dithiothreitol was included in the incubation media; activity was about 8 times higher in the dark or 15 times higher in the light than found in thalli floated in absence of the reducer.[152]

2. Differential effects obtained after light or dark treatment could be related to the amount of newly synthesized lichen phenolics which used CO_2 evolved from urea as synthetic units. When *E. prunastri* thalli were floated on ^{14}C-urea, 2.2% of radioactivity was incorporated into lichen phenols in the light, but only 0.5% in the dark.[74] Evernic acid appeared to be the primary phenolic labeled after incubations.

3. *N,N*-dimethylformamide is a nonmetabolizable urea analog (no CO_2 evolved) whereas thiourea can be metabolized (CO_2 evolved) by algae containing urease.[165] Urease could be induced in *E. prunastri* by incubation of thalli on 20 mM *N,N*-dimethylformamide or 40 mM thiourea. Filtration of cell-free extracts from thalli incubated on *N,N*-dimethylformamide through Sepharose 6B showed a main peak of urease activity which had a molecular weight of about 56×10^4. However, extracts from thalli floated on thiourea showed several peaks of similar activity, which had molecular weights of about 1.1×10^6, 0.67×10^6, and 0.14×10^6. Concentration of total lichen phenolics after 6 hr incubation on both urea analogs was 3.7 and 7.0 μmol $\times$ g^{-1} dry weight, respectively.[166]

The main problem with this possible regulation mechanism is that urease is primarily restricted to phycobiont cells[148] whereas phenolics are fungal products.[166] However, a certain amount of urease can be synthesized by the fungal partner in several lichen species,[99,147] and, in addition, crystals of exocellular products, like phenols, have been located on the cell walls of the phycobionts.[167] This suggests that lichen phenolics may be in a position to effect metabolism of the phycobiont.

5. The Relationship Between Energy Supply and Urease Activity

It has been shown that *Peltigera canina* contained urease before incubation with exogenous urea under laboratory conditions.[145] When added to dark-incubated thalli, a marked increase in urease activity was not observed and endogenous activity often declined over time. In the light, there was some increase of urease activity which was maintained for at least 13.5 hr.[145] The role of light in the maintenance of urease activity was demonstrated by light preincubations in the absence of urea followed by urea treatments. Activity was only observed when thalli were incubated in the light throughout the experiment; a light pretreatment followed by darkness during urea addition failed to induce urease activity. Illumination of *P. canina* thalli may result, by photoreductive processes, in a more reducing intracellular environment thereby permitting the extraction of more active, reduced, urease molecules. This hypothesis is in agreement with the results obtained by Legaz and Brown[153] where *E. prunastri* urease was most active when a strong reducer, dithiothreitol, was present in the extracts during both extraction procedure and enzymatic reaction.

VI. OTHER ENZYMES

Several types of esterases (arylesterase, carboxylesterase, lipase, acetylesterase, and arylsulfatase) have been demonstrated in algal preparations of *Trebouxia* from *Lasallia papulosa*.

Phosphatase activities included glucose-6-phosphatase, orthophosphoric monoester phosphohydrolase, and phosphodiesterase.[168] The relative specificity and activities of these enzymes against various substrates (e.g., ethyl acetate, ethyl butyrate, α-naphthyl acetate, fluorescein diacetate, glucose-6-phosphate) suggested the possibility that at least two functions existed for this esterase system.

The esterases might be closely related to biosynthetic and substrate transfer mechanisms in lichens.

Some lichens with cyanobionts, such as *Lichina pygmaea, L. confinis, Placynthium nigrum,* and *Peltigera canina,* contain structures similar to the bacteria mesosome. Associated autophagic vacuoles have been shown to possess acid phosphatase activity.[169]

Rennert and Gubański[170] reported the presence of ribonuclease in water extracts of *Cetraria islandica.*

Results of qualitative surveys for certain enzyme activities in a number of lichens have proved to be variable and not reproducible. Such enzymatic activities include catalase, amylase, invertase, aspartase, tyrosinase, phenol oxidase, lichenase, tannase, lipase, allantoinase, and allantoicase.[171]

VII. EQUILIBRIA OF ENZYME ACTIVITIES BETWEEN PHOTOBIONT AND MYCOBIONT CELLS

It is well known that the physiology of isolated photobionts differs dramatically from that of the same algal species in its lichenized form. There are several ways in which this may be affected. Two such molecular mechanisms include: (1) modification of cell permeability; and (2) compartmentation of both enzymes and their effectors.

A. Changes in Membrane Permeability

The action of lichen substances on cells permeability has been postulated by several workers. Follmann and Villagrán[172] found that usnic acid increased the permeability of cells of *Spyrogira, Elodea,* and *Allium* to glucose. Kinraide and Ahmadjian[173] showed that usnic acid increased the permeability of *Trebouxia* isolated from *Acarospora fuscata* and *Cladonia boryi.* Vicente and Cifuentes[174] reported that sodium L-usnate enhanced the oxidation of fumarate by bacteria by increasing the cell permeability to this organic acid. Usnic acid inhibited growth, photosynthesis, and respiration in *Trebouxia,*[173] as well as inhibiting the growth of *Chlamydomonas reinhardii* and causing loss of its flagella.[175] D-Usnic acid disorganized the photosynthetic membrane system of several algae, such as *Chlamydomonas* and *Anacystis.*[176] Even in bacteria, changes in permeability were accompanied by a polar disruption of cell membrane followed by an enlargement of the cell wall at that point. This enlargement could be explained through an activation of glucosamine phosphate isomerase by the phenol, which initiates the synthesis of cell wall material,[177,178] as well as by an inhibition of the enzyme secretion outside the cells.[179] These properties of lichen phenolics, particularly usnic acid, are probably due to their ability to act as surfactant agents.[180]

Richardson[181] has argued against any possible role of lichen phenols in the release of carbohydrates from photobionts on the grounds that modifications of algal permeability in lichens must be selective (e.g., restricted to a motile polyalcohol) rather than a generalized increase in carbohydrate flow. This argument, compounded by the great diversity of lichen phenolics (or lack of them in certain lichens), strongly suggests that phenols do not play a general role in carbohydrate transfer. However, permeability changes may greatly affect many metabolic properties of the photobiont.

B. Compartmentalization of Both Enzymes and Their Effectors

Only few studies have been made on compartmentalization of enzyme activities in lichen symbionts due to technical problems of symbiont isolation and purification.[182] An attempt

Table 8
LOCALIZATION OF THE ACTIVITIES OF NITROGENASE, PRIMARY NH$_4^+$ ASSIMILATING ENZYMES, AND SOME AMINOTRANSFERASES IN THE LICHEN *PELTIGERA CANINA*

Enzymes	Enzymic activity (nmol product formed min^{-1} mg^{-1} protein)		Probable location
	Intact thalli	Digitonin treated thalli[a]	
Nitrogenase	0.10	0.27	Cyanobiont
Alanine dehydrogenase	3.80	1.04	Cyanobiont and mycobiont
Glutamine synthetase	0.60	1.18	Cyanobiont
Glutamate dehydrogenase	195.80	4.15	Mycobiont
Glutamate synthase	0.28	0.84	Cyanobiont
Glutamate-pyruvate aminotransferase	18.12	0.86	Mycobiont
Glutamate-oxaloacetate aminotransferase	41.00	13.80	Mycobiont and cyanobiont
Alanine-oxaloacetate aminotransferase	126.00	58.60	Mycobiont and cyanobiont

[a] For digitonin treatment, whole thalli were incubated in BG-11$_o$ medium containing 0.01% (w/v) digitonin at 20°C and a photon flux density of 60 μmol m^{-2} sec^{-1}. After 24 hr, the thalli were thoroughly washed in buffer and enzymic activities determined.

From Rai, A. N. et al., *Arch. Microbiol.*, 134, 136, 1983. With permission.

was made to localize enzymes involved in ammonia assimilation by using both intact and digitonin-treated thalli of *Peltigera canina*.[183] Digitonin disrupts the structure of fungal membranes but not of cyanobacterial membranes. As shown in Table 8, glutamine synthetase and glutamate synthase were mainly located in the cyanobiont, as deduced from the increase of enzyme activity after digitonin treatment. Glutamic acid dehydrogenase was also restricted to the mycobiont, since enzyme activity was negligible after detergent treatment (see also Chapter VI.B).

Direct measurement of enzymatic activities in freshly isolated symbionts has been done for proteins involved in L-arginine metabolism. Agmatine iminohydrolase[134] and L-ornithine decarboxylase[125] were located in the mycobiont of *Evernia prunastri*, whereas arginase, L-arginine decarboxylase, and agmatine amidinohydrolase were restricted to the algal cells.[148] This implies that the mycobiont was the main source of putrescine in this lichen although a small pool of putrescine may exist in algal cells since agmatine amidinohydrolase was localized in both symbionts (Table 9).[184]

The accessibility of the substrates to their enzymes can also be regulated through the differing affinities of two enzymes for the same metabolite. Agmatine is preferentially used by the *Evernia* mycobiont, which contains the highest agmatine iminohydrolase activity. In addition, the K$_m$ value of this enzyme for agmatine is eight times higher than that of agmatine amidinohydrolase, which is located in both symbionts.[127,134] Inducible arginase of *E. prunastri* has a K$_m$ value for L-arginine of 0.2 mM[122] whereas the value of L-arginine decarboxylase is 12.5 mM.[126] This implies that the affinity of arginase for the amino acid is 62.5 times higher than the affinity of arginine decarboxylase. This may drastically limit the hydrolysis of agmatine in the phycobiont, since urea, mainly produced in algal cells, behaves as an absolute inhibitor of agmatine amidinohydrolase.[185]

Table 9
LOCATION OF ENZYMES OF L-ARGININE CATABOLISM IN *EVERINIA PRUNASTRI* THALLUS

Enzyme	Intact thalli				Mycobiont				Phycobiont			
	Specific activity (units)	%	Total activity (units)	%	Specific activity (units)	%	Total activity (units)	%	Specific activity (units)	%	Total activity (units)	%
Arginase*	0.164	100	0.656	100	0.054	32.92	0.208	88.88	0.11	67.08	0.026	11.11
L-Arginine decarboxylase *	0.72	100	2.88	100	0.17	23.61	0.64	82.90	0.55	76.39	0.132	10.10
L-Ornithine decarboxylase**	1.01	100	5.46	100	0.82	81.10	4.50	81.90	0.19	18.80	0.99	18.00
Agmatine amidinohydrolase *	0.16	100	0.64	100	0.01	6.25	0.04	51.08	0.15	93.75	0.036	48.92
Agmatine iminohydrolase ***	0.233	100	1.38	100	0.194	83.15	1.24	89.74	0.039	16.85	0.142	10.26

Note: The mycobiont contained 15.7*, 1.05**, and 1.8*** more protein than the phycobiont.

From Legaz, M. E., in *Surface Physiology of Lichens,* Vicente, C., Brown, D. H., and Legaz, M. E., Eds., Complutense University Press, Madrid, 1985, 57.

VIII. CONCLUDING REMARKS AND FUTURE PROSPECTS

Many topics in lichen physiology require a more detailed study at their enzymatic level. To facilitate diffusion of photoassimilates, which is not dependent on the conventional K^+-ATPase system, could require very specific and unique permeases, the synthesis of which could be regulated through a very complex cross-control system. Recognition by mycobionts of antigens present on the cell walls of the phycobionts[186,187] involves the synthesis, control, and attachment of a series of wall-associated proteins. The enzymological studies on the biosynthesis of secondary lichen compounds are yet only beginning. The assumed functions of structures such as concentric bodies or lysosomes, based on ultrastructural observation, are in need of biochemical analysis.

On the other hand, lichen enzymology was pioneering in the biotechnology of immobilized enzymes and cells. Mosbach and Mosbach[188] described a simple technique for the entrapment of enzymes and lichen cells using cross-linked polyacrylamide and found that orsellinic acid decarboxylase was able to produce orcinol without any significant loss of activity after 14 days at 20°C. When lichen cells were entrapped in the granules, they retained part of their decarboxylase activity after 3 months at 20°C. With this method, it should be possible to trap enzymes which appear transiently in a biosynthetic sequence, thus being able to isolate intermediates in great quantities.[189] This is very important in order to elucidate chemical structures, to produce chemicals and antibiotics, and to apply to lichens the new analytical methods in biochemistry, medicine, and pharmacy.

ACKNOWLEDGMENTS

We wish to thank sincerely the great help given by Miss Elena PerezUrria in the typewriting of this manuscript and to Adolfo Avalos for his technical assistance.

REFERENCES

1. **Arnon, D. I.**, Role of ferredoxin in photosynthesis, *Naturwissenschaften*, 56, 295, 1969.
2. **Estévez, M. P. and Vicente, C.**, Ferredoxina y ferredoxin-NADP$^+$-reductasa de *Evernia prunastri*, *Rev. Bryol. Lichenol.*, 44, 111, 1978.
3. **Vicente, C. and Requena, V.**, Purification and some properties of a new ferredoxin from the lichen *Lobaria pulmonaria*, *Photosynthetica*, 18, 57, 1984.
4. **Tagawa, K. and Arnon, D. I.**, Ferredoxins as electron carriers in photosynthesis and in the biological production and consumption of hydrogen gas, *Nature (London)*, 195, 537, 1962.
5. **Keresztes-Nagy, S. and Margoliasch, E.**, Preparation and characterization of alfalfa ferredoxin, *J. Biol. Chem.*, 241, 5955, 1966.
6. **Schürmann, P., Buchanan, B. B., and Matsubara, H.**, Ferredoxin from fern and *Amaranthus:* two diverse plants with similar ferredoxins, *Biochim. Biophys. Acta*, 223, 450, 1970.
7. **Matsubara, H.**, Purification and some properties of *Scenedesmus* ferredoxin, *J. Biol. Chem.*, 243, 370, 1968.
8. **Tagawa, K. and Arnon, D. I.**, Oxidation-reduction potentials and stoichiometry of electron transfer in ferredoxins, *Biochim. Biophys. Acta*, 153, 602, 1968.
9. **Serrano, A., Rivas, J., and Losada, M.**, Studies on one- and two-electron FAD-mediated reactions by ferredoxin-NADP$^+$ oxidoreductase from *Anabaena*, *FEBS Lett.*, 170, 85, 1984.
10. **Ramírez, R., Batzán, M. T., and Vicente, C.**, Regulation of ferredoxin-NADP$^+$ reductase from *Evernia prunastri* by several nucleotides, *Phyton (Buenos Aires)*, 37, 81, 1979.
11. **Bednar, T. W. and Smith, D. C.**, Studies in the physiology of lichens. VI. Preliminary studies of photosynthesis and carbohydrate metabolism of the lichen *Xanthoria aureola*, *New Phytol.*, 65, 211, 1966.
12. **Drew, E. A.**, Some Aspects of the Carbohydrate Metabolism of Lichens, Ph.D. thesis, University of Oxford, Oxford, England, 1966.
13. **Feige, B.**, Stoffwechselphysiologische Untersuchungen an der tropischen Basidiolichene *Cora pavonia*, *Flora*, 160A, 169, 1969.
14. **Rai, A. N., Rowell, P., and Stewart, W. D. P.**, Nitrogenase activity and dark CO_2 fixation in the lichen *Peltigera aphthosa*, *Planta*, 151, 256, 1981.
15. **Hill, D. J.**, The physiology of lichen symbiosis, in *Lichenology: Progress and Problems*, Brown, D. H., Hawksworth, D. L., and Bailey, R. H., Eds., Academic Press, New York, 1976, 457.
16. **Snelgar, W. P. and Green, T. G. A.**, Carbon dioxide exchange in lichens: low carbon dioxide compensation levels and lack of apparent photorespiratory activity in some lichens, *Bryologist*, 83, 505, 1980.
17. **Smith, D. C.**, Mechanisms of nutrient movement between the lichen symbionts, in *Cellular Interactions in Symbiosis and Parasitism*, Cook, C. B., Pappas, P. W., and Rudolph, E. D., Eds., Ohio State University Press, Columbus, 1980, 197.
18. **Richardson, D. H. S. and Smith, D. C.**, Lichen physiology. IX. Carbohydrate movement from the *Trebouxia* symbiont of *Xanthoria aureola*, *New Phytol.*, 67, 61, 1968.
19. **Richardson, D. H. S., Hill, D. J., and Smith, D. C.**, Lichen physiology. XI. The role of the alga in determining the pattern of carbohydrate movement between lichen symbionts, *New Phytol.*, 67, 469, 1968.
20. **Richardson, D. H. S. and Smith, D. C.**, The physiology of the symbiosis in *Xanthoria aureola*, *Lichenologist*, 3, 202, 1966.
21. **Pueyo, G.**, Recherches sur la nature et evolution des glucides solubles chez quelques lichens du bassin parisienne, *Annee Biol.*, 36, 117, 1960.
22. **Smith, D. C.**, The biology of lichen thalli, *Biol. Rev. Cambridge Philos. Soc.*, 37, 537, 1962.
23. **Legaz, M. E., De Torres, M., and Escribano, M. I.**, Putrescine affects mannitol metabolism in the lichen *Evernia prunastri*, *Photosynthetica*, 19, 230, 1985.
24. **Asahina, Y. and Shibata, S.**, *Chemistry of Lichen Substances*, Lubrecht & Cramer, Forestburgh, N. Y., 1972.
25. **Shibata, S.**, Lichen substances, in *Modern Methods of Plant Analysis*, Linskens, H. F. and Tracey, M. V., Eds., Springer-Verlag, Berlin, 1963, 155.
26. **Pueyo, G.**, Chromatographie de separatrion sur columne des polyalcools derivés des sucres. I, *Am. Fals. Exp. Chim.*, 749, 27, 1977.
27. **Culberson, C. F.**, Some constituents of the lichen *Ramalina siliquosa*, *Phytochemistry*, 4, 951, 1965.
28. **Lindberg, B., Wachtmeister, C. A., and Wickberg, B.**, Studies in the chemistry of lichens. II. Umbilicin, an arabitol galactoside from *Umbilicaria pustulata*, *Acta Chem. Scand.*, 6, 1052, 1952.
29. **Pueyo, G.**, Identification par chromatographie sur papier des glucides solubles des lichens, *Rev. Bryol. Lichenol.*, 32, 285, 1963.
30. **Lindberg, B. and Meier, H.**, Studies on the chemistry of lichens. XV. Siphulitol, a new polyol from *Siphula ceratites*, *Acta Chem. Scand.*, 16, 543, 1962.
31. **Lindberg, B.**, Low-molecular carbohydrates in algae. I. Investigation of *Fucus vesiculosus*, *Acta Chem. Scand.*, 7, 1119, 1953.

32. **De Lestang Laisne, G.**, Sur l'interférence due métabolisme glucidique d'un champignon ascomycéte et du métabolisme glucidique d'une algue bleue dans un lichen marin: *Lichina pygmaea* Agardh., *Rev. Bryol. Lichenol.*, 34, 346, 1966.

33. **Lewis, D. H. and Smith, D. C.**, Sugar alcohols (polyols) in fungi and green plants. I. Distribution, physiology and metabolism, *New Phytol.*, 66, 143, 1967.

34. **Ingram, J. M. and Wood, W. A.**, Enzymatic basis for D-arabitol production by *Saccharomyces rouxii*, *J. Bacteriol.*, 89, 1186, 1965.

35. **Klungsϕyr, L.**, Mannitol kinase in cell-free extracts of *Escherichia coli*, *Biochim. Biophys. Acta*, 122, 361, 1966.

36. **Yamada, H., Okamoto, K., Kodama, K., and Tanaka, S.**, Mannitol formation by *Pyricularia oryzae*, *Biochim. Biophys. Acta*, 33, 271, 1959.

37. **Yamaki, S.**, Subcellular location of sorbitol-6-phosphate dehydrogenase in protoplast from apple cotyledons, *Plant Cell Physiol.*, 22, 359, 1981.

38. **Yamaki, S.**, NADP$^+$-dependent sorbitol dehydrogenase found in apple leaves, *Plant Cell Physiol.*, 25, 1323, 1984.

39. **Negum, F. B. and Loescher, W. P.**, Detection and characterization of sorbitol dehydrogenase from apple callus tissue, *Plant Physiol.*, 64, 69, 1979.

40. **Du Toit, P. J. and Kotzé, J. P.**, The isolation and characterization of sorbitol-6-phosphate dehydrogenase from *Clostridium pasteurianum*, *Biochim. Biophys. Acta*, 206, 333, 1970.

41. **Yamaki, S.**, Localization of sorbitol oxidase in vacuoles and other subcellular organelles in apple cotyledons, *Plant Cell Physiol.*, 23, 891, 1982.

42. **Fromm, H. J. and Bietz, J.**, Ribitol dehydrogenase. IV. Purification and crystallization of the enzyme, *Arch. Biochem. Biophys.*, 115, 510, 1966.

43. **Glaser, L.**, Ribitol-5-phosphate dehydrogenase from *Lactobacillus plantarum*, *Biochim. Biophys. Acta*, 67, 525, 1963.

44. **Strobel, G. A. and Kusuge, T.**, Polyol metabolism in *Diplodia viticola* Desm., *Arch. Biochem. Biophys.*, 109, 622, 1965.

45. **Edmundowicz, J. M., and Wislon, J. C.**, Mannitol dehydrogenase from *Agaricus campestris*, *J. Biol. Chem.*, 238, 3539, 1963.

46. **Wolff, J. B. and Kaplan, N. O.**, D-Mannitol 1-phosphate dehydrogenase from *Escherichia coli*, *J. Biol. Chem.*, 218, 849, 1956.

47. **Liss, M., Horwitz, S. B., and Kaplan, N. O.**, D-Mannitol 1-phosphate dehydrogenase and D-sorbitol 6-phosphate dehydrogenase in *Aerobacter aerogenes*, *J. Biol. Chem.*, 237, 1342, 1962.

48. **Rapsch, S. and Cifuentes, B.**, Glucosamine 6-P isomerase of *Evernia prunastri* (L.) Ach., *Cryptog. Bryol. Lichenol.*, 4, 161, 1983.

49. **Sols, A. and De la Fuente, G.**, Glucosa-oxidasa en análisis, *Rev. Esp. Fisiol.*, 13, 231, 1975.

50. **Muñoz, M. L. and Rodriguez, M.**, Induced synthesis of phosphates in microalgae by antibiotic inhibitor of protein synthesis: mechanism of action, *Plant Sci. Lett.*, 13, 275, 1978.

51. **Cifuentes, B.**, Modificación por Ácido L-úsnico de Estructuras que Soportan la Permeabilidad Celular, Ph.D. thesis, Complutense University, Madrid, 1980.

52. **Phelps, C. I., Hardingham, T. E., and Winterburn, P. J.**, Studies on the control of nucleotides sugar metabolism in glucosamine-glycan biosynthesis, *Expo. Annu. Biochim. Med.*, 30, 79, 1970.

53. **Cifuentes, B. and Rapsch, S.**, N-Acetylglucosamine 6-P synthetase of *Evernia prunastri*, *Phyton (Buenos Aires)*, 43, 97, 1983.

54. **Brodo, I. M.**, Substrate ecology, in *The Lichens*, Ahmadjian, V. and Hale, M. E., Eds., Academic Press, New York, 1973, 401.

55. **Ascaso, C., González, C., and Vicente, C.**, Epiphytic *Evernia prunastri*: ultrastructural facts, *Cryptog. Bryol. Lichenol.*, 1, 43, 1980.

56. **Basham, H. G. and Bateman, D. F.**, Relationships of cell death in plant tissue treated with a homogeneous endopectate lyase to cell wall degradation, *Physiol. Plant Pathol.*, 5, 249, 1975.

57. **Bateman, D. F. and Basham, H. G.**, Degradation of plant cell walls and membranes by microbial enzymes, in *Encyclopedia of Plant Physiology*, Vol. 3, Heitefuss, R. and Williams, P. H., Eds., Springer-Verlag, Berlin, 1976, 316.

58. **Yagüe, E., Orús, M. I., and Estévez, M. P.**, Extracellular polysaccharidases synthesized by the epiphytic lichen *Evernia prunastri*, *Planta*, 160, 212, 1984.

59. **Bloomer, J. L. and Hoffman, W. F.**, On the origin of the C$_3$-unit in (+)-protolichesterinic acid, *Tetrahedron Lett.*, 50, 4339, 1969.

60. **Bloomer, J. L., Eder, W. R., and Hoffman, W. F.**, The biosynthesis of (+)-protolichesterinic acid, *Chem. Commun.*, 120, 354, 1968.

61. **Cifuentes, B. and Gómez, A.**, Purification and properties of two lypoxygenase isozymes of *Evernia prunastri*, *Z. Pflanzenphysiol.*, 109, 429, 1983.

62. **Gaucher, G. M. and Shepherd, M. G.,** Isolation of orsellinic acid synthetase, *Biochem. Biophys. Res. Commun.*, 32, 664, 1968.

63. **Mosbach, K.,** On the biosynthesis of lichen substances. I. The depside gyrophoric acid, *Acta Chem. Scand.*, 18, 329, 1964.

64. **Fox, C. H. and Mosbach, K.,** On the biosynthesis of lichen substances. III. Lichen acids as products of a symbiosis, *Acta Chem. Scand.*, 21, 2327, 1967.

65. **Packter, N. M.,** Biosynthesis of acetate-derived phenols (polyketides), in *The Biochemistry of Plants*, Vol. 4, Stumpf, P. K., Ed., Academic Press, New York, 1980, 535.

66. **Yamazaki, M., Matsuo, M., and Shibata, S.,** Biosynthesis of lichen depsides. Lecanoric acid and atranorin, *Chem. Pharm. Bull.*, 13, 1015, 1965.

67. **Yamazaki, M. and Shibata, S.,** Biosynthesis of lichen substances. II. Participation of C_1-unit to the formation of β-orcinol type lichen depside, *Chem. Pharm. Bull.*, 14, 96, 1966.

68. **Mosbach, K. and Ehrensvärd, U.,** Studies on lichen enzymes. I. Preparation and properties of a depside hydrolysing esterase and of orsellinic acid decarboxylase, *Biochem. Biophys. Res. Commun.*, 22, 145, 1966.

69. **Schultz, J. and Mosbach, K.,** Studies on lichen enzymes. Purification and properties of an orsellinate depside hydrolase obtained from *Lasallia pustulata*, *Eur. J. Biochem.*, 22, 153, 1971.

70. **González, A. and Vicente, C.,** Purification and properties of an orsellinate depside hydrolase of *Evernia prunastri*, Proc. 4th Congr. of the Federation of European Societies of Plant Physiologists, Strasbourg, July 29-August 3, 1984.

71. **González, A., Vicente, C., and Legaz, M. E.,** A simple assay demonstrating the effect of rehydration on the orsellinate depside hydrolase activity of *Evernia prunastri*, *J. Plant Physiol.*, 116, 219, 1984.

72. **Mosbach, K. and Schultz, J.,** Studies on lichen enzymes. Purification and properties of orsellinate decarboxylase obtained from *Lasallia pustulata*, *Eur. J. Biochem.*, 22, 485, 1971.

73. **Scott, G. D.,** *Plant Symbiosis*, Edward Arnold Publ., London, 1971, 58.

74. **Blanco, M. J., Suárez, C., and Vicente, C.,** The use of urea by *Evernia prunastri* thalli, *Planta*, 162, 305, 1984.

75. **Legaz, M. E.,** The relationship between seasonal variation in free L-arginine, arginine degrading enzymes and the phenolic content of *Evernia prunastri* thalli, Proc. Int. Symp. on Recent Advances in Lichen Physiology, Bristol, April 16—18, 1984.

76. **Culberson, C. F.,** Joint occurrence of a lichen depsidone and its probable depside precursor, *Science*, 143, 255, 1964.

77. **Vicente, C., Legaz, M. E., Arruda, E. C., and Xavier Filho, L.,** The utilization of urea by the lichen *Cladonia sandstedei*, *J. Plant Physiol.*, 115, 397, 1984.

78. **Taguchi, H., Sankawa, U., and Shibata, S.,** Biosynthesis of usnic acid in lichens, *Tetrahedron Lett.*, 42, 5221, 1966.

79. **Taguchi, H., Sankawa, U., and Shibata, S.,** Biosynthesis of natural products. VI. Biosynthesis of usnic acid in lichens. I. A general scheme of biosynthesis of usnic acid, *Chem. Pharm. Bull.*, 17, 2054, 1969.

80. **Shibata, S. and Taguchi, H.,** Occurrence of isousnic acid in lichens with reference to isodihydrousnic acid derived from dihydrousnic acid, *Tetrahedron Lett.*, 48, 4867, 1967.

81. **Vicente, C., Ruiz, J. L., and Estévez, M. P.,** Mobilization of usnic acid in *Evernia prunastri* under critical conditions of nutrient availability, *Phyton (Buenos Aires)*, 39, 15, 1980.

82. **Estévez, M. P., Legaz, M. E., Olmeda, L., Pérez, F., and Vicente, C.,** Purification and properties of a new enzyme from *Evernia prunastri*, which reduces L-usnic acid, *Z. Naturforsch.*, 36c, 35, 1981.

83. **Vicente, C. and González, A.,** D-Usnic acid dehydrogenase of *Evernia prunastri*, *Lichen Physiol. Biochem.*, 1, 47, 1986.

84. **Forsén, S., Nilsson, M., and Wachtmeister, C. A.,** Spectroscopic studies on enols. IV. Hydrogen bonding in usnic acid, *Acta Chem. Scand.*, 16, 583, 1962.

85. **Bandoni, R. J. and Towers, G. H. N.,** Degradation of usnic acid by microorganisms, *Can. J. Biochem.*, 45, 1197, 1967.

86. **Vainstein, E. A. and Ravinskaya, A. P.,** Biological degradation in lichen acids in the soil, *Bot. Zh. (Leningrad)*, 69, 1347, 1984.

87. **Kutney, J. P., Baarschers, W. H., Chin, O., Ebizuca, Y., Hurley, L., Leman, J. D., Salisbury, P. J., Sánchez, I. H., Yee, T., and Bandoni, R. J.,** Studies in the usnic acid series. VIII. The biodegradation of (+)-usnic acid by *Mortierella isabellina*, *Can. J. Chem.*, 55, 2930, 1977.

88. **Kutney, J. P., Leman, J. D., Salisbury, P. J., Sánchez, I. H., Yee, T., and Bandoni, R. J.,** Studies in the usnic acid series. VII. The biodegradation of (+)-usnic acid by a *Pseudomonas* species. Isolation, structure determination and synthesis of (+)-6-desacetylusnic acid, *Can. J. Chem.*, 55, 2336, 1977.

89. **Castle, H. and Kubsch, F.,** The production of usnic didymic and rhodocladonic acids by the fungal component of the lichen *Cladonia cristatella*, *Arch. Biochem.*, 23, 158, 1949.

90. **Ahmadjian, V.,** Separation and artificial resynthesis of lichens, in *Cellular Interactions in Symbiosis and Parasitism,* Cook, G. B., Pappas, P. W., and Rudolph, E. D., Eds., Ohio State University Press, Columbus, 1980, 3.

91. **Ejiri, H. and Shibata, S.,** Squamatic acid from the mycobiont of *Cladonia crispata, Phytochemistry,* 14, 2505, 1975.

92. **Komiya, T. and Shibata, S.,** Formation of lichens substances by mycobionts of lichens. Isolation of (+)-usnic acid and salazinic acid from mycobionts of *Ramalina* spp., *Chem. Pharm. Bull.,* 17, 1305, 1969.

93. **Kurokawa, S., Shibata, S., and Tomiya, T.,** Isolation of algal and fungal components of lichens and their chemical products, *Misc. Bryol. Lichenol.,* 5, 8, 1969.

94. **Umezawa, H., Shibamoto, N., Naganawa, H., Ayukawa, S., Matsuzaki, T., Takeuchi, T., Kono, K., and Sakamoto, T.,** Isolation of lecanoric acid, an inhibitor of histidine decarboxylase from a fungus, *J. Antibiot.,* 27, 587, 1974.

95. **Yamamoto, Y., Nishimura, K., and Kiriyama, N.,** Studies on the metabolic products of *Aspergillus terreus.* I. Metabolites of the strain FO 6123, *Chem. Pharm. Bull.,* 24, 1853, 1976.

96. **Paulo, M. Q., Xavier, L., Pessoa, R., and Vicente, C.,** Estudos sobre o cogumelo japonês o tea fungus. Estudo morfologico e quimico, Proc. 36th Reunao Annual Sociedade Brasileira Progresso da Ciência, Sao Paulo, 1984.

97. **Rundel, P. W.,** Clinal variation in the production of usnic acid in *Cladonia subtenuis* along light gradients, *Bryologist,* 72, 40, 1969.

98. **Fahselt, D.,** Lichen products of *Cladonia stellaris* and *Cladonia rangiferina* maintained under artificial conditions, *Lichenologist,* 13, 87, 1981.

99. **Vicente, C. and Xavier Filho, L.,** Urease regulation in *Cladonia verticillaris, Phyton (Buenos Aires),* 37, 137, 1979.

100. **Hamada, N.,** The effect of temperature on the content of medullary depsidone salazinic acid in *Ramalina siliquosa* (lichens), *Can. J. Bot.,* 60, 383, 1982.

101. **Hamada, N.,** The effect of temperature on lichen substances in *Ramalina subbreviuscula, Bot. Mag.,* 96, 121, 1983.

102. **Hamada, N.,** The content of lichen substances in *Ramalina siliquosa* cultured at various temperatures in growth cabinets, *Lichenologist,* 16, 96, 1984.

103. **Culberson, C. F., Culberson, W. L., and Johnson, A.,** Genetic and environmental effects on growth and production of secondary components in *Cladonia cristatella, Biochem. Syst. Ecol.,* 11, 77, 1983.

104. **Stephenson, N. L. and Rundel, P. W.,** Quantitative variation and ecological role of vulpinic acid and atranorin in the thallus of *Letharia vulpina, Biochem. Ecol.,* 7, 263, 1979.

105. **Culberson, C. F. and Culberson, W. L.,** Age and chemical constituents of individuals of the lichen *Lasallia papulosa, Lloydia,* 21, 189, 1958.

106. **Gandhi, A. P. and Nair, M. S.,** Role of roots, hormones and light in the synthesis of nitrate reductase and nitrite reductase in rice seedlings, *FEBS Lett.,* 40, 343, 1974.

107. **Johnson, C. B.,** Rapid activation by phytochrome of nitrate reductase in the cotyledons of *Sinapis alba, Planta,* 128, 127, 1976.

108. **Rao, L. U. M., Datta, L., Sopory, S. K., and Guha-Mukerfee, S.,** Phytochrome-mediated induction of nitrate reductase activity in etiolated maize leaves, *Physiol. Plant.,* 50, 208, 1980.

109. **Avalos, A. and Vicente, C.,** Phytochrome enhances nitrate reductase activity in the lichen *Evernia prunastri, Can. J. Bot.,* 63, 1350, 1985.

110. **Diez, J. A., Chaparro, A., Vega, J. M., and Relimpio, A.,** Studies on the regulation of assimilatory nitrate reductase in *Ankistrodesmus brownii, Planta,* 137, 231, 1977.

111. **Guerrero, M. G., Vega, J. M., and Losada, M.,** The assimilatory nitrate-reducing system and its regulation, *Annu. Rev. Plant Physiol.,* 32, 169, 1981.

112. **Rivas, J., Tortolero, M., and Panegue, A.,** Metal component of the nitrate-reducing system from the yeast *Torulopsis nitratophila, Plant Sci. Lett.,* 2, 283, 1974.

113. **Bernard, T.,** Contribution á l'etude du mètabolisme azoté des lichens. Activité de la glutamate décarboxylase de cinq espèces de la famille des Stictacées, *C. R. Acad. Sci. Paris,* 269, 823, 1969.

114. **Ramakrishnan, S. and Subramanian, S. S.,** Amino acids of *Roccella montagnei* and *Parmelia tinctorum, Indian J. Chem.,* 2, 467, 1964.

115. **Ramakrishnan, S. and Subramanian, S. S.,** Amino acids of *Peltigera canina, Curr. Sci.,* 33, 522, 1964.

116. **Ramakrishnan, S. and Subramanian, S. S.,** Amino acids of *Cladonia rangiferina, C. gracilis* and *Lobaria isidiosa, Curr. Sci.,* 34, 345, 1965.

117. **Ramakrishnan, S. and Subramanian, S. S.,** Amino acids of *Dermatocarpon moulinsii, Curr. Sci.,* 35, 284, 1966.

118. **Ramakrishnan, S. and Subramanian, S. S.,** Amino acids of *Lobaria subisidiosa, Umbilicaria pustulata, Parmelia nepalensis* and *Ramalina sinensis, Curr. Sci.,* 35, 124, 1966.

119. **Legaz, M. E., González de Buitrago, G., and Vicente, C.,** Exogenous supply of L-arginine modifies free amino acids content in *Evernia prunastri* thallus, *Phyton (Buenos Aires),* 42, 213, 1982.

120. **Jäger, H. J. and Weigel, H. J.**, Amino acids metabolism in lichen, *Bryologist*, 81, 107, 1978.
121. **Legaz, M. E. and Vicente, C.**, Arginase regulation in *Evernia prunastri* (L.) Ach., *Cryptog. Bryol. Lichenol.*, 1, 407, 1980.
122. **Legaz, M. E. and Vicente, C.**, Two forms of arginase in *Evernia prunastri* thallus, *Biochem. Biophys. Res. Commun.*, 104, 1441, 1982.
123. **Weiss, R. L. and Davis, R. H.**, Control of arginine utilization in *Neurospora*, *J. Bacteriol.*, 129, 866, 1977.
124. **Martín-Falquina, A. and Legaz, M. E.**, Purification and properties of the constitutive arginase of *Evernia prunastri*, *Plant Physiol.*, 76, 1065, 1984.
125. **Escribano, M. I. and Legaz, M. E.**, L-Ornithine decarboxylase from *Evernia prunastri*, *Phyton (Buenos Aires)*, 44, 171, 1984.
126. **Vicente, C. and Legaz, M. E.**, Purification and properties of L-arginine decarboxylase of *Evernia prunastri*, *Plant Cell Physiol.*, 22, 1119, 1981.
127. **Vicente, C. and Legaz, M. E.**, Purification and properties of agmatine amidinohydrolase of *Evernia prunastri*, *Physiol. Plant.*, 55, 335, 1982.
128. **Sanzey, B. and Ullman, A.**, Urea, a specific inhibitor of catabolite sensitive operons, *Biochem. Biophys. Res. Commun.*, 71, 1062, 1976.
129. **Ullman, A.**, Are cyclic AMP effects related to real physiological phenomena?, *Biochem. Biophys. Res. Commun.*, 57, 348, 1974.
130. **Vicente, C. and Legaz, M. E.**, Repression of arginase and agmatine amidinohydrolase by urea in the lichen *Evernia prunastri*, *Phytochemistry*, 24, 217, 1985.
131. **Smith, T. A. and Garraway, J. L.**, *N*-Carbamylputrescine — an intermediate in the formation of putrescine by barley, *Phytochemistry*, 3, 23, 1964.
132. **Smith, T. A.**, Agmatine iminohydrolase in maize, *Phytochemistry*, 8, 2111, 1969.
133. **LeRudulier, D. and Goas, G.**, Biogenèse de la *N*-carbamylputrescine et de la putrescine dans les plantulas de *Glycine max* (L.) Merr., *Physiol. Veg.*, 18, 609, 1980.
134. **Legaz, M. E., Iglesias, A., and Vicente, C.**, Regulation of agmatine iminohydrolase of *Evernia prunastri* by L-arginine metabolites, *Z. Pflanzenphysiol.*, 110, 53, 1983.
135. **Bernard, T. and Goas, G.**, Contribution à l'étude du métabolisme azoté des lichens. Charactérisation et dosage des méthylamines de quelques espèces de la famille des Stictacées, *C. R. Acad. Sci. Paris*, 267, 622, 1968.
136. **Bernard, T. and Larher, F.**, Contribution à l'etude du métabolisme azoté des lichens: rôle de la glycine [14]C-2 dans la formation des méthylamines chez *Lobaria laetevirens* Zahlbr., *C. R. Acad. Sci. Paris*, 272, 568, 1971.
137. **Bernard, T. and Goas, G.**, Biosynthèse de la sticticin chez le lichen *Lobaria laetevirens*, *Physiol. Plant.*, 53, 71, 1981.
138. **Bernard, T., Goas, G., Hamelin, J., and Jouda, M.**, Characterization of DOPA betaine, tyrosine betaine and *N*-dimethyltyrosine from *Lobaria laetevirens*, *Phytochemistry*, 20, 2325, 1981.
139. **Smith, T. A.**, Further properties of the polyamine oxidase from oat seedlings, *Phytochemistry*, 16, 1647, 1977.
140. **Fracassini, D. S., Bagni, N., Cionini, P. G., and Bennici, A.**, Polyamines and nucleic acids during the first cell cycle in *Helianthus tuberosus* tissue after the dormancy break, *Planta*, 148, 332, 1980
141. **Altman, A. and Bachrach, V.**, Involvement of polyamines in plant growth and senescence, in *Advances in Polyamines Research*, Calderea, C. M., Zappia, V., and Bachrach, U., Eds., Raven Press, New York, 1981, 365.
142. **Kaur-Sawhney, R., Shih, L., Flores, H. E., and Galston, A. W.**, Relation of polyamine synthesis and titer to aging and senescence in oat leaves, *Plant Physiol.*, 69, 411, 1982.
143. **Fishbein, W. N., Winter, T. S., and Davidson, J. D.**, Urease catalysis. I. Stoichiometry, specificity and kinetics of a second substrate: hydroxyurea, *J. Biol. Chem.*, 240, 2402, 1965.
144. **Shapiro, I. A.**, Urease activity in lichens, *Fiziol. Rast.*, 24, 1135, 1977.
145. **Brown, D. H., Vicente, C., and Legaz, M. E.**, Urease activity in *Peltigera canina*, *Cryptog. Bryol. Lichenol.*, 3, 33, 1982.
146. **Massé, L.**, Quelques aspects de l'uricolyse enzymatique chez les lichens, *C. R. Acad. Sci. Paris Ser. D*, 268, 2896, 1969.
147. **Xavier Filho, L. and Vicente, C.**, Exo- and endourease from *Parmelia roystonea* and their regulation by lichen acids, *Bol. Soc. Broteriana*, 52, 55, 1978
148. **Legaz, M. E. and Vicente, C.**, Location of several enzymes of L-arginine catabolism in *Evernia prunastri* thallus, *Z. Naturforsch.*, 36c, 92, 1981.
149. **Brown, D. H., Legaz, M. E., and Feest, A.**, Observations on bacterial contamination and urease activity of *Evernia prunastri*, *Cryptog. Bryol. Lichenol.*, 4, 263, 1983.
150. **Solberg, Y. J.**, Studies on the chemistry of lichens. VII. Chemical investigations of the lichen species *Lecanora (Aspicilia) myrinii*, *Z. Naturforsch.*, 24b, 447, 1969.

151. **Solberg, Y. J.,** Studies on the chemistry of lichens. IX. Quantitative determination of monosaccharides and amino acids in hydrolizates of several Norwegian lichen species, *Lichenologist,* 4, 283, 1970.

152. **Cifuentes, B., Estévez, M. P., and Vicente, C.,** *In vivo* protection of urease of *Evernia prunastri* by dithiothreitol, *Physiol. Plant.,* 53, 245, 1981.

153. **Legaz, M. E. and Brown, D. H.,** Factors affecting urease activity in the lichen *Evernia prunastri, Ann. Bot. (London),* 52, 261, 1983.

154. **Dixon, N. E., Gazzola, C., Blakeley, R. L., and Zerner, B.,** Jack bean urease. A metalloenzyme. A simple biological role for nickel?, *J. Am. Chem. Soc.,* 97, 4131, 1975.

155. **Polacco, J. C.,** Is nickel a universal component of plant ureases?, *Plant Sci. Lett.,* 10, 249, 1977.

156. **Winkler, R. G., Polacco, J. C., Eskeur, D. L., and Welch, R. M.,** Nickel is not required for apourease synthesis in soybean seeds, *Plant Physiol.,* 72, 262, 1983.

157. **García, I., Cifuentes, B., and Vicente, C.,** L-Usnate-urease interactions: binding sites for the ligand, *Z. Naturforsch.,* 35c, 1098, 1980.

158. **Avalos, A., Legaz, M. E., Perez Urria, E., and Vicente, C.,** L-Arginine induces a thermostable repressor of urease in *Pseudevernia furfuracea, Endocyt. Cell Res.,* 2, 15, 1985.

159. **Vicente, C., Azpiroz, A., Estévez, M. P., and González, M. L.,** Quaternary structure changes and kinetics of urease inactivation by L-usnic acid in relation to the regulation of nutrient transfer between lichen symbionts, *Plant Cell Environ.,* 1, 29, 1978.

160. **Vicente, C. and Cifuentes, B.,** Reversal by L-cysteine of the inactivation of urease by L-usnic acid, *Plant Sci. Lett.,* 15, 165, 1979

161. **Legaz, M. E., Cifuentes, B., and Vicente, C.,** Mecanismos de síntesis e inactivación urease en *Evernia prunastri,* in *Estudios sobre Biología,* Vicente, C. and Municio, A. M., Eds., Complutense University Press, Madrid, 1982, 101.

162. **Avalos, A. and Cifuentes, B.,** Urease inactivation by evernic acid, Proc. Int. Symp. Recent Advances in Lichen Physiology, Bristol, April 16—18, 1984.

163. **Gorin, G., Fuschs, E., Butler, L. G., Chopra, S. L., and Hersh, R. T.,** Some properties of urease, *Biochemistry,* 1, 911, 1962.

164. **Cifuentes, B., García, I., and Vicente, C.,** L-Usnate-urease interactions: binding sites for polymerization, *Z. Naturforsch.,* 38c, 273, 1983.

165. **Syrett, P. J. and Bekheet, I. A.,** The uptake of thiourea by *Chlorella, New Phytol.,* 79, 291, 1977.

166a. **Vicente, C., Nieto, J. M., and Legaz, M. E.,** Induction of urease by urea analogues in *Evernia prunastri* thallus, *Physiol. Plant.,* 58, 325, 1983.

166b. **Mosbach, H.,** Biosynthesis of lichen substances, in *The Lichens,* Ahmadjian, V. and Hale, M. E., Eds., Academic Press, New York, 1973, 523.

167. **Ahmadjian, V. and Jacobs, J. B.,** Algal-fungal relationships in lichens: recognition, synthesis and development, in *Algal Symbiosis,* Goff, L. J., Ed., Cambridge University Press, London, 1983, 147.

168. **Roberts, T. L. and Rosenkrantz, H.,** Esterase activity in the green alga *Trebouxia, Tex. Rep. Biol. Med.,* 25, 432, 1967.

169. **Boissiere, M. C.,** Un mécanisme possible d'absorption des glucides d'origine cyanophytique pur les hyphas de quelques lichens, *Rev. Bryol. Lichenol.,* 43, 19, 1977.

170. **Rennert, A. and Gubański, M.,** Ribonuclease in *Cetraria islandica, Naturwissenschaften,* 47, 18, 1960.

171. **Moissejeva, E. N.,** Biochemical properties of lichens and their practical importance, *Akad. Nank. SSSR Bot. Inst. V. L. Komawa,* Moscow, Leningrad, 1961.

172. **Follmann, G. and Villagrán, V.,** Flechtenstoffe und Zellpermeabilität, *Z. Naturforsch.,* 20b, 723, 1965.

173. **Kinraide, W. T. B. and Ahmadjian, V.,** The effects of usnic acid on the physiology of two cultured species of the lichen alga *Trebouxia, Lichenologist,* 4, 234, 1970.

174. **Vicente, C. and Cifuentes, B.,** L-Usnate and permeability, *Cryptog. Bryol. Lichenol.,* 2, 213, 1981.

175. **Schimmer, O. and Lehner, H.,** Untersuchungen zur Wirkung von Usninsäure auf die Grünalge *Chlamydomonas reinhardii, Ark. Mikrobiol.,* 93, 145, 1973.

176. **Muñoz-Calvo, M. L., Rodriquez-López, M., and Villaroya-Sánchez, M.,** The response of prokaryotic algae to treatment with D-usnic acid, Proc. 2nd Congr. of the Federation of European Societies of Plant Physiologist, Santiago, August 2—6, 1980.

177. **Cifuentes, B. and Vicente, C.,** Binding studies on L-usnic acid to D-fructose-6-P aminotransferase, *Biochem. Biophys. Res. Commun.,* 95, 1550, 1980.

178. **Cifuentes, B. and Vicente, C.,** Purification and properties of glucosamine phosphate isomerase of *Proteus mirabilis, Z. Naturforsch.,* 37c, 381, 1982.

179. **Cifuentes, B. and Vicente, C.,** Action of L-usnic acid on glucosamine phosphate isomerase activity of *Proteus mirabilis, Cryptog. Bryol. Lichenol.,* 4, 255, 1983.

180. **Natori, S.,** Antibacterial effect of lichen substances and related compounds. VII. The structure-activity relationship observed in compounds related to dibenzofuran and an approach to the elucidation of the mode of action, *Pharm. Bull.,* 5, 553, 1957.

181. **Richardson, D. H. S.**, The surface physiology of lichens with particular reference to carbohydrate transfer between the symbionts, in *Surface Physiology of Lichens,* Vicente, C., Brown, D. H., and Legaz, M. E., Eds., Complutense University Press, Madrid, 1985, 25.

182. **Richardson, D. H. S.**, Lichens, in *Methods in Microbiology,* Vol. 4, Booth, C., Ed., Academic Press, New York, 1971, 267.

183. **Rai, A. N., Rowell, P., and Stewart, W. D. P.**, Interactions between cyanobacterium and fungus during $^{15}N_2$-incorporation and metabolism in the lichen *Peltigera canina, Arch. Microbiol.,* 134, 136, 1983.

184. **Legaz, M. E.**, The regulation of urea biosynthesis, in *Surface Physiology of Lichens,* Vicente, C., Brown, D. H., and Legaz, M. E., Eds., Complutense University Press, Madrid, 1985, 57.

185. **Vicente, C. and Legaz, M. E.**, Regulation of urea production in *Evernia prunastri:* effects of L-arginine metabolites, *Z. Pflanzenphysiol.,* 111, 123, 1983.

186. **Galun, M. and Bubrick, P.**, Physiological interactions between the partners of the lichen symbiosis, in *Encyclopedia of Plant Physiology,* Vol. 17, Linkskens, H.-F. and Heslop-Harrison, J., Eds., Springer-Verlag, Berlin, 1984, 362.

187. **Ahmadjian, V., Russell, L. A., and Hildreth, K. C.**, Artificial reestablishment of lichens. I. Morphological interactions between the phycobionts of different lichens and the mycobionts *Cladonia cristatella* and *Lecanora chrysoleuca, Mycologia,* 72, 73, 1980.

188. **Mosbach, K. and Mosbach, R.**, Entrapment of enzymes and microorganisms in synthetic crosslinked polymers and their application in column techniques, *Acta Chem. Scand.,* 20, 2807, 1966.

189. **Mosbach, K.**, The potential in biotechnology of immobilized multistep enzyme-coenzyme systems, *Philos. Trans. R. Soc. London Ser. B,* 300, 366, 1983.

190. **Rai, A. N., Rowell, P., and Stewart, W. D. P.**, Mycobiont-cyanobiont interactions during dark nitrogen fixation by the lichen *Peltigera aphthosa, Physiol. Plant.,* 57, 285, 1983.

191. **Bernard, T. and Goas, G.**, Glutamate deshydrogénases du lichen *Lobaria laetevirens* (Lightf.) Zahlbr. Charactéristiques de l'enzyme du champignon, *Physiol. Veg.,* 17, 535, 1979.

192. **Rai, A. N., Rowell, P., and Stewart, W. D. P.**, Glutamate synthase activity in symbiotic cyanobacteria, *J. Gen. Microbiol.,* 126, 515, 1981.

193. **Millbank, J. W.**, Nitrogenase and hydrogenase in cyanophilic lichens, *New Phytol.,* 92, 221, 1981.

194. **Rai, A. N., Rowell, P., and Stewart, W. D. P.**, $^{15}N_2$ incorporation and metabolism in the lichen *Peltigera aphthosa, Planta,* 152, 544, 1981.

195. **Stewart, W. D. P., Rai, A. N., Reed, R. H., Creach, E., Codd, G. A., and Rowell, P.**, Studies on the N_2-fixing lichen *Peltigera aphthosa,* in *Current Perspectives in Nitrogen-Fixation,* Gibson, A. H. and Newton,, W. E., Eds., Elsevier, Amsterdam, 1981, 237.

196. **Horstmann, J. L., Denison, W. C., and Silvester, W. B.**, $^{15}N_2$ fixation and molybdenum enhancement of acetylene reduction by *Lobaria spp., New Phytol.,* 92, 235, 1982.

197. **Huss-Danell, K.**, The cephalodia and their nitrogenase activity in the lichen *Stereocaulon paschale, Z. Pflanzenphysiol.,* 95, 431, 1979.

198. **Huss-Danell, K.**, The influence of light and oxygen on nitrogenase activity in the lichen *Stereocaulon paschale, Physiol. Plant.,* 47, 269, 1979.

199. **Bernard, T. and Goas,, G.**, Contribution à l'étude due métabolisme azoté des lichens. Mise en évidence de quelques transaminases; activité de la glutamate-oxaloacetate transaminase dans cinq espèces de la famille des Stictacées, *C. R. Acad. Sci. Paris,* 269, 1957, 1969.

APPENDIX
CATALOG OF ENZYMES DESCRIBED IN LICHENS

Oxidoreductases

EC	Recommended name	Reaction	Source	Ref.
1.4.1.1.	Alanine dehydrogenase	L-Alanine + H_2O + NAD^+ = Pyruvate + NH_3 + NADH	Peltigera aphthosa	190
			P. canina	183
1.4.1.4.	Glutamate dehydrogenase	L-Glutamate + H_2O + $NADP^+$ = 2-oxoglutarate + NH_3 + NADPH	Pseudevernia furfuracea	120
			Lobaria laetevirens	191
			Peltigera canina	183
			P. aphthosa	190
			Hypogymnia physodes	120
1.4.7.1.	Glutamate synthase (Fd)	2-L-Glutamate + 2Fd(ox) L-glutamine + 2-oxoglutarate + 2-Fd(red)	P. canina	183, 192
			P. aphthosa	190, 192
1.6.6.1.	Nitrate reductase	NO_3^- + NADH = NO_2^- + NAD^+ + H_2O	Evernia prunastri	109
1.13.11.12.	Lipoxygenases	Linoleate + O_2 = 13-hydrogenoxy-octadeca-9,11-dienoate	E. prunastri	61
1.18.1.2.	Ferredoxin-$NADP^+$ oxidoreductase	Fd(red) + $NADP^+$ = Fd(ox) + NADPH	E. prunastri	2, 10
1.18.2.1.	Nitrogenase	6Fd(red) + $6H^+$ + N_2 + nATP = 6Fd(ox) + $2NH_3$ + nADP + nPi	Parmelia membranacea	193
			Peltigera polydactyla	193
			P. aphthosa	14, 194, 195
			P. canina	183
			Lobaria pulmonaria	193, 196
			L. oregana	196
			Stereocaulon paschale	197, 198
1.18.3.1.	Hydrogenase	2Fd(red) + $2H^+$ = 2Fd(ox) + H_2	Parmelia membranacea	193
			Peltigera polydactyla	193
			L. pulmonaria	193
	DL-Usnic acid dehydrogenase	DL-usnic acid + NADH = NAD + X	E. prunastri	82
	D-Usnic acid dehydrogenase	D-Usnic acid + NADH = NAD + X	E. prunastri	83

Transferases

EC	Recommended name	Reaction	Source	Ref.
2.3.1.4.	Glucosamine-P acetyltransferase	Acetyl-CoA + 2-amino-2-deoxy-D-glucose = CoA + 2-acetamido-2-deoxy-D-glucose	E. prunastri	53

2.6.1.1.	L-Aspartate:2-oxoglutarate aminotransferase	L-Aspartate + 2-oxoglutarate = 2-oxaloacetate + L-glutamate	*L. pulmonaria*	199
			L. laetevirens	199
			Sticta sylvatica	199
			S. limbata	199
			S. fuliginosa	199
			P. aphthosa	190
			P. canina	183
			P. aphthosa	190
2.6.1.2.	L-Alanine:2-oxoglutarate aminotransferase	L-Alanine + 2-oxoglutarate = pyruvate + L-glutamate		
2.6.1.12.	L-Alanine:oxoacid aminotransferase	L-Alanine + a 2-oxoacid = pyruvate + an L-amino acid		

Hydrolases

3.1.1.40.	Orsellinate depside hydrolase	Orsellinate depside + H_2O = 2 orsellinate	*Lasallia pustulata*	68, 69
3.2.1.4.	Cellulase	Endohydrolysis of 1,4-β-D-glucosidic linkage in cellulose and lichenin	*E. prunastri*	70, 71
			E. prunastri	58
3.2.1.15.	Polygalacturonase	Random hydrolysis of 1,4-α-D-galacturiduronic linkages	*E. prunastri*	58
3.5.1.5.	Urease	Urea + H_2O = CO_2 + $2NH_3$	*E. prunastri*	74,149,152
			Lobaria pulmonaria	148, 153
			Parmelia roystonea	147
			Pseudevernia furfuracea	158
			Cladonia verticillaris	99
			C. sandstedei	77
			C. rangiferina	144
			Cetraria islandica	144
			Hypogymnia physodes	144
			Peltigera canina	144, 145
3.5.3.1.	Arginase	L-Arginine + H_2O = L-ornithine + urea	*E. prunastri*	121, 122, 124, 130
3.5.3.11.	Agmatine amidinohydrolase	Agmatine + H_2O = putrescine + urea	*E. prunastri*	184, 185
				127, 130
3.5.3.12.	Agmatine iminohydrolase	Agmatine + H_2O = N-carbamoylputrescine + NH_3	*E. prunastri*	184, 185
				134, 184

APPENDIX (continued)
CATALOG OF ENZYMES DESCRIBED IN LICHENS

EC	Recommended name	Reaction	Source	Ref.
		Lyases		
4.1.1.15.	Glutamate decarboxylase	L-Glutamate = 4-aminobutyrate + CO_2	*L. pulmonaria*	113
			L. laetevirens	113
			S. sylvatica	113
			S. limbata	113
			S. fuliginosa	113
4.1.1.17.	L-Ornithine decarboxylase	L-Ornithine = putrescine + CO_2	*E. prunastri*	125, 184
4.1.1.19.	L-Arginine decarboxylase	L-Arginine = agmatine + CO_2	*E. prunastri*	126, 185
4.1.1.31.	Phosphoenolpyruvate carboxylase	Pi + oxaloacetate = H_2O + phosphoenolpyruvate + CO_2	*P. aphthosa*	14
4.1.1.38.	Phosphoenolpyruvate carboxyphosphotransferase (pyrophosphate)	Pi + oxaloacetate = orthophosphate + phosphoenolpyruvate + CO_2	*P. aphthosa*	14
4.1.1.39.	Ribulose-bisphosphate carboxylase	D-Ribulose-1,5-bisP + CO_2 = 2,3-phosphoglycerate	*P. aphthosa*	14
4.1.1.49.	Phosphoenolpyruvate carboxykinase(ATP)	ATP + oxaloacetate = ADP + phosphoenolpyruvate + CO_2	*P. aphthosa*	14
4.1.1.58.	Orsellinate decarboxylase	2,4-Dihydroxy-6-methylbenzoate = orcinol + CO_2	*Lasallia pustulata*	68, 72
		Isomerases		
5.3.1.19.	Glucosamine phosphate isomerase (glutamine forming)	2-Amine-2-deoxy-D-glucose-6-P + L-glutamate = D-fructose-6-P + L-glutamine	*E. prunastri*	48, 51
		Ligases (synthetases)		
6.3.1.2.	Glutamine synthetase	ATP + L-glutamate + NH_3 = ADP + Pi + L-glutamine	*P. canina*	183, 192
			P. aphthosa	190, 192, 194
6.3.4.16.	Carbamoylphosphate synthetase (ammonia)	2ATP + NH_3 + CO_2 + H_2O = 2ADP + Pi + carbamoyl-P	*P. aphthosa*	14
6.3.5.5.	Carbamoylphosphate synthetase (glutamine hydrolysing)	2ATP + glutamine + CO_2 + H_2O = 2ADP + orthophosphate + glutamate + carbamoylphosphate	*P. aphthosa*	14

Index

INDEX